T0392615

BIODEGRADATION AND DETOXIFICATION OF MICROPOLLUTANTS IN INDUSTRIAL WASTEWATER

BIODEGRADATION AND DETOXIFICATION OF MICROPOLLUTANTS IN INDUSTRIAL WASTEWATER

Edited by

IZHARUL HAQ
Department of Civil Engineering, Indian Institute of Technology, Guwahati, Assam, India

AJAY S. KALAMDHAD
Department of Civil Engineering, Indian Institute of Technology Guwahati, Assam, India

MAULIN P. SHAH
Senior Environmental Microbiologist, Applied and Environmental Microbiology, Bharuch, India

ELSEVIER

Elsevier
Radarweg 29, PO Box 211, 1000 AE Amsterdam, Netherlands
The Boulevard, Langford Lane, Kidlington, Oxford OX5 1GB, United Kingdom
50 Hampshire Street, 5th Floor, Cambridge, MA 02139, United States

ISBN: 978-0-323-88507-2

For Information on all Elsevier publications visit our website at
https://www.elsevier.com/books-and-journals

Publisher: Susan Dennis
Acquisitions Editor: Anita Koch
Editorial Project Manager: Howi M. De Ramos
Production Project Manager: Kumar Anbazhagan
Cover Designer: Mark Rogers

Typeset by Aptara, New Delhi, India

Contents

Contributors

Md. Anwaruzzaman
Department of Applied Chemistry and Chemical Engineering, Bangabandhu Sheikh Mujibur Rahman Science and Technology University, Gopalganj, Bangladesh

Christy K Benny
Department of Civil Engineering, Indian Institute of Technology, Guwahati, Assam, India

Sonali Bhardwaj
Department of Microbiology, Lovely Professional University, Phagwara, Punjab, India

Shovendu Biswas
Department of Applied Chemistry and Chemical Engineering, Bangabandhu Sheikh Mujibur Rahman Science and Technology University, Gopalganj, Bangladesh

Saswati Chakraborty
Department of Civil Engineering, Indian Institute of Technology, Guwahati, Assam, India

Sunipa Deb
Department of Chemical Engineering, National Institute of Technology Agartala, Tripura, India

Daljeet Singh Dhanjal
Department of Biotechnology, Lovely Professional University, Phagwara, Punjab, India

Sougata Ghosh
Department of Microbiology, School of Science, RK University, Rajkot, Gujarat, India

Aparna Gupta
Department of Environmental Science, Central University of South Bihar, India

Md. Lawshan Habib
Department of Applied Chemistry and Chemical Engineering, Bangabandhu Sheikh Mujibur Rahman Science and Technology University, Gopalganj, Bangladesh

Izharul Haq
Department of Civil Engineering, Indian Institute of Technology, Guwahati, Assam, India

Md. Irfanul Haque
Department of Applied Chemistry and Chemical Engineering, Bangabandhu Sheikh Mujibur Rahman Science and Technology University, Gopalganj, Bangladesh

M. Mehedi Hasan
Department of Applied Chemistry and Chemical Engineering, Bangabandhu Sheikh Mujibur Rahman Science and Technology University, Gopalganj, Bangladesh

Mohd. Tanzir Hossain
Department of Applied Chemistry and Chemical Engineering, Bangabandhu Sheikh Mujibur Rahman Science and Technology University, Gopalganj, Bangladesh

Md Didarul Islam
National Institute of Textile Engineering and Research, Dhaka, Bangladesh

J Aravind
Dhirajlal Gandhi College of Technology, Omalur, Tamil Nadu, India

Ajay S. Kalamdhad
Department of Civil Engineering, Indian Institute of Technology, Guwahati, Assam, India

Md. Kamruzzaman
Department of Applied Chemistry and Chemical Engineering, Bangabandhu Sheikh Mujibur Rahman Science and Technology University, Gopalganj, Bangladesh

Dhriti Kapoor
Department of Botany, Lovely Professional University, Phagwara, Punjab, India

Vijay Kumar
Regional Ayurveda Research Institute for Drug Development, Gwalior, MP, India

M Kamaraj
Department of Biotechnology, College of Biological and Chemical Engineering, Addis Ababa Science and Technology University, Addis Ababa, Ethiopia

Meem Muhtasim Mahdi
Environment and Natural Resources, School of Engineering and Natural Sciences, Háskóli Íslands, Sæmundargötu 2, Reykjavík, Iceland

Krishna Chaitanya Maturi
Department of Civil Engineering, Indian Institute of Technology, Guwahati, Assam, India

Soma Nag
Department of Chemical Engineering, National Institute of Technology Agartala, Tripura, India

P Snega Priya
Department of Medical Microbiology, SRM Medical College Hospital and Research Center, Kattankulathur, Tamil Nadu, India

Nidhi Pareek
Microbial Catalysis and Process Engineering Laboratory, Department of Microbiology, School of Life Sciences, Central University of Rajasthan Bandarsindri, Kishangarh, Ajmer, Rajasthan, India

Subhankar Paul
Structural Biology and Nanomedicine Laboratory, Department of Biotechnology and Medical Engineering, National Institute of Technology, Rourkela, Odisha, India

Akshay Pawar
Department of Biotechnology, Lovely Professional University, Phagwara, Punjab, India

Praveen C Ramamurthy
Interdisciplinary Centre for Water Research (ICWaR), Indian Institute of Sciences, Bangalore, India

S Sudhakar
Department of Biotechnology, PGP College of Arts and Science, Namakkal, Tamil Nadu, India

Md. Nahidul Islam Sajol
Department of Applied Chemistry and Chemical Engineering, Bangabandhu Sheikh Mujibur Rahman Science and Technology University, Gopalganj, Bangladesh

Bishwarup Sarkar
Department of Microbiology and Biotechnology Centre, The Maharaja Sayajirao University of Baroda, Vadodara, Gujarat, India

Simranjeet Singh
Interdisciplinary Centre for Water Research (ICWaR), Indian Institute of Sciences, Bangalore, India

Joginder Singh
Department of Microbiology, Lovely Professional University, Phagwara, Punjab, India

Shweta Singh
Department of Civil Engineering, Indian Institute of Technology, Guwahati, Assam, India

Prathap Somu
Structural Biology and Nanomedicine Laboratory, Department of Biotechnology and Medical Engineering, National Institute of Technology, Rourkela, Odisha, India

Twinkle Soni
Microbial Catalysis and Process Engineering Laboratory, Department of Microbiology, School of Life Sciences, Central University of Rajasthan Bandarsindri, Kishangarh, Ajmer, Rajasthan, India

V Vivekanand
Centre for Energy and Environment, Malaviya National Institute of Technology, Jaipur, Rajasthan, India

Preface

The book titled Biodegradation and Detoxification of Micropollutants in Industrial Wastewater mainly covers various aspects of biodegradation of micropollutants while discussing their toxic effects on humans and animals health and state-of-the-art treatment technologies.

As evident globally, the industries are an imperative economic sector in several countries around the world but, also notorious polluters of the environment. The industry uses wide range of technology, which varies among industries as well as within the industry. These industries use large volume of water and different chemicals during various operational and manufacturing processes and generate a huge amount of toxic wastewater which is loaded with many micropollutants. The wastewater generated in industries is released into the environment with or without proper treatment and affect aquatic as well as terrestrial ecosystem and human being also. Thus, there is an urgent need to cope with this challenge and develop sustainable methods for treating/detoxifying wastewater before its release into the environment.

This book mainly focuses on treatment of industrial wastewater and its toxic effects on living organism. The individual chapters will provide detailed information about the occurrence, source, characteristics, toxicity and effects of industrial micropollutants on living organism and environment and its biodegradation and detoxification through various advanced treatment technologies. The book is addressing new and emerging work in the field of wastewater treatment system. Treatment of micropollutants is the toughest task but it is necessary to remove these pollutants from environment for the safety of humans and animals.

This book will boost the knowledge of students, researchers, scientists, professors, engineer, and professionals who aspire to work in the field of environmental science, environmental biotechnology, environmental microbiology, environmental engineering, eco-toxicology, and other relevant area of industrial waste management for the safety of environment. In addition of this, the readers of the book can also get the valuable information related to various environmental problems and their solutions.

Editors
Izharul Haq
Ajay S. Kalamdhad
Maulin P. Shah

Acknowledgments

We are grateful to the authors in compiling the pertinent information required for chapter writing, which we believe will be a valuable source for both the scientific community and the audience in general. We are thankful to the expert reviewers for providing their useful comments and scientific insights, which helped shape the chapter organization and improved the scientific discussions, and overall quality of the chapters. We sincerely thank the Elsevier production team for their support in publishing this book. Last but not least, I'd want to express my gratitude to my mother, a strong and compassionate spirit, who taught me to trust in God, believe in hard effort, and that a lot can be accomplished with a little, and my father, for always encouraging and morally supporting me throughout my life.

CHAPTER 1

Evaluation of micro-pollutants removal from industrial wastewater using conventional and advanced biological treatment processes

Md. Didarul Islam[a], Meem Muhtasim Mahdi[b]
[a]National Institute of Textile Engineering and Research, Dhaka, Bangladesh
[b]Environment and Natural Resources, School of Engineering and Natural Sciences, Háskóli Íslands, Sæmundargötu 2, Reykjavík, Iceland

1.1 Introduction

The industrial revolution started from the mid-18th century onwards has brought about ease and flexibility in our lifestyle and provided versatile employment to earn a daily livelihood. The profits earned from industrialization have fulfilled our desire for having a happy life but led the environment towards a successive degradation. The disposal of solid wastes and discharge of wastewaters from different household chores and industrial activities result in a hazardous impact on natural resources and human health. Scholars have shown that the wastewaters released from domestic, agricultural, and industrial streams contain a generous amount of toxic pollutants and hazardous substances that causes a shortage of pure drinking waters by decreasing the aesthetic water quality, accelerate water pollution in many countries and also responsible for imbalancing the aquatic ecosystem when they mix up with groundwater and aquifers.

Among harmful pollutants, micro-pollutants are considered as an emerging type pollutant because they contaminate natural elements like soil matrices, water bodies at an integral rate, and shows adverse effect in aquatic living species and human health through multigenerational interactions (Kanaujiya et al., 2019). Generally, micro-pollutants (MPs) are categorized as those anthropogenic chemicals that exist in the marine environment at a very lower concentration/trace amount (ranged μg/L-ng/L), released from both point and non-point sources due to human activities and finally end up in the water streams almost at an undetectable level (≤1 ppb) and hence these chemicals are also referred to as 'Trace organic contaminants' (TrOCs) (Chavoshani et al., 2020). Most of the MPs chemicals undergo biotransformation during their treatment processes and form new chemical species (persistent metabolites) that react with humic agents in sunlight, turn bioactive, and might get bioaccumulated in the plant organs. Previous studies have stated that MPs are not fully metabolized by the animals

Biodegradation and Detoxification of Micropollutants in Industrial Wastewater
DOI: https://doi.org/10.1016/B978-0-323-88507-2.00011-7

and their parent molecules can show deleterious effects like mutagenesis, estrogenicity, cytotoxic and genotoxic syndromes in prolonged exposure. Therefore, several international environmental and health organizations like the United States Environmental Protection Agency (USEPA), United Nations Environment Programme (UNEP), Water Framework Directive (WFD) under the EU authorization, and World Health Organization (WHO) have suggested legit regulations and guidelines to eliminating these micro-pollutants from industrial wastewaters before discharging them into the water streams for water conservation while taking proper precautionary steps (Kanaujiya et al., 2019).

In recent years, attempts have been taken by using several wastewater treatment methods viz. advanced oxidation process, adsorption, filtration, ozonization, activated sludge process, etc. to investigate their feasibility on micro-pollutant removal from municipal and industrial wastewaters (Karaolia et al., 2017). Despite having no established standard limit and a less detectable amount of MPs on effluents, it is very difficult to mitigate these harmful chemicals on large-scale effluent treatment. Previous researches have shown that biological wastewater treatment processes (BWWTP) have a significant influence on MPs removal. Scholars have published a few review papers on the effectiveness of conventional and advanced BWWTPs on micro-pollutant elimination from water bodies but only a few of them have outlined a systematic summary and approaches to large-scale wastewater treatment (Grandclément et al., 2017; Obotey Ezugbe and Rathilal, 2020).

Regarding these facts, we attempted a synchronized approach to sketch a summary on the feasibility of the conventional and advanced/hybrid biological wastewater treatment methods to remove some common micro-pollutants from industrial wastewater. Besides, the occurrence and fate of these pollutants in wastewater, their behavior during biological wastewater treatment, and the pros and cons of the treatment techniques along with their future perspectives are also described in this chapter.

1.2 Types of micro-pollutants (MPs) in industrial wastewaters and their toxic health effects

The extensive application of chemical products and mechanical equipment in our daily life has resulted in a huge micro-pollutant load into the environment. The wastewaters released from sewage lines, agricultural leachates, stormwater runoff and industries like pharmaceuticals, textiles, tanneries, paper and pulp, food processing, packaging, etc. are the prominent sources to generate MPs into the aquatic streams (Chen et al., 2020). These MPs are found mostly in the drainage systems (detergents, surfactants, additives), medicinal residues (steroids, hormones, antibiotics, protein supplements) exerted with human feces/urine, household products (plastics, flame retardants), commodities used for personal hygiene and sanitation (cosmetics and toiletries); and agricultural leachates from crop harvesting sites (pesticides) (Margot et al., 2015). MPs are mainly organic compounds with a few inorganic anions and metalloids, and classified into distinguished categories according to their molecular structure, behavior, applications.

1.2.1 Organic micro-pollutants

Organic micro-pollutants are generally classified as polycyclic aromatic hydrocarbons (PAHs), polychlorinated biphenyls (PCBs), Endocrine disrupters (EDCs), and pesticide residues. Polycyclic aromatic hydrocarbons are fused benzene rings (ranged 2–8) may be either homocyclic or heterocyclic in chemical configuration.

PAHs are often nondegradable, show carcinogenic even at a very lower concentration, and have a tendency to get accumulated in aquatic food chain from wastewaters and sediments (diagenetic transformation). These PAHs occurs from automobile tar, oil spillages, herbicides, coal and petroleum combustion (pyrogenic). Anthracene (dye, plastic, pharmaceutical industry), fluorine (pharmaceutical, thermoset plastic, and pesticide industry), fluroanthene (pharmaceutical industry), pyrene (pigment industry), 1-naphthalene, etc. are some common PAHs found in different industrial wastewaters. PAHs show adverse effect on both marine species and mammals that depends on exposure dose and exposure duration. The short term health effect of PAHs on human health is not described properly in pervious literatures but researchers assume that PAHs can results in some acute diseases skin and eye irritation, short term memory loss, diarrhea, and lead to many more chronic diseases like inflammation, breathing problems, and red blood cell (RBC) cessation.

Polychlorinated biphenyls (PCBs) are special type of persistent organic pollutants consist of 209 chemical compounds possessing characteristics of high toxicity, non-biodegradability, high bioaccumulation capacity and transportation. These chemicals are now almost banned from using in insulator and lubricants but still PCBs contamination exists in many parts of the globe due to waste incineration and secondary by-products of chemical manufacturing industries. Researchers showed that the concentration of several PCBs in treated effluents is present at ranged 1.01–8.12 ng/L (dissolved media) and 2.03–31.2 ng/L (suspended media) that causes sicknesses like neurological malfunction, reproductive toxicity, cancer, and skeletal deformity if PCBs are entered into the human body upon inhalation/swallowed (Yao et al., 2014).

Endocrine disruptors (EDCs) are those exogenous synthetic chemicals that imitate as hormones and interrupt the normal functions (hormonal secretion) of endocrine receptors in animals. These chemicals and their derivatives are widely used as industrial solvents in producing plasticizers, plastics, medicinal agents, antimicrobial agents, and other commercial products. Industrial effluents originated from paper and pulp, leather tanning, fabrics manufacturing, distilleries are reported as the vital source of various EDCs. Dioxins, alkylphenols, bisphenol A (BPA), triclosan, phthalates, nonylphenol (NP) are the common EDCs that are transferred to human body via different possible routes (food/water/air) and accumulated in adipose tissues and show synergetic effects in human physiopathology such as cancer in several organs (lungs, liver, prostate), metabolic disorder, obesity, problems in reproductive system, thyroids, along with cardiovascular, gastrointestinal, neurological and muscular infections.

1.2.2 Inorganic micro-pollutants

Heavy metals are considered as a significant inorganic micro-pollutant. The metals with equal or higher density of 6 g/cm^3 are called heavy metals. Chromium (Cr), Cadmium (Cd), Mercury (Hg), Lead (Pb), and Manganese (Mn) are the most well-known heavy metals with extensive industrial applications (Duffus, 2002). Moreover, some of these metals are often considered as potential micronutrients for plant growth. Despite having several benefits, heavy meals typically show a negative influence on environment and human health.

Chromium (Cr(III) and Cr(VI)) is the most toxic of all heavy metals that has the most deleterious health symptoms. It has widespread usages in leather tanning, metal and alloy finishing, textile, paper and pulp, and paint and dyeing industries. These industries generate tons of unreacted chromium with wastewaters. Researchers showed that the consumption of chromium at an excess rate with food and water can cause severe diseases like dermatitis, skin allergy, vomiting and nausea, cancer in different visceral organs and circulatory systems. Among the two oxidation states of chromium, Cr(VI) is 1000 times more carcinogenic and corrosive than Cr(III). Cr(VI) can easily enter inside of living cells and coverts into reactive intermediates in presence of enzymes and other human antibodies that can reduce cell division rate and functional properties of cellular organs, degrades DNA configuration and essential biomolecules like proteins, lipids. Although Cr(III) can accelerate the metabolic rate during digestion (glycogenolysis) process, the excess Cr(III) uptake in human body can show adverse health syndromes.

Cadmium (Cd(II)) and its isotopes are released from dyes and pigments, alloy and metal finishing, electroplating, coatings in stabilizers, and battery industries. The Cd deposition in human body can cause noxious diseases like cancer, endocrine malfunction, and damages in peripheral nerves, bones, respiratory, gastrointestine and excretory systems due to metal intake by inhalation/ingestion.

Mercury (Hg(II)) appears as semiconductors in electronic equipment, in sphygmomanometer, thermometer, amalgams used for dentistry, barometer, paints and coatings and can damage renal cortex, liver, brain upon inhalation and it is listed as the third most toxic element on earth by the US Government Agency for Toxic Substances and Disease Registry.

Lead (Pb(II)) is mostly found in surface water and gets transmitted into water bodies from municipal water supply pipes, leaded gasoline production, pottery and boat making, mordant dye synthesis, calligraphy and in many more industrial processes. Toxicological reports showed that children are mostly get affected by the toxicity of lead than adults for their premature muscular and neural systems. Pb(II) ions at a very lower limit can cause physiological disorder, IQ deficiency and learning problems in child, and can raise anemia, blood pressure, cardiovascular diseases, brain hemorrhage, kidney failure, miscarriages during pregnancy, and even led towards death if the lead concentration is higher than its lethal dose.

Manganese (Mn(II)) exists in mining and volcanic residues and has widespread usage in producing electrical equipment, chemical reagents, metallurgical amalgams, and batteries. However, manganese predominately pollutes water by making foul odors, changes water chromaticity, and mostly affects human central neural and endocrine system in prolonged exposure.

Apart from heavy metals, some inorganic anions like nitrates (NO_3^-), and halides (viz. chlorate, bromate) are also distinguished as inorganic micro-pollutants. These anions mostly occur in municipal wastewater obtained from household surfactants and cleaning agents, but also noticeable in pharmaceutical wastewaters (Ahmed et al., 2017). These inorganic anions are slightly difficult to remove for their high solubility, forming chemically stable water solubilized products, and complete dissociation into the solution. Among these anions, nitrates are considered very harmful to babies, excess nitrate deposition affects the blood vessels and arteries that hampers total blood circulation system in infants and cause methemoglobinemia (blue baby syndrome) and many more severe sicknesses i.e. diabetes mellitus, expansion in thyroids, homeostasis (stiffness) of veins. Besides, halides are responsible for stunned growth, weakening of bones, fibromyalgia, and Alzheimer's disease.

1.3 Conventional biological treatment

1.3.1 Activated sludge process

The activated sludge process (ASP) is one of the most extensively used aerobic process typically primarily removal of solid and organic contaminants from wastewater. In this process a suspension of microorganisms such as bacterial, fungal or algal biomass that leads to degrade and decompose organic substances present in the wastewater (McCarty, 2018). Micro-organisms are grown-up in an aeration tank by consuming dissolved oxygen and organic materials from wastes and causing floc formation. Moreover, hydrophobic compounds get absorbed on the microbe's surface and form larger solid floc particles, which slowly settle at the bottom of the tank and clear liquid is obtained at the top. This treatment process is comparatively better than physical and chemical treatment processes such as adsorption, membrane separation, ion exchange, electrodialysis etc. in the point of cost, secondary pollutant generation, large scale treatment and applicability of the method in certain conditions. Now a day, ASP has been used in the developing country for sewage treatment plants, industrial wastes treatment and combined effluent treatment processes. However, this process possesses several drawbacks such as this process cannot accomplish the discharge limit standards, low quality of treated effluent is obtained, require comparatively larger area for setup, need a tertiary treatment system with ASP to remove sludge. Moreover, ASP requires intensive power supply due to active aeration of the wastewater, which is almost 50–65 percent of the total operational cost (Zhou et al., 2013).

In 2018 Maroua et al., studied a full scale textile wastewater treatment process with ASP to remove dye wastes which achieved 95 percent dye degradation at optimum operating condition of 5 days HRT and sludge recycling rates (SRR) of 0.22 (Haddad et al., 2018). Later, modified form of ASPs known as integrated fixed film activated sludge process (IFASP) has been introduced, in which microorganisms are both suspended and attached on the surface of the reactor whereas in case of conventional ASP microorganisms are in suspended state (Jabari et al., 2014). IFASP is effective in removing dissolved organics and offers a significant degree of nitrification-denitrification. Moreover, it offers several advantages such as enhanced nutrients removals, enhanced anthropogenic composites and longer solids retention time compared to conventional one.

1.3.2 Biological trickling filter

Trickling filter sometimes known as biological filter in which micro-organisms are attached or grown on supporting materials mostly on large rocks or plastic carrier that allows air ventilation and prevent clogging as well. In this type of treatment processes oxygen may be supplied by natural ventilation or by forced aeration, to control aerobic condition. In biological trickling filter (BTF), influent wastewater is entered in to the reactor from the top through a proper distribution system and flows to the downward direction of the biofilm. Where microorganisms degrade the organic substances and finally treated water is disposed from the bottom of the reactor. In some cases, recirculation of the partially treated wastewater may be carried out to complete removal of micro-pollutants. However, TFs processes are not effective for the removal of nitrogen because during degradation of organic substances it produces CO_2 and leaves limited residual carbon that act as electron donor for the conversion of nitrate/nitrite to nitrogen gas (denitrifcation). For that reason, TFs processes are designed in those treatment plants where COD is main focused (high COD and BOD values of industrial effluent treatment plant) as a pre-treatment step. In 2017 Liao et al. scrutinized a stand-alone down-flow hanging sponge (DHS) system with a two-stage configuration to treat wastewater from soft the drink industry for 700 days with COD of 3000 mg/L (Liao et al., 2017). This study reported a COD removal efficiency of over 98 percent at an organic loading rate (OLR) of 4.9 Kg-COD/m^3-sponge/day and hydraulic retention time (HRT) of 15.3 h Kornaros and Lyberatos in 2006, established a TF for the treatment of organic dyes and varnishes wastewater with an initial COD of 40,000 mg/L (Kornaros and Lyberatos, 2006). The system showed over 70 percent COD removal efficiency at an OLR of 1.1 m^3/m^2 day.

The TF reactor can be used as a post treatment step for the removal of nutrients where discharged effluents characteristics are strictly regulated. For example, in 2020 Hunter and Deshusses used trickling nitrification filters and submerged denitrifcation filters for the post-treatment of fecal waste digestate with an influent COD and nitrogen levels of about 4500 and 3000 mg/L (Hunter and Deshusses, 2020). This study achieved 84 percent, 69 percent, and 89 percent removal of COD, nitrogen and phosphorus respectively. In another study, Forbis and Stokes used laboratory-scale TF for the removal of nitrogenous substances from swine waste anaerobic digester with in influent COD and total ammonium nitrogen (TAN) (Forbis-Stokes et al., 2018). The process achieved over 79 percent COD and 90 percent TAN removal due to the activity of nitrifying microorganisms. Moreover, the nitrate generated in the nitrification step was further treated in a subsequent submerged filter.

1.3.3 Biological passive aeration nitrification/denitrification reactor

Nitrification/denitrification is another biological process reported for the removal of micro-pollutants from the wastewater with proper nitrifying and denitrifying conditions. In this biological treatment process, nitrogen has been removed from the wastewater by two steps: aerobic nitrification and anaerobic denitrification process. In the first step (nitrification) oxidation of ammonia/ammonium takes place and converted into nitrite and nitrate in presence of nitrifying micro-organism, while in the second step (denitrification) reduction of nitrite and nitrate takes place to form nitrogen gas which is discharged in to the environment or may be used for industrial purpose with or without any treatment (Kanaujiya et al., 2019). These two reaction conditions and operating conditions are completely different for that reason, conventional activated sludge reactor require separate reactor for the removal of nitrogen. The nitrification step is the most temperature sensitive step in biological treatment process. Several studies showed that the oxidation of ammonia/ammonium to nitrite/nitrate in presence of nitrifying biomass may be slowed down due to temperature decrease (Grandclément et al., 2017). Along with temperature other operating conditions such as sludge retention time, hydraulic retention time, and effluent pHmay influence overall treatment efficiency. The main drawbacks of nitrification/denitrification process are it requires a pretreatment step to remove any solids present in the wastewater, large area requirement in case of ASP based nitrogen removal process (which is almost 50–60 percent of the total operating cost of ASP based wastewater treatment plant (Zhou et al., 2013)) and cannot able to remove phosphorus (Abdelfattah et al., 2020). The large area problem has been almost solved by using biofilm based bioreactor where both oxidation and reduction reaction zone can be created in the same reactor (Abdelfattah et al., 2020).

In 2015, Flavigny and his coauthors developed glycogen accumulating organism (GAP) based aerated biofilm for the removal of poly-hydroxy-alkanoate (as a source of acetate an organic carbon) from the wastewater (Flavigny and Cord-Ruwisch, 2015).

This study showed this biofilm can remove over 95 percent organic carbon based waste but not effectively eliminate nitrogen and phosphorous based organic substances. But the main advantage of this method it can treat wastewater at a rate of 128 g-BOD/m^3/h with 60 percent energy saving compared to ASP method. Later another study showed that GAO dominated aerated biofilm based can minimize excess sludge production rate 0.05 g VSS g^{-1} BOD (more than 9 times lower than other wastewater treatment processes) (Hossain et al., 2018). However, the drawbacks of biofilm based treatment processes were not effective for the reduction of nitrogenous organic substances that limits the application of these treatment processes in real scenarios.

In 2017 Ahmed critically reviewed EDCs treatment by denitrification technique (Ahmed et al., 2017). In his study, it was found that denitrification technique can be best way for the removal of several EDCs such as estrone (E1), 17-ethinylestradiol (EE2), 17β-estradiol (E2), estriol (E3), bisphenol A, 4-tertbutylphenol and 4-tert-octylphenol and PCPs (such as benzophenone, galaxolide, oxybenze, salicyclic acid and tonalide) with in the range of 82–100 percent. Denitrification technique can also effective for the removal of pharmaceutical waste such as ibuprofen, metronidazole and ketofenac within the range of 82–97 percent but less effective for the removal of carbamazepine, diclofenac, clofibric acid, gembrozil, erythromycin and roxythromycin (Ahmed et al., 2017).

1.4 Advanced biological treatment

1.4.1 Two-phase partitioning bioreactor

Two phase partitioning bioreactors (TPPB) are composed of two separate phases, one containing immiscible, non-biodegradable and biocompatible organic phase containing organic pollutant and second is aqueous phase containing micro-organism. Although this system is so called biphasic culture, actually there are at least four phases present in the system: hydrophobic, aqueous, gas and micro-organism. The main purpose of organic phase is to remove the end-products or residue, to control the aqueous substance (low polar substance) and enhance the biotransformation of water-insoluble substrates by increasing its bioavailability to biocatalysts. In TPPB system, high speed agitation creates small droplet of organic pollutants into the aqueous phase that may contact and consumed by micro-organism for their livelihood. However, this treatment process possess some disadvantages such as this system require high agitation speed for higher droplet production and connection between micro-organism and organic pollutant, which is highly consuming at the same time may lead to damage of micro-organism cells. TPPB demonstrates a promising prospect system for the removal of highly hydrophobic natured micro-pollutants, heavy metals which are poisonous to micro-organism. Several studies reveals that TPPBs can be successfully be used for the removal of micro-pollutants such as polycyclic aromatic hydrocarbons (PAH) (e.g. pyrene, styrene, volatile organic compound (VOC) such as (toluene)) and phenolic compounds from wastewater (Baskaran et al., 2020). Recently, Baskaran and his team evaluated the performance of a TPPB reactor system to remove trichloroethylene by *Rhodococcus opacus* for microbial treatment (Baskaran et al., 2020). During the operation, 100 percent TCE removal was obtained for when TCE concentration was within the range of 1.12–2.62 g/m^3. Tomei et al. (2018) employed a continuous TPPB system that can degrade more than 96 percent 2,4-dimethylphenol (initial concentration of ~1200 mg/L) (Tomei et al., 2018). In 2017, Domenica and his co-author use TPPB system for the removal of synthetic tannery wastewater containing 4-chlorophenol (concentration in the range of 1000–2500 mg/L) and potassium dichromate (100 mg/L as Cr (VI)) (Angelucci et al., 2017). They achieved more than 98 percent reduction of organic load with almost 100 percent removal of Cr(VI). In 2019, the same author established another TPPB technology for the treatment of pentachlorophenol (PCP) the most toxic micro-pollutant among phenolic group compounds (Angelucci et al., 2019). This study demonstrated almost complete biodegradation was achieved if the rate of effluent intake/release is between 4.0–7.8 mg/Lh with PCP concentrations up to 100 mg/L and biomass concentration of 1 gVSS/L.

1.4.2 Membrane based reactor

Membrane biofilm reactor (MBR) system is a combination of two separate units, membrane unit responsible for physical separation (gas permeable and separation) and delivers air or oxygen that requires for the oxidation purpose and biological reactor unit where degradation of micro-pollutant arises. These systems work generally in two main arrangements: a) external or side stream and b) submerged/immersed. In external or side stream system the membrane is kept outside of the bioreactor and requires an additional pump for transporting feed from the biofilm reactor to the membrane. On the other hand, membrane unit is inserted inside the bioreactor that allows the feed to flow through between two units without any external or internal pump. In MBR system almost 100 percent oxygen transfer efficiency can be achieved due to use of membrane unit, which is more than 2.5 times higher than conventional activated sludge process. Another advantage of this treatment process is it can physically separate larger molecular weight micro-pollutants that are than membrane pore size. However, the overall efficiency of this reactor mostly depends on sludge retention time (SRT), hydraulic retention time (HRT) composition and types of effluents, organic loading rate, operating condition (e.g. temperature, pH, turbidity and conductivity), concentration, and existence of anoxic and anaerobic compartments.

For the last few decades, MBR technology has been used for the treatment of municipal, domestic and industrial wastewater due to its versatile use (removal of numerous amounts of micro-pollutants) with higher efficiency. Recent study showed that removal efficiency by MBRs and showed better removal efficiency for several pollutants such as pharmaceuticals wastes, fragrances, endocrine disrupting compounds such as ranitidine, gemfibrozil, diclofenac, ketoprofen, ofloxacin, ibuprofen, pravastatin, naproxen, bezafibrate, acetaminophen, paroxetine, and hydrochlorothiazide (Grandclément et al., 2017). In 2016 Prasertkulsak develop a MBR system for the removal of pharmaceutical compounds from hospital wastewater operated with short HRT of 4 h. Within short 4 h HRT this treatment achieved 100 degradation of ibuprofen (IBP), 17b-Estradiol (E2), triclosan (TCS) (Prasertkulsak et al., 2016). However, fouling on the membrane surface is one of the most common and important disadvantages of using MBRs system which is mainly caused by the deposition of organic and inorganic substances during operation. This problem can be almost resolved by changing the operational parameter (aeration which is costly almost 70 percent of the total operating cost), using adaption membrane cleaning method and pre-treating the micro-organism to limit its fouling properties (e.g. addition of adsorbent).

1.4.3 Moving bed biofilm reactor

Compare to the trickling filter to, moving bed biofilm reactor (MBBR) microorganisms grow both suspended and attached growth processes which are supported on plastic carriers. The movement of the support material within the biofilm reactor is achieved either by the agitation (aeration) or by mechanical means (anoxic/anaerobic process). The main advantages of using MBBR over other biofilm reactors is that the whole volume of the reactor can be used for microorganisms growth and the carriers includes existence of both aerobic outer layer and anoxic/anaerobic inner layers that reduces the land area requirement. Moreover, MBBR treatment plants offer more pollutant removal efficiency with reduced energy consumption. Irankhah et al. set up a batch moving bed biofilm reactor for the removal of phenolic compounds with several activated sludge microorganisms and phenol-degrading bacteria (Irankhah et al., 2018). This study showed under optimum operating condition, more than 90 percent COD (initial load of 2795 mg/L) was reduced at 24 hydraulic retention time (HRT). In 2019, Chen et al. integrated an anaerobic MBBR and a microbial fuel cell (MFC) to produce bioelectricity and to treat wastewater from pulp/paper industry (Chen et al., 2020). Over 66 percent of COD reduction from wastewater was observed with an influent of 6300–6500 mg/L at 72 HRT with a maximum power density of 94.5 mW/m^2. Schneider et al. investigate a hybrid system combination of MBBR, ozone reactor and a column of biological activated carbon for the treatment of oil refinery wastewater (Schneider et al., 2011).

1.4.4 Cell-immobilized bioreactor

Cell-Immobilization bioreactors are sometimes known as semi-fluidized bed bioreactors in where microbial cell mobility is constrained in a particular area of the reactor. Cell immobilization system can be divided into two categories, a) active immobilization system in where micro-organisms are entrapped and/or encapsulated in a solid supportive material or carrier by physical and or chemical bonding, and b) passive immobilization system in where micro-organisms grows and forms multilayer in the supportive material or carrier and form a biological film without any bond formation. Passive immobilization sometimes known as free cell (FC) systems are most commonly used method in the biofilm based reactor, moving bed bioreactor, trickling filter system which are discussed in the earlier sections. In case of active cell immobilization (ACI) system selection of carrier/supporting materials types, porosity and concentration and micro-organism are most critical and reactor efficiency mostly depends on that parameter. Porous alginate, agar, polyacrylamide, polyvinyl alcohol, chitosan, collagen and gelatin are the most widely used carriers along with some additives to enhance the cell performance. In technical and economical point of view ACI treatment offers enormous advantages over FC systems such as this system can be applicable in a wide range of operating conditions (pH, temperature, and in the presence of highly toxic substances while FC system should be optimized (Kanaujiya et al., 2019). Moreover, micro-organism that used during ACI processes can be reuse, prevent to cell washout and protect against shear damage that leads to reduce operating cost as well. After all of these advantages, ACI have some limitations such as diffusion rate in substrate can be reduced over time, lower viability, durability due to little evidence in literature (practical application for the wastewater treatment). More investigation should be carried out to improve mass transfer as well as to explain the observed effects associated with substrate diffusion into the beads and cell release from the beads.

Sarma et al. in 2010 investigated a low aqueous phased immobilized cell bioreactor system for the removal of pyrene a polycyclic aromatic hydrocarbon (PAH at initial concentration of 1000 mg/L) (Sarma and Pakshirajan, 2011). The biodegradation batch profile revealed complete degradation of PAH within 10 days without any lag phase formation. While another study by Mahanty et al. found similar results for the reduction of PAH within 200 h by a FC system but micro-pollutant concentration level was 50 mg/L (Mahanty et al., 2008). Mahanty in his earlier investigation reported that over 90 percent pyrene reduction in 6 days was achieved for an initial concentration of 50 mg/L. (Mahanty et al., 2007). While Pagnout reported almost complete pyrene (Initial concentration of 202 mg/L) biodegradation within 4 days by an alkaliphilic Mycobacterium sp. strain but the drawback of this ACI system was formation of lag phase (Pagnout et al., 2006).

1.4.5 Hybrid methods

Most of the conventional and advanced biological treatment processes possess several disadvantages such as lower removal of micro-pollutants, higher operating and maintenance cost, higher SRT shown in the Tables 1.1 & 1.2. Those problems can be mostly solved by the implementation of hybrid treatment processes. In recent years varieties of biological hybrid treatment processes are reported and their application achieved

Table 1.1 Micro-pollutant degradation efficiency from wastewater (real/synthetic) by conventional biological treatment processes and their advantage, limitations.

Treatment process	Advantages	Disadvantages	Wastewater types	Operating conditions	Micro-pollutants (mg/L)	Micro-organism	Removal efficiency (%percent)	References
Activated sludge process	Co-metabolism by consortia	Cannot accomplish discharge limit ECs cannot effectively remove where COD levels are high.	Pharmaceutical	Effective volume: 4 LC:N:P of 100:5:1 pH: 6.8–7.5 HRT: 8h Temp: 25±1 °C	COD: 240–410 NH_4^+ –N: 30–50	NA	COD: 88.85 NH_4^+ -N: 69.0	(Wang et al., 2020)
			Textile wastewater	HRT: 5 d Reactor volume: 2.5 L Dye/Biomass ratio of 0.72 g CODS/g VSS pH: 7.5–8.5	Indigo dye (304.5 kg/day)	NA	89	(Haddad et al., 2018)
Biological trickling filter	Require a complex water and air distribution system. Lower maintenance cost.	Sludge may clog trickling filter. Long setup time. Require backwash after certain operating periods.	Dye manufacturing wastewater	pH: 5.5–8.0 Hydraulic loading: 0.6 m^3/m^2 d	COD (40,000)	NA	80–85	(Kornaros and Lyberatos, 2006)
			High strength human waste	Urine, feces and water: 6.8:2.7:12.8 HRT: 25 d Temp: 20±2 °C Loading rate: 1.73 kgCOD/m^3/day & 0.80 kgN/m^3/d	COD: 4500 mgCOD/L TN: 3000 mgN/L PO_4-P: 1370 mgp/L	NA	COD: 84 TN: 69 PO_4-P: 89	(Hunter and Deshusses, 2020)
Simultaneous Nitrification/ denitrification	Small plant footprint. Lower aeration energy and lower operating time.	Should be including additional pretreatment step to remove solid particles. Not able to remove phosphorus for the wastewater.	Sewage with low carbon/ nitrogen	C/N: 3.4 Operation Time: 266 d Volume: 11 L HRT: 14–18 h Temp: 24±2 °C	NH_4^+-N Total inorganic nitrogen Phosphorus	Denitrifying glycogen-accumulating organisms (N removal), Denitrifying polyphosphate accumulating organisms (P Removal)	NH_4^+-N: 96.3 Total inorganic nitrogen: 81.4 Phosphorus: 91.0	(Yuan et al., 2020)

Table 1.2 Micro-pollutant degradation efficiency from wastewater (real/synthetic) by advanced biological treatment processes and their advantage, limitations.

Treatment process	Advantages	Disadvantages	Wastewater types	Operating conditions	Micro-pollutants (mg/L)	Micro-organism	Removal efficiency (%percent)	References
Membrane-based bioreactor (MBR)	High and complete separation of micro-pollutants. Small foot print. No space required for aeration tanks and sludge storages	Membrane fouling problem. Higher aeration cost. Resistance in mass transfer due to thick biofilm.	Hospital wastewater	FR: 500 L/h Operated under 2 phase, first phase: 0–42 d and second phase: 43–76 d HRT: 4h SRT: first phase: 27.7 d and second phase: 27.0 d WF: 107 L/d Temp: 18±2 °C	IBP (0.0365) E2 (0.128) TCS (0.04) GFZ (0.051) NPX (0.052) TMD (0.102) TMP (0.032) SMX (0.028)	Microbial consortia	IBP: 100 E2: 100 TCS: 100 GFZ: 45.8 NPX: 82.3 TMD: 14.4 TMP: 80.1 SMX: 78.5	(Prasertkulsak et al., 2016)
			Pharmaceutical wastewater	Volume of 8.5 L SRT: 30 d HRT: 24 h pH: 7.3 ± 0.2 Temp: 25±1 °C Pressure: 200 mbar	Etodolac (511mg/L)	NA	66.8	(Kaya et al., 2013)
Two phase partitioning bioreactor	Higher molecular wt. hydrophobic micro-pollutant removal. Higher oxygen transfer efficiency.	Require higher agitation speed	Synthetic tannery wastewater	Organic load: 19–94 mg/h HRT: 3–6 h pH: 7.5 Temp: 27±0.5 °C Volume: 4 L	4-chlorophenol (1000–2500)	NA	89–95	(Angelucci et al., 2017)
			Hypersaline wastewater	pH: 7.5 Temp: 28 °C Reaction Volume: 2.0 L Wastewater flow rate: 0.001–0.24 L/h	2,4-dimethylphenol (1200)	NA	> 96	(Tomei et al., 2018)

(continued)

Table 1.2 (Cont'd)

Treatment process	Advantages	Disadvantages	Wastewater types	Operating conditions	Micro-pollutants (mg/L)	Micro-organism	Removal efficiency (%percent)	References
Immobilized cell bioreactor	Effective removal of highly toxic substances. Reuse of biomass.	Scale formation over time.	NA	Contact time: 36 h Carrier: Bagasse pH: 7.0 Temp: 25 °C	Tetradecane (400)	*Acinetobacter venetianus*	93.3	(Lin et al., 2015)
			NA	Contact time: 24 h Carrier: Sugarcane bagasse pH: 6.0 Temp: 30 °C Agitation speed: 120 rpm	Phenol (2400)	*Candida tropicalis* PHB5	97.67	(Basak et al., 2014)
			NA	Contact time: 24 h Carrier: alginate bead pH: 6.0 Temp: 30 °C Agitation speed: 90 rpm	Phenol (2400)	*Candida tropicalis* PHB5	99.71	(Basak et al., 2014)
			NA	Contact time: 240 h pH: 7.0 Temp: 28 °C Agitation speed: 150 rpm	Pyrene (1000)	*Mycobacterium frederiksbergense*	>90	(Sarma and Pakshirajan, 2011)

Treatment process	Advantages	Disadvantages	Wastewater types	Operating conditions	Micro-pollutants (mg/L)	Micro-organism	Removal efficiency (%percent)	References
Moving bed biofilm Reactor (MBBR)	Compact, simple and robust process Reduced sludge production and no need to sludge treatment. Lower foot print.	Long setup period. Higher installation and operating cost	Pesticide industrial wastewater	Biofilm carrier: Kaldnes K1 media C/F: 1:1 HRT: 6 h pH: 6.7 & 6.8 Temp: 26±4 °C	NH_4—N: 17–42	NA	95	(Pinto et al., 2018)
			Pharmaceutical wastewater	Biofilm carrier: AnoxKaldnes™ C/F: 2:3 HRT: 10 h pH: neutral Temp: 37 °C	Diclofenac (10) Ibuprofen(10)	NA	Diclofenac: 61 Ibuprofen: 37	(Fatehifar and Borghei, 2018)
			Petroleum refinery wastewater	Biofilm carrier: Ceramsite C/F: 1:1 3 stage (Anaerobic nad aerobic unit) pH: 7.3 HRT: 72 h	NH_4—N (31–35)	Photobacterium phosphoreum	80–85	(Lu et al., 2013)

huge degradation/removal capacity comparatively conventional and advanced biological treatment processes which may be due to synergistic effects. Biological hybrid methods can be categorized into three sections: (a) two similar or different biological treatment processes, (b) combination of biological and chemical treatment processes and (c) Combination of biological and physiological treatment process. Table 1.3 summarizes several hybrid techniques, reported in previous literature that has been used for the treatment of micro-pollutants from wastewater. In case of combination of biological and chemical treatment process biological treatment process is followed by chemical treatment process mostly oxidation process. Now today, advanced oxidation processes (AOP) such as Fenton and photo-Fenton processes, electro-Fenton processes, ozonation (catalytic), wet peroxide/air oxidation (catalytic), heterogeneous photocatalysis, electrochemical oxidation, or a combination of these AOPs are the most widely used oxidizing agent. Reports from the literature suggested that pretreatment by AOP before biological processes or hybrid system combination between AOP and biological processes can enhance its biodegradability and efficiency. Moreover, these systems are comparatively more sustainable, environmental friendly. However, these hybrid processes have some limitations such as higher operating cost, higher energy consumption, generation of toxic secondary by-product and prone to damage the reactor system due to contact between reactor surface and radical scavenging compounds. On the other side in case of combination of biological and physical hybrid processes, biological systems are followed by ultrafiltration (UF) or microfiltration (MF) or reverse osmosis (RO) or nanofiltration (NF).

The reduction of micro-pollutants such as EDCs, pesticides, pharmaceuticals, beta blockers by biological hybrids has been studied. It was found from the literature that more than 90 percent of EDCs based micro-pollutants such as E1, E2, EE2, E3, 17 β-estradiol 17-acetate, bisphenol A, 4-n-nonylphenol and 4-tert-butylphenol can be removed by a hybrid system use of combination of biological treatment along with some physical treatment techniques such as reverse osmosis, ultrafiltration or nanofiltration. Nguyen et al. in 2013 demonestrated single MBR, UV oxidation, nanofiltration, reverse osmosis and hybrid biological treatment process combination of MBR with UV oxidation or high pressure membrane filtration processes such as nanofiltration (NF) or reverse osmosis (RO) for the removal more than 22 micro-pollutants (trace organic contaminants (TrOC)) (Nguyen et al., 2013). This study showed that hybrid systems possess comparatively higher removal efficiency over single treatment process (single UV, NF, RO), of them combination of MBR and RO results best performance (more than 96 percent removal capacity of total TrOC). It is noteworthy that the higher efficiency of MBR-RO treatment process is due to absorption of hydrophobic natured TrOC on the surface of membrane polymeric matrix and separation of charged and hydrophilic natured TrOC by RO results higher removal capacity. Later in 2019 Moser et al., developed a hybrid UF-Osmotic MBR and compared it with conventional

Table 1.3 The removal efficiency of micro-pollutants achieved by biological hybrid treatment processes and their operating condition.

Method	Wastewater types	Operating conditions	Micro-pollutants	Removal efficiency (%percent)	References
Two biological treatment processes					
MBBR+MBR	Synthetic wastewater	HRT: (MBBR: 24 h; MBR: 6 h) Flow rate: 27.8 ml/min pH: 7	(Ketoprofen, Carbamazepine, Primidone, Bisphenol A and Estriol) Each of them 5 μg/L	25.5–99.5	(Luo et al., 2015)
ASP+Biofilm reactor	Pharmaceutical wastewater	Maximum flow rate: 0.2 ml/min Temp: 20 °C	Propranolol (0.055 μg/L) Diclofenac (0.24 μg/L) Propiconazole (0.11 μg/L) Tebuconazole (0.022 μg/L) Iohexol (3.28 μg/L) Iomeprol (20.8 μg/L) Iopromide (2.9 μg/L)	Propranolol (98 percent) Diclofenac (82 percent) Propiconazole (21 percent) Tebuconazole (59 percent) Iohexol (91 percent) Iomeprol (93 percent) Iopromide (91 percent)	(Casas and Bester, 2015)
Anaerobic SBR+Aerobic SBR	Textile wastewater	HRT: 48 h for anaerobic and 6 h for aerobic SBR SRT: Infinite for anaerobic and 20 d for aerobic SBR MLSS/MLVSS Ratio: 0.6–0.7 pH: 6.8–7.2 Temp: 30–35 °C	Dye (Disperse, Reactive, Pigment, Vat)	78.4 percent	(Shoukat et al., 2019)
Combination of biological and chemical treatment processes					
MBR+UV	Synthetic wastewater	HRT: 24 h Active volume: 4.5 L Temp: 22±0.1 °C pH: 7.2–7.5	(17β-Estradiol-17-acetate, Bisphenol A, 4-n-Nonylphenol, 4-tert-Butylphenol, Fenoprop, Pentachlorophenol) Each of them 5 μg/L	>85 percent each of the component	(Nguyen et al., 2013)

(continued)

Table 1.3 (Cont'd)

Method	Wastewater types	Operating conditions	Micro-pollutants	Removal efficiency (%percent)	References
ACF+ Ozonation	WWTP	NA	Atrazine (0.001 μg/L) 2,4-D (0.102 μg/L) Diazinon (0.778 μg/L) Diuron (0.046 μg/L) Metolachlor (0.002 μg/L) Praziquantel (0.003 μg/L) Triclopyr (0.115 μg/L) Atenolol (0.598 μg/L) Metoprolol (0.919 μg/L) Propranolol (0.061 μg/L)	Atrazine (70) 2,4-D (92.9) Diazinon (99.3) Diuron (99.1) Metolachlor (69.0) Praziquantel (97.1) Triclopyr (87.4) Atenolol (99.8) Metoprolol (99.97) Propranolol (99.7)	(Reungoat et al., 2012)
MBR+AOP (Solar Fenton)	Sewage	MBR treatment capacity: 10 m^3/day MBRSRT: 30 d MBRHRT: 9 h MBR pH: 2.8 AOP reactor capacity: 21.4 L AOP Flow rate: 150 L/h	Sulfamethoxazole (0.540 μg/L) Erythromycin (92 ng/L) Clarithromycin (43 ng/L)	Sulfamethoxazole (100) Erythromycin (100) Clarithromycin (84)	(Karaolia et al., 2017)
IBR+AOP (Solar Fenton)	Sewage	IBR treatment capacity: 20 L AOP reactor capacity: 44.6 L AOP Temp: 35 °C	Diuron (1080–9280 ng/L) 4-FAA (4620–5830 ng/L) Gemfibrozil (2020 ng/L) 4-AAA (132–6730 ng/L) Bisphenol A (820–910 ng/L) Ibuprofen (726–11,280 ng/L) Ciprofloxacin (450–1170 ng/L) Ofloxacin (1610–7500 ng/L) Paraxanthine (874–17,750 ng/L) Nicotine (910–1000 ng/L) Caffeine (3840–15,460 ng/L	Overall 95 percent	(Prieto-Rodríguez et al., 2017)

Method	Wastewater types	Operating conditions	Micro-pollutants	Removal efficiency (%percent)	References
Combination of biological and physiological treatment process					
MBR+NF	Synthetic wastewater	HRT: 24 h Active volume: 4.5 L Temp: 22±0.1 °C pH: 7.2–7.5	(17β-Estradiol-17-acetate, Bisphenol A, 4-n-Nonylphenol, 4-tert-Butylphenol, Fenoprop, Pentachlorophenol) Each of them 5 μg/L	>90 percent each of the component	(Nguyen et al., 2013)
MBR+RO	Synthetic wastewater	HRT: 24 h Active volume: 4.5 L Temp: 22±0.1 °C pH: 7.2–7.5	(17β-Estradiol-17-acetate, Bisphenol A, 4-n-Nonylphenol, 4-tert-Butylphenol, Fenoprop, Pentachlorophenol) Each of them 5 μg/L	>95 percent each of the component	(Nguyen et al., 2013)

NF: Nano-filtration; RO: Reverse osmosis; BACF: Biological activated carbon filtration; AOP: Advanced oxidation process; IBR: Immobilized biomass reactor; SBR: Sequential Bioreactor; ASP: Activated sludge process; OMBR: Osmotic membrane bioreactor; UF: Ultrafiltration.

MBR for the removal of dissolved organic carbon (DOC) from oil refinery wastewater (Moser et al., 2019). More than 99.6 percent of DOC was reduced by using hybrid UF-Osmotic MBR system which was more than 0.51 times higher than the conventional MBR treatment process.

In 2015 Casas and Bester reported a hybrid method combination of activated sludge process and suspended biomass and biofilm-type bioreactor for the removal of seven micro-pollutants (propranolol, diclofenac, propiconazole, tebuconazole, iohexol, iomeprol and iopromide) from pharmaceutical wastewater (Casas and Bester, 2015). This hybrid system possesses several advantages such as lower or almost no sign of clogging, easy to control in terms of hydraulics and sorption of micro-pollutants and can remove most of the micro-pollutant over 88 percent. But the major challenge of this treatment process was slow sand filtration with low-operating loading rates. Later, Luo et al. reported a MBBR and MBR hybrid system and compare with conventional MBR system for the removal of micro-pollutants (Luo et al., 2015). The result showed that the MBBR-MBR system could effectively reduce most of the selected micro-pollutants. Moreover, membrane fouling problem can be reduced because some micro-pollutant and altered extracellular polymeric substances were subtract on MBBR system that leads to increase shelf-life of overall system. On the other side degradation products of micro-pollutants generated in MBBR were effectively removed along in the MBR system.

In summary, biological hybrid treatment processes can be highly effective for the removal of a wide range of micro-pollutants (EDCs, antibiotics and other pharmaceuticals etc.) from wastewater (raw and synthetic). The most common advantage of all hybrid systems is higher removal efficiency, reduction of sludge generation, easily be modified with other conventional and advanced treatment process and in case of MBR based hybrid system fouling and clogging problem comparatively conventional biological treatment processes. But hybrid systems may involve higher operating and maintenance cost, and need to optimize operating condition such as HRT, SRT, pollutant load, temperature, pH. Furthermore, combination of biological treatment with nanomaterial based materials or sequential biological treatment processes can be alternatives and needs more attention.

1.5 Pros and cons of biological treatment process on MPs removal over conventional ones

The remediation of bioactive chemicals from water bodies is one of the vital challenges for researchers in this new millennium. Research reports described the presence of versatile chemical and biological contaminants in water and wastewaters that requires a constant effort to set up a suitable water treatment strategy. Biological wastewater treatment methods has always been a point of interest for many industrial and domestic wastewater treatment since the methods are efficient for their high removal efficiency for remediation of persistent organic pollutants and to eliminate suspended solids of larger quantity (Table 1.2). The research of Contreras et al. in 2019 showed that the pilot treatment plant of horizontal subsurface flow (HSSF) in constructed wetlands by activated sludge method can remove 80 percent of the MPs like naproxen, ibuprofen, diclofenac, triclosan, bisphenol-A, 2-hydroxyl ibuprofen, etc. which was 10–95 percent higher than the removal efficiency of other physicochemical treatments of these chemicals (Reyes Contreras et al., 2019). Technologies like chemical precipitation, coagulation/flocculation, and chemical oxidation are simple processes to operate but possess noticeable demerits including excess chemical consumption, maintenance cost, and sludge/residual disposal problem (Crini and Lichtfouse, 2019). Although BWWTTs have numerous benefits on large-scale MPs containing wastewater treatment, they have some disadvantages. Both aerobic and anaerobic treatments can become inactive at a regressed condition, and generate huge amount of inactive and active microorganisms with sludge that can cause detrimental health hazards and may raise the treatment cost higher for extra sludge treatment. Therefore, sludge control, important treatment tools, zero biomass production is required to overcome these dilemmas.

1.6 Future aspects of BWWTPs of large-scale MPs remediation

In present days, industrial authorities do not show particular interest to provide extra financial expenditure on remediation processes for the global economic crisis, lack of skilled manpower, high labor cost, sophisticated machine maintenance, and other terrestrial factors. This reluctance gradually affects the environmental matrices and brings about detrimental impacts on living biota. Though the overall treatment cost of these harmful industrial wastewaters is high than the capital production cost in some cases, the development of efficient, advanced, and low-cost treatment strategies and legalized policies can make this extra investment profitable for environmental conservation and industrial sustainability.

The articles reviewed in the previous sections of this chapter aim to highlight the bioremediation of hazardous micropollutants to reduce their negative effects. The previous researches have driven generous information on sources of MPs, their probable accumulation gateways to living cells, and possible remediation by conventional and advanced treatment methods. However, studies have shown that the complete remediation of MPs by biological treatment processes is not possible at all and requires profound knowledge to upgrade these processes. (Kanaujiya et al., 2019) suggested that the improvement of conventional BBWWTPss like activated sludge process/trickling filters integrating with advanced bioreactors can increase the treatment efficiency of micropollutant enriched wastewaters. In 2016, Singhal and Perez-Garcia explained that MPs are required to be cometabolized with intermediates (to become concentrated and bioactive) for better removal efficiency and proposed to use enzyme stimulated techniques to degrade them under a redox environment (Singhal and Perez-Garcia, 2016). They also opined that the degradation of organic MPs can be accelerated in this technique by using enthused enzymes such as laccase, oxidoreductases in presence of nondegradable substrates that help to retain the enzyme production constant. Therefore, sufficient research is required to foucs on highlighting the potential of BWWTTs for bulk implementation.

1.7 Conclusion

The sustainability of water resources has become a vital challenge because of the growing population, freshwater scarcity, climate changes, industrial activities, and so on. Emerging contaminants like micro-pollutants released with untreated municipal and industrial wastewaters along with other effluents discharged from WWTPs have detrimental influences on environmental matrices and living beings throughout their life cycle. Occurrences of MPs in industrial wastewaters are even measured after treatment and occasionally the threshold limit of some chemicals is found quite higher and their ecotoxicological threats, characteristics, possible health risks, and disposals are also missing in previous literature. Therefore, proper policy making, arise of public awareness, standard guidelines and remediation methods with precautionary approaches, set up of threshold limits, proper monitoring and maintenance are required to mitigate MPs in bulk level water treatment.

Present wastewater treatment technologies viz. adsorption, ozonization, advanced oxidation process (AOP), use of nanomaterials may have shown satisfactory removal efficiency towards MPs despite having several demerits such as high treatment and maintenance expenditure, generation of secondary pollutants, poor remediation of bulk quantity pollutants. As a result, developing a compatible, efficient and low-cost treatment process for MPs mitigation is getting more priority. Studies have shown that BWWTTs can reduce the concentration on some specific pollutants up to 12.5–100 percent despite these techniques are not beneficial in the case of reducing the polar persistent micro-pollutants. However, the optimization of treatment procedures and functional parameters (pH, pK_a, SRT), choice of proper microorganisms, application of modern BBWTTs (MBR, CIBR, enzymatic approaches) can overcome these barriers up to certain extents and be proved as cost-effective too. Taking account of the above discussion it can be concluded that the integration of the conventional and advanced BWWTTs along with physicochemical methods can become a prominent approach for decomposing micro-pollutants to increase the treatment efficiency and facilitate a prospective bioremediation tactic on large scale MPs containing wastewater treatment.

References

Abdelfattah, A., Hossain, M.I., Cheng, L., 2020. High-strength wastewater treatment using microbial biofilm reactor: a critical review. World. J. Microbiol. Biotechnol. 36, 1–10.

Ahmed, M.B., Zhou, J.L., Ngo, H.H., Guo, W., Thomaidis, N.S., Xu, J., 2017. Progress in the biological and chemical treatment technologies for emerging contaminant removal from wastewater: a critical review. J. Hazard. Mater. 323, 274–298.

Angelucci, D.M., Piscitelli, D., Tomei, M.C., 2019. Pentachlorophenol biodegradation in two-phase bioreactors operated with absorptive polymers: box-Behnken experimental design and optimization by response surface methodology. Process Saf. Environ. Prot. 131, 105–115.

Angelucci, D.M., Stazi, V., Daugulis, A.J., Tomei, M.C., 2017. Treatment of synthetic tannery wastewater in a continuous two-phase partitioning bioreactor: biodegradation of the organic fraction and chromium separation. J. Cleaner Prod. 152, 321–329.

Basak, B., Bhunia, B., Dey, A., 2014. Studies on the potential use of sugarcane bagasse as carrier matrix for immobilization of Candida tropicalis PHB5 for phenol biodegradation. Int. Biodeterior. Biodegrad. 93, 107–117.

Baskaran, D., Sinharoy, A., Paul, T., Pakshirajan, K., Rajamanickam, R., 2020. Performance evaluation and neural network modeling of trichloroethylene removal using a continuously operated two-phase partitioning bioreactor. Environ. Technol. Innov. 17, 100568.

Casas, M.E., Bester, K., 2015. Can those organic micro-pollutants that are recalcitrant in activated sludge treatment be removed from wastewater by biofilm reactors (slow sand filters)? Sci. Total Environ. 506, 315–322.

Chavoshani, A., Hashemi, M., Amin, M.M., Ameta, S.C., 2020. Conclusions and future research. Micropollutants and Challenges: Emerging in the Aquatic Environments and Treatment Processes, 249.

Chen, F., Zeng, S., Luo, Z., Ma, J., Zhu, Q., Zhang, S., 2020. A novel MBBR–MFC integrated system for high-strength pulp/paper wastewater treatment and bioelectricity generation. Sep. Sci. Technol. 55, 2490–2499.

Contreras, R., C., L., D., L., M., A., Domínguez, C., Bayona, J.M., et al., 2019. Removal of organic micropollutants in wastewater treated by activated sludge and constructed wetlands: a comparative study. Water (Basel) 11, 2515.

Crini, G., Lichtfouse, E., 2019. Advantages and disadvantages of techniques used for wastewater treatment. Environ. Chem. Lett. 17, 145–155.

Duffus, J.H., 2002. Heavy metals a meaningless term. Pure Appl. Chem. 74, 793–807.

Fatehifar, M., Borghei, S.M., 2018. Application of moving bed biofilm reactor in the removal of pharmaceutical compounds (diclofenac and ibuprofen). J. Environ. Chem. Eng. 6, 5530–5535.

Flavigny, R.M.-G., Cord-Ruwisch, R., 2015. Organic carbon removal from wastewater by a PHA storing biofilm using direct atmospheric air contact as oxygen supply. Bioresour. Technol. 187, 182–188.

Forbis-Stokes, A.A., Rocha-Melogno, L., Deshusses, M.A., 2018. Nitrifying trickling filters and denitrifying bioreactors for nitrogen management of high-strength anaerobic digestion effluent. Chemosphere 204, 119–129.

Grandclément, C., Seyssiecq, I., Piram, A., Wong-Wah-Chung, P., Vanot, G., Tiliacos, N., et al., 2017. From the conventional biological wastewater treatment to hybrid processes, the evaluation of organic micropollutant removal: a review. Water Res. 111, 297–317.

Haddad, M., Abid, S., Hamdi, M., Bouallagui, H., 2018. Reduction of adsorbed dyes content in the discharged sludge coming from an industrial textile wastewater treatment plant using aerobic activated sludge process. J. Environ. Manage. 223, 936–946.

Hossain, M.I., Paparini, A., Cord-Ruwisch, R., 2018. Direct oxygen uptake from air by novel glycogen accumulating organism dominated biofilm minimizes excess sludge production. Sci. Total Environ. 640, 80–88.

Hunter, B., Deshusses, M.A., 2020. Resources recovery from high-strength human waste anaerobic digestate using simple nitrification and denitrification filters. Sci. Total Environ. 712, 135509.

Irankhah, S., Ali, A.A., Soudi, M.R., Gharavi, S., Ayati, B., 2018. Highly efficient phenol degradation in a batch moving bed biofilm reactor: benefiting from biofilm-enhancing bacteria. World. J. Microbiol. Biotechnol. 34, 164.

Jabari, P., Munz, G., Oleszkiewicz, J.A., 2014. Selection of denitrifying phosphorous accumulating organisms in IFAS systems: comparison of nitrite with nitrate as an electron acceptor. Chemosphere 109, 20–27.

Kanaujiya, D.K., Paul, T., Sinharoy, A., Pakshirajan, K., 2019. Biological treatment processes for the removal of organic micropollutants from wastewater: a review. Curr. Pollut. Rep. 5, 112–128.

Karaolia, P., Michael-Kordatou, I., Hapeshi, E., Alexander, J., Schwartz, T., Fatta-Kassinos, D., 2017. Investigation of the potential of a Membrane BioReactor followed by solar Fenton oxidation to remove antibiotic-related micro contaminants. Chem. Eng. J. 310, 491–502.

Kaya, Y., Ersan, G., Vergili, I., Gönder, Z.B., Yilmaz, G., Dizge, N., et al., 2013. The treatment of pharmaceutical wastewater using in a submerged membrane bioreactor under different sludge retention times. J. Membr. Sci. 442, 72–82.
Kornaros, M., Lyberatos, G., 2006. Biological treatment of wastewaters from a dye manufacturing company using a trickling filter. J. Hazard. Mater. 136, 95–102.
Liao, J., Fang, C., Yu, J., Sathyagal, A., Willman, E., Liu, W.-T., 2017. Direct treatment of high-strength soft drink wastewater using a down-flow hanging sponge reactor: performance and microbial community dynamics. Appl. Microbiol. Biotechnol. 101, 5925–5936.
Lin, J., Gan, L., Chen, Z., Naidu, R., 2015. Biodegradation of tetradecane using Acinetobacter venetianus immobilized on bagasse. Biochem. Eng. J. 100, 76–82.
Lu, M., Gu, L.-P., Xu, W.-H., 2013. Treatment of petroleum refinery wastewater using a sequential anaerobic–aerobic moving-bed biofilm reactor system based on suspended ceramsite. Water Sci. Technol. 67, 1976–1983.
Luo, Y., Jiang, Q., Ngo, H.H., Nghiem, L.D., Hai, F.I., Price, W.E., et al., 2015. Evaluation of micropollutant removal and fouling reduction in a hybrid moving bed biofilm reactor–membrane bioreactor system. Bioresour. Technol. 191, 355–359.
Mahanty, B., Pakshirajan, K., Dasu, V.V., 2008. Biodegradation of pyrene by Mycobacterium frederiksbergense in a two-phase partitioning bioreactor system. Bioresour. Technol. 99, 2694–2698.
Mahanty, B., Sarma, S., Pakshirajan, K., 2007. Evaluation of different surfactants for use in pyrene biodegradation by Mycobacterium frederiksbergense. Int. J. Chem. Sci. 5, 1505–1512.
Margot, J., Rossi, L., Barry, D.A., Holliger, C., 2015. A review of the fate of micropollutants in wastewater treatment plants. Wiley Interdisciplinary Reviews: Water 2, 457–487.
McCarty, P.L., 2018. What is the best biological process for nitrogen removal: when and why? ACS Publications.
Moser, P.B., Bretas, C., Paula, E.C., Faria, C., Ricci, B.C., Cerqueira, A.C.F., et al., 2019. Comparison of hybrid ultrafiltration-osmotic membrane bioreactor and conventional membrane bioreactor for oil refinery effluent treatment. Chem. Eng. J. 378, 121952.
Nguyen, L.N., Hai, F.I., Kang, J., Price, W.E., Nghiem, L.D., 2013. Removal of emerging trace organic contaminants by MBR-based hybrid treatment processes. Int. Biodeterior. Biodegrad. 85, 474–482.
Obotey Ezugbe, E., Rathilal, S., 2020. Membrane technologies in wastewater treatment: a review. Membranes 10, 89.
Pagnout, C., Rast, C., Veber, A.-M., Poupin, P., Férard, J.-F., 2006. Ecotoxicological assessment of PAHs and their dead-end metabolites after degradation by Mycobacterium sp. strain SNP11. Ecotoxicol. Environ. Saf. 65, 151–158.
Pinto, H.B., de Souza, B.M., Dezotti, M., 2018. Treatment of a pesticide industry wastewater mixture in a moving bed biofilm reactor followed by conventional and membrane processes for water reuse. J. Cleaner Prod. 201, 1061–1070.
Prasertkulsak, S., Chiemchaisri, C., Chiemchaisri, W., Itonaga, T., Yamamoto, K., 2016. Removals of pharmaceutical compounds from hospital wastewater in membrane bioreactor operated under short hydraulic retention time. Chemosphere 150, 624–631.
Prieto-Rodríguez, L., Oller, I., Agüera, A., Malato, S., 2017. Elimination of organic micro-contaminants in municipal wastewater by a combined immobilized biomass reactor and solar photo-Fenton tertiary treatment. J. Adv. Oxid. Technol. 20.
Reungoat, J., Escher, B., Macova, M., Argaud, F., Gernjak, W., Keller, J., 2012. Ozonation and biological activated carbon filtration of wastewater treatment plant effluents. Water Res. 46, 863–872.
Sarma, S.J., Pakshirajan, K., 2011. Surfactant aided biodegradation of pyrene using immobilized cells of Mycobacterium frederiksbergense. Int. Biodeterior. Biodegrad. 65, 73–77.
Schneider, E., Cerqueira, A., Dezotti, M., 2011. MBBR evaluation for oil refinery wastewater treatment, with post-ozonation and BAC, for wastewater reuse. Water Sci. Technol. 63, 143–148.
Shoukat, R., Khan, S.J., Jamal, Y., 2019. Hybrid anaerobic-aerobic biological treatment for real textile wastewater. J. Water Process. Eng. 29, 100804.
Singhal, N., Perez-Garcia, O., 2016. Degrading Organic Micropollutants: the Next Challenge in the Evolution of Biological Wastewater Treatment Processes. Front. Environ. Sci. 4.

Tomei, M.C., Stazi, V., Angelucci, D.M., 2018. Biological treatment of hypersaline wastewater in a continuous two-phase partitioning bioreactor: analysis of the response to step, ramp and impulse loadings and applicability evaluation. J. Cleaner Prod. 191, 67–77.
Wang, G., Wang, D., Xu, Y., Li, Z., Huang, L., 2020. Study on optimization and performance of biological enhanced activated sludge process for pharmaceutical wastewater treatment. Sci. Total Environ. 739, 140166.
Yao, M., Li, Z., Zhang, X., Lei, L., 2014. Polychlorinated biphenyls in the centralized wastewater treatment plant in a chemical industry zone: source, distribution, and removal. J. Chem. 2014.
Yuan, C., Wang, B., Peng, Y., Li, X., Zhang, Q., Hu, T., 2020. Enhanced nutrient removal of simultaneous partial nitrification, denitrification and phosphorus removal (SPNDPR) in a single-stage anaerobic/micro-aerobic sequencing batch reactor for treating real sewage with low carbon/nitrogen. Chemosphere 257, 127097.
Zhou, Y., Zhang, D.Q., Le, M.T., Puah, A.N., Ng, W.J., 2013. Energy utilization in sewage treatment–a review with comparisons. J. Water Clim. Change 4, 1–10.

CHAPTER 2

Fate and occurrence of micro- and nano-plastic pollution in industrial wastewater

Simranjeet Singh[a], Vijay Kumar[b], Dhriti Kapoor[c], Sonali Bhardwaj[d], Daljeet Singh Dhanjal[e], Akshay Pawar[e], Praveen C Ramamurthy[a], Joginder Singh[d]
[a]Interdisciplinary Centre for Water Research (ICWaR), Indian Institute of Sciences, Bangalore, India
[b]Regional Ayurveda Research Institute for Drug Development, Gwalior, MP, India
[c]Department of Botany, Lovely Professional University, Phagwara, Punjab, India
[d]Department of Microbiology, Lovely Professional University, Phagwara, Punjab, India
[e]Department of Biotechnology, Lovely Professional University, Phagwara, Punjab, India

2.1 Introduction

In 1869, John Wesley Hyatt invented the first synthetic polymer from cotton fiber, halted the slaughtering of elephants which were used to produce ivory and relieved people from economic restrictions levied by the shortage of natural resources. Later in the 1970's, Leo Baekeland invented a completely synthetic substance made up of polymers and named it Bakelite. Both Hyatt's and Baekeland's inventions, due to their amendable nature and outstanding characteristics at such a low cost revolutionized the era. Plastic products played an imperative role during World War II. However, this unblemished optimism for plastic and plastic products didn't last long as people became aware of the environmental issues imposed by plastic debris, especially the non-biodegradable nature of these plastics which allowed them to sustain in the environment for years.

Although the natural degradation of plastic material is not well-understood, it is possible to mineralize it into simpler forms by different methods. However, the mineralization process doesn't resolve its endurance in the environment; instead, it becomes a pollutant of soil, water and air. The contamination of natural resources by these pollutants has led to the development of various diseases in animals and humans. According to Rhodes study, from 1950 to 2015, approximately 6.3 billion plastics was produced, out of which only 9% was recycled, 12% was incinerated and rest 79% still persists on earth as a pollutant and affecting the environment (Rhodes, 2018). Plastic is widely used in different countries for manufacturing different items as it can be moulded into any form under pressure and heat. With the increasing use of plastic, its improper management and negligible degradation potential have made it a serious issue for water, land and environment. It can get accumulated anywhere and get associated with other contaminants alone or in association with other contaminants, in both ways, it affects the human health and environment. Plastic polymer on degradation gets converted into

Biodegradation and Detoxification of Micropollutants in Industrial Wastewater
DOI: https://doi.org/10.1016/B978-0-323-88507-2.00008-7

Microplastic (MP) (100 nm-5 mm size) and Nanoplastic (NP) (<100 nm). These MPs and NPs getting dissolved in water become toxic to various living forms as they get easily ingested by aquatic organisms.

Globally, micro- and nano-plastics are found as pollutant everywhere, and their pollution varies according to the geographical location. Hydrodynamic, anthropogenic, meteorological, and geographical factors either alone or in combination influence the behaviour of micro-and nano-plastics in the environment. The similarity of their size with sediments makes them bioavailable for ingestion by aquatic organisms which leads to accumulation in their bodies and eventually causing harm to these organisms. Various physical and chemical methods have been developed to remove these micro- and nano-plastics from the environment. Due to highly stable nature of plastic, it takes millions of years to get completely degraded by natural means, but with the help of physical, chemical and biological methods, it can be fragmented into micro- and nano-plastics. The effective remediation of micro- and nano-plastics can be done by two major approaches viz. technological and biotechnological approach. The technological approach involves wastewater treatment plants that can degrade the particles before disposal into the environment. This approach has been divided into many methods like membrane bioreactor, activated sludge, rapid gravity sand filters and dissolved air floatation. The biotechnological approach involves the use of biodegradable materials to ensure complete removal of plastic materials from the environment through the use of microorganisms. Therefore, this book chapter aims to cover all aspects of structural differences between micro- and nano-plastics. Various sections and subsections are dedicated to discussing the toxic effects of micro-and nano-plastics on different life forms.

2.2 Micro- and nano-plastics in the environment

The presence of plastic pollutants in the form of micro- and nano-plastics were first reported in the 1970s and were found to have direct access into the environment through various commodities of commercial importance like a) cosmetics and hygiene products b) industrial raw materials used to manufacture different plastic items c) plastic polymers in air cannons d) plastic nanoparticles released during 3D printing on textiles e) paints and adhesives etc. (da Costa et al., 2016). In terms of coverage in kilograms per square kilometres, large-sized plastic items are among the major contributors while micro- and nano-plastics top the charts with respect to their numbers when measured in items per square kilometres (GESAMP, 2015). These two even surpass physical boundaries and cause large-scale multiple economic and environmental fallouts. Developmental spree in industrialisation and globalisation has consequently led to increased production of plastic containing articles that get piled up in both land and aquatic ecosystems after being discarded owing to their impervious nature and extremely slow chemical and thermal degradation. Besides, manufacturing processes also involve adding certain substances to plastic articles like plasticizers, colouring materials, and other chemicals that impart softness, peculiar colours, UV- radiation, and fire resistance. However, these small-sized additives are not bound to polymeric constituents of plastic via any chemical linkages. Upon degradation, these micro- and nano-plastics get leached into the environment in the vicinity.

2.3 Structural difference between micro- and nano-plastics

Although both micro- and nano-plastics share similar chemical composition microplastics have some distinctive characteristics as compared to nano-plastics. The major difference lies in their size, i.e. plastic fragments of 0.1–5000 μm are considered as microplastics, whereas nanoplastics lie in size range of 1–100 nm (Andrady, 2011). This size difference is considered responsible for the transport of relatively small-sized nanoplastics to far remote parts of the globe as compared to microplastics. Moreover, type of interaction display by microplastics with biota is also distinct as microorganisms are able to grow as biofilms on them, whereas nanoplastics with either same or smaller size than that of microorganisms do not support the growth of latter on them (Levin and Angert, 2015).

Upon degradation, microplastics can have the shape of fibres and even smaller fragments, however, nanoplastics may assume the shape of fragments, granules and spheres but never fall in the respective length of fibres. Microplastics exhibit vertical motion in water due to their density while nano plastics are comparatively small and display Brownian motion. Further, the differences in genesis between them lead to their categorization into primary and secondary types. Primary micro- and nano-plastics are those which are manufactured on purpose in their respective size range. In contrast, secondary micro- and nano-plastics are formed via gradual abiotic and biotic degradation of primary plastic articles over time (Lambert and Wagner, 2016).

2.4 Characterisation of micro- and nano-plastics

For the purpose of characterization of microplastics and nanoplastics, different techniques belonging to chromatographic, optical and spectroscopic analysis have been in use which brings to the fore their peculiar qualitative and quantitative characteristics (Mintenig et al., 2018). Three main steps constituting the plan of action for identifying these plastic fragments involve- a) isolation from the heterogeneous polymeric matrix; b) quantification and sorting and c) adoption of suitable characterization method (Correia and Loeschner, 2018). All these are summed up in a simplified manner in Fig. 2.1.

The very first step after sample collection and pre-separation procedures utilize methods like density, size or hydrophobicity-based approaches for the separation of microplastics. In comparison, methods such as magnetic field flow fractionation (MFFF), gel electrophoresis and size-exclusion chromatography (SEC) are used to separate nanoplastics (Robertson et al., 2016). Further, micro- and nano-plastics' quantification and characterization of micro- and nano-plastics include visual identification based on

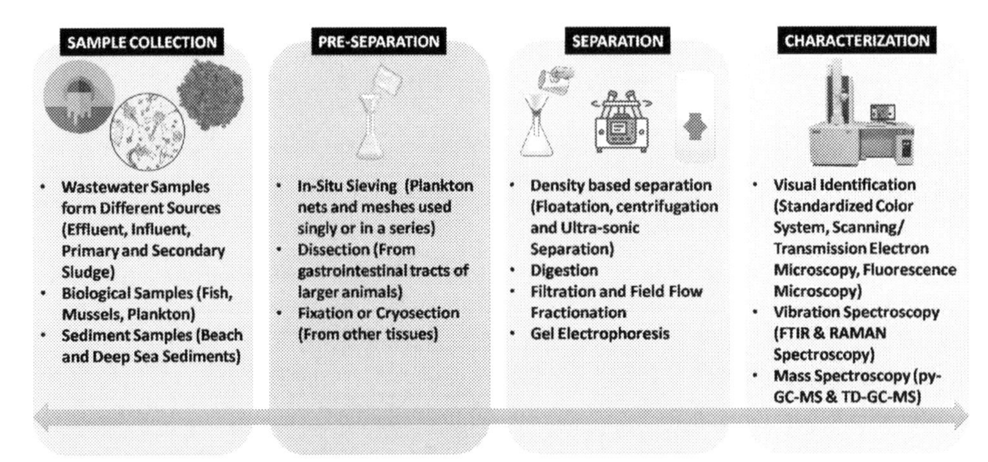

Fig. 2.1 ***Simplified presentation of the process of isolation, identification and characterization of Micro and nano plastics.***

morphology, namely standardized size and color sorting (SCS) system, fluorescence microscopy, scanning and transmission electron microscopy with the confirmation of their composition through vibrational spectroscopy and mass spectrometry methods.

2.5 Techniques used for characterization of micro- and nano-plastics from wastewater

Following is a brief account of major characterization techniques adopted for the characterization of micro- and nano-plastics (Fig. 2.2):

2.5.1 Visual characterization

The primary method used in the identification of microplastics is via visual recognition. Lack of high-priced analytical tools and limited resources make this technique apt for large volumes of samples, despite the fact that it consumes a lot of time. Correct visual examination of microplastic samples requires the absence of organic or cellular structures, uniformity in color and consistency in length and diameter without any bends in case the plastic sample is in the form of fibres. Thus, visual sorting has a chance of occurrence of error at the individual level. Therefore, microscopic and spectroscopic conformation are at times further supplemented with high-magnification fluorescence microscopy or scanning electron microscopy to improve visual identification outcomes and minimize the errors. Some major visual identification procedures are:

2.5.1.1 The size and color sorting (SCS) system

It is an efficient characterization technique which provides distinct codes to heterogeneous plastic samples, especially microplastics. The procedure involves i) collecting a

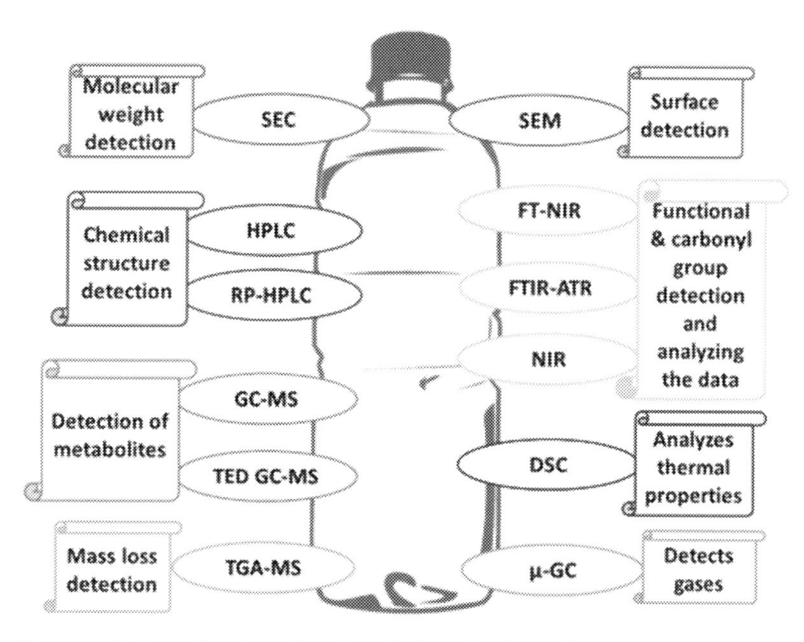

Fig. 2.2 ***Different techniques for the estimation of microplastics and nano plastics in wastewater.***

sample from the environment, ii) sorting of plastic fragments on the basis of their size in their longest dimensions iii) providing an abbreviated categorical name to differently sized plastic fragments such as MP to microplastics and NP to nanoplastics, iv) further sorting and abbreviated coding by their shape, size and constitution such as pellets (PT), microbeads (MBD), fragments (FR), fibres (FB), films (FI) and foam (FM) v) giving an individual color code to these sorted plastic pieces like all opaque (AO), all transparent (AT), any color (ALL), blue (BL), brown (BN), etc., vi) determination of the type of polymer by adopting appropriate analytical tools like polyamide (PA), polyethylene (PE), expanded polystyrene (EPA) etc., vii) quantification of each plastic polymer sorted so far. Finally, each plastic piece is assigned a unique SCS code which characterizes its size, shape, color, type of polymer and quantity.

2.5.1.2 Scanning electron microscopy (SEM)

A scanning electron microscope (SEM) performs the raster scanning of microplastics' surface in a zig-zag manner by firing a beam of high-intensity electrons to produce a high-resolution image. Surface features of microplastic samples collected from the environment along with other physical characteristics, can be efficiently analyzed with SEM. This technique helps in differentiating these plastic items from those non-plastic items which have a similar appearance. SEM's efficiency to characterize the chemical composition of microplastics and its additives can be further enhanced if it is coupled with EDX, i.e., Energy dispersive X-ray micro analyzer (Goldstein et al., 2017).

2.5.2 Vibrational spectroscopy

Though dependent on particle size and purity of the sample for analysis, this technique is an important tool to confirm the visible and internal characteristics, particularly when coupled with optical microscopic methods. Some of the major methods based upon vibrational spectroscopy are:

2.5.2.1 Fourier transform infrared spectroscopy (FTIR)

It is one of the most common and efficient technique to characterize microplastics on the basis of their peculiar IR-spectra accurately. The signal originates from an alteration in dipole moments of chemical linkages, so that the polar functional groups in the plastic polymeric matrix can be easily identified. The requirements like minimum sample thickness of about 150 nm along with its coating on IR- transparent substrate material, given that the spatial resolution is 5 μm make this method more suitable to particle size usually bigger than 20 μm and films or aggregations of smaller sized particles (Hernandez, Yousefi and Tufenkji, 2017).

2.5.2.2 Raman spectroscopy

This technique is suitable to identify microplastic fragments smaller than 20 μm in size. It utilizes molecular vibrations coming out from sample plastic materials which bring about the scattering of polarized light. Features like sample thickness and size of particles to be examined, the spatial resolution of about 1 um and irradiance of UV rays does not hinder the measurements and enhance the sensitivity and efficiency of this technique as compared to FTIR. However, coloring materials and additives of organic and inorganic nature in the plastic samples do greatly affect the outcomes. Therefore, this technique can be used in combination with FTIR, as both have been reported to complement each other.

2.5.3 Mass spectrometry

This method can carry out the analysis of samples in bulk and has enhanced sensitivity and suitability for nanoplastics' qualitative detection. Here, the signal for detection from the samples depends on the mass of the sample to be identified and analyzed. Important mass spectroscopy-based methods are:

2.5.3.1 Thermal desorption coupled with gas chromatography and mass spectrometry (TDS-GC–MS)

In this, the sample is first thermally degraded at 1000 °C by placing it on a thermogravimetric balance followed by adsorption of products onto a solid phase (Dümichen et al., 2017). After thermally desorbing the latter inside the designated thermal desorption unit, separation of different particles is brought about by a gas chromatographic column followed by analysis via mass spectrometry. Though it can analyze samples with relative mass as high as 100 mg, this technique's main disadvantage lies in its appropriateness for qualitative analysis only (Dümichen et al., 2015).

2.5.3.2 Pyrolysis coupled with gas chromatography and mass spectrometry (py-GC–MS)

Here, the sample is first pyrolysed at high temperature in different types pyrolyzers like quartz tubes, preheated furnaces and ferromagnetic foils. After pyrolysis, the degraded material particles are separated by a gas chromatographic column followed by their analysis via mass spectrometry. The final measurements rely heavily on the preparation of samples, type of pyrolysis adopted and transfer of degraded pyrolysates (Dworzanski and Meuzelaar, 2016). For instance, the quartz tube and heated coil in electrically heated filament pyrolysis, preheated furnaces and a carrier gas in furnace pyrolysis while ferromagnetic foils and induction pyrolyzers in curie point pyrolysis are used to place and transfer the sample pyrolysates. However, out of all these methods of pyrolysis, curie point pyrolysis is advantageous over others because it is much rapid, accurate, does not alter the composition of pyrolysate and leaves no un-pyrolysed residue behind. py-GC–MS is comparatively more sensitive and efficient than TDS-GC–MS for characterization of nanoplastics of mass as small as 50 μg.

2.6 Health hazards of microplastics and nanoplastics

Heavy metals like Pb, Ni, Cd and Zn, hydrophobic organic chemicals such as polychlorinated biphenyls (PCBs), organochlorine pesticides (DDT, Eldrin etc.), polycyclic aromatic hydrocarbons (PAHs) and other substances such as persistent, accumulative and toxic contaminants (PBTs) get adsorbed and amassed over hydrophobic surfaces of microplastics to a great extent (Avio, Gorbi and Regoli, 2017). All these turn micro- and nano-plastics into potential carriers of other pollutants of primary nature.

2.6.1 Effects on terrestrial environment

The impact of anthropogenic pollution on the terrestrial environment is far less studied vis-à-vis the marine environment. However, it has now been projected to be 4–23 times greater on land than on the oceanic reservoirs, the main contributor being agricultural soil alone (Horton et al., 2017). Physico-chemical attributes of soil such as its texture, structure, chemical composition and function are extensively altered due to microplastic contamination. In turn, a number of ecosystem services like biogeochemical cycling, hydrological balance, plant growth etc. are negatively affected (Bergmann et al., 2016). Modulations in optimum water content of soil due to microplastics also hamper the principal beneficial mutual interactions prevailing the terrestrial ecosystems like plant-fungi (Mycorrhizal) and plant-bacteria/cyanobacteria interaction (Nitrogen fixation) by harming microbial biodiversity of the soil. Soil fauna, mainly earthworms, are largely affected by microplastic soil pollution because these particles accumulate in their body and hamper their growth (Huerta Lwanga et al., 2016). Earthworms are also found to percolate these contaminants in soil both horizontally and vertically due to their movements. Furthermore, accumulation levels of microplastics and nanoplastics in food webs of terrestrial and continental ecosystems are estimated to be at par or sometimes higher than in the marine environment, though still needs sufficient substantial evidence.

2.6.2 Effects on aquatic environment

75-90% of the plastic polluting aquatic environment has been reported to have landfills, industrial and other land-based waste collection areas as its source whiles the remaining 10-25% comes from sea sources (Duis and Coors, 2016). The direct access of secondary micro- and nano-plastics and other PBTs to the aquatic environment has resulted in their ingestion and accumulation inside aquatic organisms, putting them at risk and marring the safety of seafood items via invasion in the food chain. In aquatic animals, these plastic materials have been reported to clog alimentary tract and gills, hindering major organ systems' activity and altering the growth and behavioural patterns. Nanoplastics are more toxic as their bioaccumulation potential is quite high due to their small size and ability to invade the nervous system, alter cell physiology, and cause membrane disintegration.

2.6.3 Major consequences on human health

Detection of microplastics and nanoplastics in drinking water, air, food provisions (seafood, beer, salt, honey, sugar etc.) and cosmetics has escalated the probability of human exposure to these substances. The otherwise chemically inert nature of polymeric constituents of microplastics and nanoplastics present in the environment, upon entering the food chain finally leads to serious health problems in human beings (Besseling et al., 2014). Microplastics cause the destruction of immune cells in particular, whereas nanoplastics have been proven fatal as they cause disruption of cell membranes. These microparticles may turn out to be carcinogenic and cast a number of other adverse effects on getting ingested or inhaled like internal blockages, reduced fecundity, abrasions, the fallacious feel of satiation and cytotoxicity.

Besides, some of the major sources and toxicity symptoms of microplastics and nanoplastics upon interacting with a biotic component in the ecosystem as the likes of plants, fungi and animals are summarised in the Table 2.1.

Literature analysis reveals that the addition of these plastic materials in our environment is continuously rising and is expected to double in near-decade (Wu, Yang and Criddle, 2017). Although multiple efforts are presently being carried out to handle and eliminate them effectively, these two plastic pollutants of cosmopolitan occurrence are yet to end up in better recycling mechanisms.

Table 2.1 Microplastics and Nanoplastics - Manufactured (Primary) and degradation (Secondary) plastic products.

➢ Major sources: Products of cosmetics and cleaning; Industrial emissions of plastic raw materials; Discharge of ship-generated plastic waste; Biomedical plastic waste; Plastic containing wastewater in irrigation; Thermal cutting of Styro foam and Polystyrene foam	
➢ Interaction of these plastic pollutants with living biota and production of toxicity symptoms:	
In Plants	• Microplastics indirectly affect plant growth via changing soil properties and functioning of soil enzymes; immobilizing nutrients and beneficial microbial community (McCormick et al., 2016). • Microplastics are also reported to affect plant-pollinator relationship because of their presence in inflorescences of diverse species as they get detected in commercially available honey produced locally and industrially (Liebezeit and Liebezeit, 2015). • Reactive oxygen species-induced morphotoxic, cytotoxic and genotoxic effects, and affects the cell cycle regulator cdc2 gene expression due to the presence of PS microbeads (100 nm) on *Allium cepa* (Maity et al., 2020). • Both the time required for germination of *Lepidium sativum* seeds and the germination rate have been found to be significantly reduced with the increasing particle size of plastics (50, 500, and 4,800 nm) (Bosker et al., 2019). • Nanoplastics cause growth inhibition in green algae (*Chlorella* sp. *and Scenedesmus sp.*) (Besseling et al., 2014).
In Fungus	• Lethal toxicity to fungi as observed in case of yeast cells (*Saccharomyces cerevisiae*) exposed to polystyrene nanobeads (50 and 100 nm, 10–15 mg/L) in 5 mM NaCl culture media (Miyazaki et al., 2015).
In Animals	• Microplastic accumulation causes inflammation and damage to the vital organs of the body like kidneys, liver, lungs, lymphatic system and gut as reported in mammals like dogs, mice and rats (Deng et al., 2017). • In fish crucian carp, behavioural, physiological and metabolic alterations have been observed due to deposition of nanoplastics (*Carassiuscarassius*) (Mattsson et al., 2015). • Severe developmental defects in sea urchin (*Paracentrotuslividus*) embryos (Della Torre et al., 2014) and decreased fertility and reproduction rates in crustaceans *Daphnia magna* (Besseling et al., 2014). • Reduction in the growth of earthworms (Huerta Lwanga et al., 2016).

2.7 Impact of micro- and nano-plastics in industrial wastewater

Chemicals of organic and inorganic origin found in wastewater released from different industries raise the grievousness of the problem of water pollution. Out of all the other sources, the plastic industry discharge contributes to both biodegradable and non-biodegradable compounds of monomeric or polymeric nature such as polyacrylics, polystyrene, polyethylene, polyacrylonitrile, acrylonitrile-butadiene-styrene copolymers etc. into this wastewater. Solutions, emulsions or suspensions present in this industrial wastewater have been reported to contain a large number of waste raw materials, intermediates, by-products and end-products with the presence of micro- and nano-plastics (both primary and secondary) in them. Further, microplastic scrubbers which are used to scour paint and rust from machines in industries get reduced in size and acquire heavy metals like cadmium, chromium and lead after repetitive and get washed off in industrial wastewater.

The sewage water generated by industries having a significant amount of micro- and nano-plastics is responsible for polluting rivers even in magnitudes higher than the marine environments (Basel, 2016). When this wastewater is made to pass through a treatment plant, plastic pollutants have still been found to be persistent in wastewater. Thence, the treated effluents can further act as a major source of micro- and nanoplastics into the environment, as found by a study carried in the USA, where wastewater treatment plants have been reported to release eight trillion microbeads in a single day into the water resources (Rochman et al., 2015).

Around 80%–90% of the incoming microplastics in the wastewater treatment plant has been reported to be retained in the sludge (Talvitie et al., 2017). When this sludge is put into use to increase soil fertility, microplastics present in it are retained in the soil for durations longer than the essential nutrients and that too, without even undergoing any change in their properties (Horton et al., 2017).

2.8 Conclusions

Microplastics and nano particles' occurrence and fate mostly depend on the physico-chemical properties of soil, water, and plastics. The major concern related to nano- and micro-plastics is their ability to get transferred from one system to another without mineralization or degradation. They can easily enter the food chain and get accumulated in various cells and tissues of an organism. With increased worldwide production and uncontrolled disposal of plastic materials, it is essential to develop effective detection methods for these materials. Plants are also reported to uptake micro- and nano-plastics and accumulate them from contaminated soils around the root zone. The major challenge concerning the use of biodegradable plastics is the assurance of accurate results, for which optimal conditions and required strain of microbes that can carry out the process effectively must be maintained well. Based on constraints and mechanisms stated above, the mineralization of plastics in various ecosystems should be limited within the lifetime of human scale. Future studies must be conducted to mitigate the liberation of nano- and micro-plastics in different ecosystems.

Acknowledgements

Dr. Simranjeet Singh is thankful to the Interdisciplinary Centre for Water Research (ICWaR), Indian Institute of Sciences Bangalore for the financial assistance IOE-IISc Fellowship (Sr. No: IE/REAC-20-0134).

References

Andrady, A.L., 2011. Microplastics in the marine environment. Mar. Pollut. Bull. 62 (8), 1596–1605. http://doi.org/10.1016/j.marpolbul.2011.05.030.

Avio, C.G., Gorbi, S., Regoli, F., 2017. Plastics and microplastics in the oceans: from emerging pollutants to emerged threat. Mar. Environ. Res. 128, 2–11. http://doi.org/10.1016/j.marenvres.2016.05.012.

Basel, et al., 2016. News: rhine one of the most microplastic polluted rivers worldwide. Mar. Pollut. Bull. 102 (1), 4–8.

Bergmann, J., et al., 2016. The interplay between soil structure, roots, and microbiota as a determinant of plant–soil feedback. Ecol. Evol. 6 (21), 7633–7644. http://doi.org/10.1002/ece3.2456.

Besseling, E., et al., 2014. Nanoplastic affects growth of S. obliquus and reproduction of D. magna. Environ. Sci. Technol. 48 (20), 12336–12343. http://doi.org/10.1021/es503001d.

Bosker, T., et al., 2019. Microplastics accumulate on pores in seed capsule and delay germination and root growth of the terrestrial vascular plant Lepidium sativum. Chemosphere 226, 774–781. http://doi.org/10.1016/j.chemosphere.2019.03.163.

Correia, M., Loeschner, K., 2018. Detection of nanoplastics in food by asymmetric flow field-flow fractionation coupled to multi-angle light scattering: possibilities, challenges and analytical limitations. Anal. Bioanal. Chem. 410 (22), 5603–5615. http://doi.org/10.1007/s00216-018-0919-8.

da Costa, J.P., et al., 2016. (Nano)plastics in the environment - Sources, fates and effects. Sci. Total Environ. 566–567, 15–26. http://doi.org/10.1016/j.scitotenv.2016.05.041.

Della Torre, C., et al., 2014. Accumulation and embryotoxicity of polystyrene nanoparticles at early stage of development of sea urchin embryos Paracentrotus lividus. Environ. Sci. Technol. 48 (20), 12302–12311. http://doi.org/10.1021/es502569w.

Deng, Y., et al., 2017. Tissue accumulation of microplastics in mice and biomarker responses suggest widespread health risks of exposure. Sci. Rep. 7 (April), 1–10. http://doi.org/10.1038/srep46687.

Duis, K., Coors, A., 2016. Microplastics in the aquatic and terrestrial environment: sources (with a specific focus on personal care products), fate and effects. Environmental Sciences Europe 28 (1), 1–25. http://doi.org/10.1186/s12302-015-0069-y.

Dümichen, E., et al., 2015. Analysis of polyethylene microplastics in environmental samples, using a thermal decomposition method. Water Res. 85, 451–457. http://doi.org/10.1016/j.watres.2015.09.002.

Dümichen, E., et al., 2017. Fast identification of microplastics in complex environmental samples by a thermal degradation method. Chemosphere 174, 572–584. http://doi.org/10.1016/j.chemosphere.2017.02.010.

Dworzanski, J.P., Meuzelaar, H.L.C., 2016. Pyrolysis mass spectrometry, methods, 3rd edn, Encyclopedia of Spectroscopy and Spectrometry, 3rd edn. Elsevier Ltd. http://doi.org/10.1016/B978-0-12-409547-2.11686-5.

GESAMP: Sources, fate and effects of microplastics in the marine environment: a global assessment. Reports and Studies 90. London: IMO/FAO/UNESCO-IOC/UNIDO/WMO/IAEA/UN/UNEP/UNDP Joint Group of Experts on the Scientific Aspects of Marine Environmental Protection; 2015.

Goldstein, J.I., et al., 2017. Scanning Electron Microscopy and X-Ray Microanalysis. Springer, New York.

Hernandez, L.M., Yousefi, N., Tufenkji, N., 2017. Are there nanoplastics in your personal care products?. Environ. Sci. Technol. Lett. 4 (7), 280–285. http://doi.org/10.1021/acs.estlett.7b00187.

Horton, A.A., et al., 2017. Microplastics in freshwater and terrestrial environments: evaluating the current understanding to identify the knowledge gaps and future research priorities. Sci. Total Environ. 586, 127–141. http://doi.org/10.1016/j.scitotenv.2017.01.190.

Huerta Lwanga, E., et al., 2016. Microplastics in the Terrestrial Ecosystem: implications for Lumbricus terrestris (Oligochaeta, Lumbricidae). Environ. Sci. Technol. 50 (5), 2685–2691. http://doi.org/10.1021/acs.est.5b05478.

Lambert, S., Wagner, M., 2016. Characterisation of nanoplastics during the degradation of polystyrene. Chemosphere 145, 265–268. http://doi.org/10.1016/j.chemosphere.2015.11.078.

Levin, P.A., Angert, E.R., 2015. Small but mighty: cell size and bacteria. Cold Spring Harb. Perspect. Biol. 7 (7), 1–11. http://doi.org/10.1101/cshperspect.a019216.

Liebezeit, G., Liebezeit, E., 2015. Origin of synthetic particles in honeys. Polish J. Food Nutr. Sci. 65, 143–147.

Maity, S., et al., 2020. Cytogenotoxic potential of a hazardous material, polystyrene microparticles on Allium cepa L. J. Hazard. Mater. 385, 121560. http://doi.org/10.1016/j.jhazmat.2019.121560.

Mattsson, K., et al., 2015. Altered behavior, physiology, and metabolism in fish exposed to polystyrene nanoparticles. Environ. Sci. Technol. 49 (1), 553–561. http://doi.org/10.1021/es5053655.

McCormick, A.R., et al., 2016. Microplastic in surface waters of urban rivers: concentration, sources, and associated bacterial assemblages. Ecosphere 7 (11). http://doi.org/10.1002/ecs2.1556.

Mintenig, S.M., et al., 2018. Closing the gap between small and smaller: towards a framework to analyse nano-and microplastics in aqueous environmental samples. Environ. Sci.: Nano 5 (7), 1640–1649.

Miyazaki, J., et al., 2015. Cytotoxicity and behavior of polystyrene latex nanoparticles to budding yeast. Colloids Surf. A 469, 287–293. http://doi.org/10.1016/j.colsurfa.2015.01.046.

Rhodes, C.J., 2018. Plastic pollution and potential solutions. Sci. Prog. 101 (3), 207–260. http://doi.org/10.3184/003685018X15294876706211.

Robertson, J.D., et al., 2016. Purification of Nanoparticles by Size and Shape. Sci. Rep. 6, 1–9. http://doi.org/10.1038/srep27494.

Rochman, C.M., et al., 2015. Scientific Evidence Supports a Ban on Microbeads. Environ. Sci. Technol. 49 (18), 10759–10761. http://doi.org/10.1021/acs.est.5b03909.

Talvitie, J., et al., 2017. How well is microlitter purified from wastewater? – A detailed study on the stepwise removal of microlitter in a tertiary level wastewater treatment plant. Water Res. 109, 164–172. http://doi.org/10.1016/j.watres.2016.11.046.

Wu, W.M., Yang, J., Criddle, C.S., 2017. Microplastics pollution and reduction strategies. Frontiers of Environmental Science and Engineering 11 (1), 1–4. http://doi.org/10.1007/s11783-017-0897-7.

CHAPTER 3

Biosensors as an effective tool for detection of emerging water and wastewater pollutants

Twinkle Soni[a], V Vivekanand[b], Nidhi Pareek[a]
[a]Microbial Catalysis and Process Engineering Laboratory, Department of Microbiology, School of Life Sciences, Central University of Rajasthan Bandarsindri, Kishangarh, Ajmer, Rajasthan, India
[b]Centre for Energy and Environment, Malaviya National Institute of Technology, Jaipur, Rajasthan, India

3.1 Introduction

Water is among the three key components that are vital for the survival of mankind. As per the WHO (World Health Organization) report, 829,000 people are estimated to die every year from the water borne illnesses i.e. diarrhoea (WHO, 2021). WHO also published its fourth edition of guidelines for drinking-water quality (GDWQ) which builds on 50 years of recommendations by organization on drinking-water quality. The rising industrialization and urbanization has led to enhanced pollution in water bodies due to the enhancement of both quantity and quality of pollutants. The increased consumption of pharmaceutical products, pesticides, hormones, personal care products, etc. had led to increments in the level of these materials into the water bodies. As majority of these chemicals and compounds are synthetic in nature so there is no natural degradation mechanism exist for their complete degradation. The covalent bonds present in these compounds are very resistant to degradation and hence, contributes to the recalcitrance of the pollutants e.g. plastics.

The first step in the direction to clean the water and make it hygienic for variety of applications, is to diagnose the types and levels of contaminants present (NCBI, 2021) (Table 3.1). However, in many cases the polluted water does not exhibit the altered physical appearance in terms of colour, odour and texture. Hence, due to the lack of physical characteristics in polluted water, an urgent need has been aroused to check the physico-chemical parameters of water in order to analyze the pollution level and pollutant load. The assessment of chemical, physical and bacteriological parameters is required to check the quality of water. These measurements not only required trained staff but also customized apparatus. The physical assessment of water includes pH, conductivity and temperature, while estimation of alkalinity, oxygen, phosphorous and nitrogen compounds present an idea about the chemical parameters.

The traditional methods to detect the pollution in waste water includes are culture based techniques, estimation of BOD and COD along with chromatographic procedures. Limited sensitivity and specificity in addition to long and tedious processes

Biodegradation and Detoxification of Micropollutants in Industrial Wastewater
DOI: https://doi.org/10.1016/B978-0-323-88507-2.00010-5

Table 3.1 Range of compounds present in the waste water and detection limits.

S.No.	Compounds	Source of origin	Guideline value (μg L^{-1})	Method of detection	Limit of detection (μg L^{-1})
1	Acrylamide	Coagulant	0.5	GC	0.032
2	Alachlor	Herbicide	20	GC	0.1
3	Aldicarb	Pesticide	10	HPLC	1
4	Antimony	Alloys	20	AAS	0.01
5	Arsenic	Earth crust	10	ICP-MS	0.1
6	Atrazine	Herbicide	100	GC–MS	0.01
7	Barium	Industrial application	1300	ICP-MS	0.004
8	Bentazone	Herbicide	500	GC	0.1
9	Benzene	Vehicular emissions	10	GC	0.2
10	Boron	Manufacture of glass	2400	ICP-MS	0.15
11	Bromate	Textile	10	IC	0.2
12	Cadmium	Steel and plastic industries	3	ICP-MS	0.01
13	Carbofuran	Pesticide	7	GC	0.1
14	Carbon tetrachloride	CFC production	4	GC–MS	0.1
15	Chloramines	Chlorination process	3000	Colorimetric	10
16	Chlordane	Insecticide	0.2	GC	0.014
17	Chlorine	Disinfectant and bleach	5000	HPLC	0.01
18	Chlorotoluron	Herbicide	30	HPLC	0.1
19	Chlorpyrifos	Insecticide	30	GC	1
20	Chromium	Essential nutrient	50	AAS	0.05
21	Copper	Essential nutrient	2000	ICP-MS	0.02
22	Cyanazine	Herbicide	0.6	GC–MS	0.01
23	2,4-D	Plant hormone	30	GC	0.1
24	2,4-DB	Herbicide	90	GC	1
25	DDT	Insecticide	1	GC	0.011
26	Dibromoethane	Lead scavenger	0.4	GC–MS	0.01
27	Dichloroacetic acid	Chlorination	50	GC	0.1
28	Dichloromethane	Paint stripping	20	GC	0.3
29	DEHP	Plasticizer	8	GC–MS	0.1
30	Endrin	Insecticide	0.6	GC	0.002
31	Ethylbenzene	Petroleum product	300	GC	0.002
32	Fenoprop	Herbicides	9	GC	0.2
33	Fluoride	Dental	1500	IC	0.01
34	HCBD	Pesticide	0.6	GC–MS	0.01

S.No.	Compounds	Source of origin	Guideline value (μg L^{-1})	Method of detection	Limit of detection (μg L^{-1})
35	Isoproturon	Herbicide	9	HPLC	0.01
36	Lead	Battery	10	AAS	1
37	Lindane	Insecticide	2	GC	0.01
38	Mecoprop	Herbicide	10	GC–MS	0.01
39	Mercury	Electrical appliance	6	AAS	0.05
40	Metolachlor	Herbicide	10	GC	0.75
41	Molinate	Herbicide	6	GC–MS	0.01
42	Nickel	Steel	70	ICP-MS	0.1
43	NTA	Detergents	200	GC	0.2
44	Pendimethalin	Herbicide	20	GC–MS	0.01
45	Selenium	S-containing mineral	40	AAS	0.5
46	Simazine	Herbicide	2	GC–MS	0.01
47	Styrene	Plastic	20	GC	0.3
48	2,4,5-T	Herbicide	9	GC	0.02
49	Terbuthylazine	Herbicide	7	HPLC	0.1
50	Tetrachlorethene	Dry cleaning industry	40	GC	0.2
51	Toluene	Blending in petrol	700	GC	0.13
52	Trichloroacetic acid	Chlorination	200	GC–MS	1
53	Trichloroethene	Metal degreasing	20	GC	0.01
54	Trifluralin	Herbicide	20	GC	0.05
55	Uranium	Mineral	30	ICP-MS	0.01

*Electro-thermal atomic absorption spectrometry (AAS).
Inductively coupled plasma mass spectrometry (ICP-MS).
Gas chromatography (GC).
Gas chromatography – mass spectroscopy (GC–MS).
High performance liquid chromatography (HPLC).
Ion chromatography (IC).

restricts the application of these methods and raised the demand for development of more sophisticated, fast and specific methods and technologies to improve the detection. Development of biosensors has provided one such efficient alternative. The assessment of biosensors depends upon various features. Some of the key features are:

1. Accuracy: The signal produced by biosensors is accurate, precise, linear over desirable range and reproducible (Turner et al., 1987).
2. Biocompatibility: Biocompatibility of the biosensor is highly desired.
3. Selectivity: Bio-elements are stable under normal storage environment, specific for the analyte and compatible over many assays.

4. Contamination: The chemicals and bio-elements should be trapped efficiently into the sensor in order to avoid any leakage from the system.
5. Cost: Biosensor must be cheap, portable, small and user friendly.
6. Stability: The biosensor must be stable for a period of at least six months (Brecht, 2005).

3.2 Conventional methods of pollutants' detection in wastewater

The detection of pollutants in wastewater is an important aspect of wastewater management. The quality of treated water is primarily measured by BOD, COD, TSS, chromatography, etc. (Table 3.2) (Peyravi et al., 2020). BOD refers to the biological oxygen demand i.e. the amount of oxygen required to remove organic matter by aerobic bacteria present in water. The permissible limit of BOD in water is 1.5 mg L^{-1} and the method of measurement involves azide modification at 20 °C for 5 days (WEPA-DB, 2021). COD is chemical oxygen demand i.e. amount of oxygen needed to chemically oxidize the organic waste found in water. The COD of drinking water should be equal to or less than 5 mg L^{-1} (WEPA-DB, 2021). TSS are the total suspended solids, it's the dry weight of total insoluble suspended solids in the water. It's maximum concentration varies from type of water. Maximum TSS in waste water generated from sugar cane factory is 100 mg L^{-1} whereas in textile and garment factories it is 40 mg L^{-1}.

The customary methods to estimate water pollution also involves the use of chromatography techniques. Chromatographic separations are based upon the distribution

Table 3.2 Conventional method of pollutant detection vs biosensors.

Category	Pollutants	Conventional analytics	Biosensors
Organic material	Industrial waste water Persistent organic pollutants Organic pesticides Organic herbicides Others	• BOD • COD	• Optical biosensor • Electrochemical • Thermal biosensor
Heavy metals	Pb Cr Cd Hg As Sb Other	• Atomic absorption spectroscopy • Ultraviolet-visible spectroscopy • Chromatography	• Optical biosensor • Electrochemical • Piezoelectric • Microfluidic
Microorganisms	Pathogenic Non-pathogenic	• Culture techniques • Microscopic methods	• Optical biosensor • Electrochemical biosensor • Aptamer-based biosensor

ration of analyte over stationary and mobile phase. Gas chromatography (GC) and high performance liquid chromatography (HPLC) are among the top two chromatographic techniques employed for the detection of pollutants in water. The advancements in the detection methods had led to the use of atomic spectroscopy and fluorescence based detection of pollutants in waste water. Atomic spectroscopy is based upon the principle that atoms in ground state absorb the light of a particular wavelength. This characteristic wavelength is specific for each atom. The quantity of light absorb is used to measure the quantity of atom in the analytical sample. The three major atomic spectroscopy used in detection of pollutants are- atomic absorption spectroscopy, inductively coupled plasma (emission spectroscopy) and atomic mass spectroscopy. Fluorescence based detection of pollutants in waste water is an excellent technique for the detection of organic materials particularly pigments (chlorophyll) in the water (Foley et al., 2019). The basic working principle of fluorescence methods lies in the detection of emitted wavelength. The sample is expose to a range of wavelength, a part of wavelength is absorbed by the sample contaminants and rest are emitted from the sample. This emitted light is detected by the detector providing an illustration of contaminants' concentration in the sample. Additionally, DNA amplification, multiple fermentation tube technique, mass spectroscopy, fluorescence in-situ hybridization, field-flow fractionation and capillary electrophoresis can be used for detection of pollutant load in waste water. However, these standard procedures are associated with a range of constraints viz. time consuming, costly, high sample volume requirement, low sensitivity etc. Furthermore, these limitations could be addressed by employment of biosensors with better selectivity and sensitivity.

3.3 Biosensors

Biosensor is an analytical device used to detect chemical and biological moiety. The key component of a biosensor includes five elements - bioreceptor, interface, transducer elements, computer software and user interface. Bioreceptor is the first and most important element in the biosensor signalling pathway. Bioreceptor comes in direct contact with the analyte of interest. Bioreceptor binds selectively to the analyte which trigger the downstream signalling pathway. The second element is interface. It is a platform where biological interactions between analyte and key component occur. Key component here implies to the enzyme, pH change, temperature variation etc. Interface detect the analyte and produces a measurable signal out of it. The third component varies with the type of biosensor used. Transducer element measures the signal generated from the interface. Apart from detecting the signal, transducer also amplifies it. In case of electrochemical biosensor, the transducing element is electrochemical transducer which convert the transducing signal into the electrochemical signal. Fourth component in the signalling is computer software and programmes. These computer programmes accept the signals from transducing element and convert them into the physical parameters which itself is used to interpret the analysis. Last component in the biosensing pathway is user interface. This is used to present the results obtained from computer programmes. The user interface is generally based upon graphical interface in place of command line interfaces. Antibiotics, enzymes, single stranded DNA and phages are the most common biomolecules that are used for the detection as an analyte of interest (Rackus et al., 2015; Bahadir and Sezgintürk, 2015; Pullano et al., 2018).

The biosensors can be broadly categorized in two types on the basis of their interaction with the element of interest-affinity and catalytic biosensors, where affinity biosensor has permanent or semi-permanent binding the catalytic one have non-permanent interactions. Further, on the basis of transducing element biosensors can be classified as optical, electrochemical, thermal, enzymatic, paper based biosensor etc.

3.3.1 Optical biosensors

Optical biosensors are so far the best type of biosensors employed due to their compact nature and high signal transduction efficiency. The transducing element in these biosensors is the optical transducer. Glucometer employed for the blood glucose detection is an example of optical biosensor. Optical biosensing can be broadly divided into two modes: label free and label based. The label free mode detects the signal directly from the interaction between analyte and transducer whereas, label based mode use a label and later the optical signal is generated. The optical transducer use in these biosensors are majorly of 5 types: SPR and LSPR based, interferometers, resonators, gratings and refractometers (Damborský et al., 2016). The SPR (Surface plasmon resonance) is a physical phenomenon which was first observed in 1902. According to this phenomenon the surface and plasmon resonance occurs at the interface of two media i.e. on the surface of conducting material when illuminated at a specific angle by polarized light. This intensity of light and the angle is proportional to the mass on the surface. In this way SPR offers a direct label free detection of analyte. LSPR (Localized surface plasmon resonance) is based on the fact that metallic nanoparticles like gold, silver possess specific and unique optical properties which are not there in the larger metal structures.

Bioluminescent optical fibre biosensors uses bioluminescent recombinant cells of *E.coli*. This strain of *E.coli* in the presence of genotoxic agent emits a luminescent signal. In addition, the recombinant cells are immobilized on fibre optic which later produces a optrode response with a detection limit of 10 pg L^{-1} in the presence of genotoxin atrazine (Jia et al., 2012). Optical waveguide interferometric biosensors is a combination of optical phase difference measurement methods and evanescent field sensing. This technique is also known as resonance wavelength grating (RWG) and is suitable for the detection of redistribution of cellular response, cellular processes and cellular contents (Zaytseva et al., 2011).

3.3.2 Electrochemical biosensors

This class of biosensors deals with the electrochemical transductance. These biosensors can detect not only biological materials viz. enzyme, tissue, ligands etc. but also non-biological materials like gases. The key advantages of electrochemical biosensors are sensitivity, size, selectivity and real time analysis. There are three major classes of electrochemical biosensors namely-potentiometric, amperometric and conductometric.

The potentiometric biosensors involve the detection of change in potential of a cell. The potential here corresponds to the log value of analyte (Eggins, 2002). That mean change in potential occurs with the change in analyte concentration and change is potential is detected for the analysis. The amperometric or voltametric biosensors measures the current flow in between the two electrodes. The glucose biosensor is a type of amperometric biosensor in which the enzyme glucose oxidase catalyse the reaction of oxygen with glucose and form hydrogen peroxide and gluconolactone. The current generated by the release of hydrogen peroxide is measured and is directly proportional to the glucose in the sample. The amperometric biosensors generally work in synergy with the enzyme based biosensors. The conductometric biosensors are designed in such a way that they measure change in electrical conductivity. This is done by connecting the drain and source electrode with a film of conjugated polymers (Zhang et al., 2017). The resistivity or conductivity so generated is measured against the concentration of analyte. Table 3.3 presents a summarized list of biosensors, analyte detected and the element of detection.

3.3.3 Thermal biosensors

As the name suggests, thermal biosensor measures the heat release. Conventionally, thermal biosensors were used to detect COD of the water. The organic compounds present in water are oxidized in the presence of strong oxidizing agents i.e. permanganates and

Table 3.3 Biosensors: types and detection.

S.no.	Type of biosensor	Analyte	Element	Reference
1	Optical	Mercury (II)	DNA	Knecht et al. (Forzani et al., 2004)
		Lead (II)	DNA	Knecht et al. (Hoa et al., 1992)
		Paraoxon	Alkaline phosphatase	Mostafa (Knecht and Sethi, 2009)
2	Electrochemical	Zinc	*Pseudomonas* sp. B4251	Gruzina et al. (Claude et al., 2007)
		Cobalt	*Bacillus cereus* B4368	Gruzina et al. (Claude et al., 2007)
		Copper	*E.coli* 1257	Gruzina et al. (Claude et al., 2007)
3	Amperometric	Phenol	Mushroom tissue	Silva et al. (Gruzina et al., 2007)
		catechol	DNA	Claude et al. (Esfahani et al., 2019)
		Parathion	Parathion hydrolase	Mostafa (Rogers, 1995)
4	Potentiometric	Simazina	Peroxidase	Salgado et al. (Silva et al., 2010)

dichromates. The oxidation reactions are exothermic in nature, hence release energy in the form of heat. The heat so released was measured using thermal biosensor. The high detection efficiency of thermal biosensors can be assumed with the fact that thermal biosensors are insensitive to the optical and electrochemical properties of water. The other type of thermal biosensor is MEMS i.e. micro electro-mechanical system technology. The key advantage of thermal biosensors is there long term stability as there is no direct contact between transducer and sample, less cost and high accuracy. Whereas, the disadvantage lies in its complexity, poor reputation and sensitivity (Dharshni et al., 2019). However, there are a range of applications possible for the thermal biosensors due to their universal detection principle. The recent application of thermal biosensor relay on its easy to maintain and operate.

3.3.4 Enzymatic biosensors

The enzymatic biosensor is an analytical device used in collaboration of enzyme and transducer to product signal proportional to the concentration of target analyte (Fig. 3.1). The signal generated can be due to proton change, gaseous exchange, temperature fluctuation, light emission etc. Enzymatic biosensors are so diverse in use that a range of transducers are available for them viz. potentiometric transducer, amperometric transducer, thermal transducers etc. Enzymatic biosensors when fabricated should maintain two important qualities- long-term use and is operational stability. Both of these

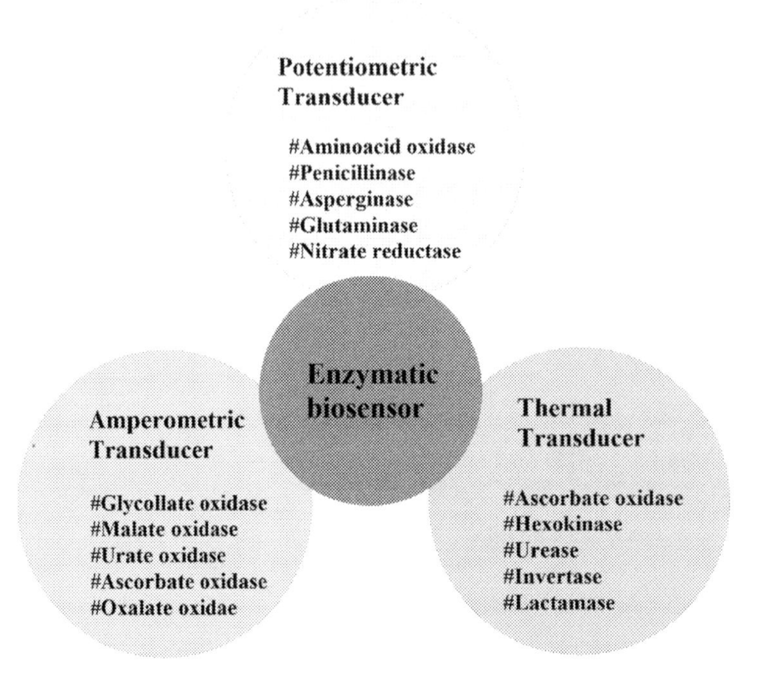

Fig. 3.1 *Enzyme in synergy with various transducers.*

qualities are directly or indirectly dependent upon the immobilization of enzyme on the matrix. Enzyme immobilization could be performed either by physical and/or chemical methods. The most common method of enzyme immobilization involves entrapment, covalent binding, cross linking and adsorption. All these methods possess some major advantages and disadvantages. Entrapment when use for immobilization have advantage of being gentle and high diffusion barrier is the key disadvantage. Covalent binding has drawback of being single time matrix whereas the advantage is to its low diffusion resistance. Cross linking immobilization possess drawback of involving harsh chemicals while, the adsorption possess drawback of having weak bonds. The application of enzyme biosensor lies in its flexibility to employ in medical and clinical diagnostics and research (Mulchandani and Rogers, 1998).

3.3.5 Paper based biosensors

The paper based biosensor are high in demand due to their high liquid wicking rate, portability, real time monitoring, fibre surface affinity and porosity. They are highly employed in various sectors viz. environmental, clinical and nutraceuticals (Liana et al., 2012). The detection of phosphate via paper based biosensor has been made possible by Cinti et al. (2016). They combined the two cutting edge technologies of screen printing and wax-printing (Cinti et al., 2016). Paper based biosensors are highly in trend for the detection of lead, nickel, Copper, chromium, cadmium, gold and iron (Rattanarat et al., 2014). With the help of paper based biosensors, the detection limit of chromium was estimated to be 0.25 ng. Cinti et al. recently developed ready to use, fully integrated paper based biosensor for the detection of paraoxon (nerve agent simulant) in the environment with a limit of detection of 3 $\mu g\ L^{-1}$. In order to improve the sensitivity Cinti et al. used the carbon black/Prussian blue nanocomposite was used as a working electrode (Cinti et al., 2016).

3.3.6 Other biosensors

The extensive application of biosensor has led to extensive and intense research in the direction of fabrication and designing of a range of biosensors. There are various other range of biosensors available for various applications.

3.3.6.1 Piezoelectric biosensors

Piezoelectric biosensor are based upon the piezoelectric effect which originated in 20th century. Piezoelectric effect was given by two physicist Pierre curie and Jacques curie. When anisotropic materials i.e. material which don't have a centre of symmetry can generate electric dipole when squeezed mechanically. The electric dipole so generate is known as piezoelectricity. The phenomenon reverse upon the voltage imposition. Although a lot of crystals possess these piezoelectric effects but only a few can be employed for the biosensor. Quartz crystal is one of them due to its good availability and high reliability. Apart from organic and inorganic materials various biological molecules also possess these effects. Protein ion channels, cellulose and collagen are excellent examples of piezoelectric effect in biological molecules (Rattanarat et al., 2014; Ravi et al., 2012).

3.3.6.2 Microfluidic biosensors

Microfluidics is a branch of science that deal with the small amount of fluid (10^{-9} to 10^{-18} L) with the use of channel in dimension of 10–100 um. The most common material employed for the fabrication of microfluidic device is PDMS (Poly-dimethylsiloxane) due to its transparent nature. Other materials used for fabrication are glass, silicon, polyolefin and polycarbonates. The major advantage of microfluidic devices lies in the consumption of small amount of liquid sample and compact size. Microfluidic biosensors provide a stable and close environment for the assay hence enhance the performance parameters. Microfluidic channels significantly reduce the sensing area and therefore shortened the sensing time with higher sensitivity (Rajala et al., 2016).

3.3.6.3 Immunochemical biosensors

Immunochemical biosensors are solid state devices which use immunochemical reactions along with the transducer. Along with the antibiotics and aptamers the immunoreactions form highly sensitive and selective biosensors. Currently, a range of biomolecules more specifically biomarkers are available which express the presence of pathological conditions, has been in use for immunoassay. Depending upon the transducing elements the immunosensors can be further divided into four types: electrochemical, optical, thermometric and micro-gravimetric. All these types of biosensors can be run as non-labelled (direct) or labelled (indirect). The most commonly labelled immunosensors employed are enzymes i.e. alkaline phosphates, glucose oxidase, luciferase etc. (INTECHOPEN, 2021).

3.4 Advantages of biosensors over conventional detection methods

The quality of water is the key component that effects the human life by influencing food, health, economy and energy. A range of wastewater monitoring systems have been set up, however the traditional methods possess some major drawbacks of being costly, time consuming and low efficiency. Biosensors can be considered as an alternative method for both in-situ and real time monitoring of wastewater. Biosensors are highly effective as compare to the conventional methods of detection. The effectiveness of biosensors is mainly due to their lower limits of detection and dynamic range. Limit of detection is an intuitive quantity which indicates the lowest concentration which can be measured and is well differentiated from the noise. Signal to noise ratio below one is not accepted by any type of sensors. The data obtained by biosensors can be of two type i.e. transient (kinetic) and end point (equilibrium). On the contrary, the traditional methods provide only end-point assay. The transient data obtained from the biosensors can be used to study the kinetic parameters of the assay.

3.5 Nanotechnology for biosensors

Nanotechnology had opened up the new dimensions in the field of pollutant detection. Nanomaterials are unique in their structure, electrical and catalytic properties (INTECHOPEN, 2021). Nanomaterials possess some major advantages over the standard regular methods owing to their small size, fast signal response, portability, high sensitivity and high surface to volume ratio. The use of nanomaterial as a detector has increased rapidly due to their compact size, excellent electrical and catalytic properties, portable and fast signal response (Mao et al., 2015; Willner and Vikesland, 2018). There are three major ways to employ nanomaterials for pollutant detection and treatment of waste water.

3.5.1 Nanophotocatalysis

Nanophotocatalysis is the decomposition of compounds in the presence of light (UV/Sunlight/Visible). The photocatalyst involves in this process induce the chemical transformation in compounds without involving itself directly. The high surface ratio and shape dependent feature of nanocatalyst enhances the reactivity of catalyst involved in waste water treatment (Chen et al., 2019). The chemistry behind the action of nanophotocatalyst lies in its flexible oxidation ability. Nanophotocatalyst produces the oxidizing species on the surface of material which in turn help in the degradation of pollutants in water (Gómez-Pastora et al., 2017). There are various photocatalystviz. SiO_2, ZnO, Al_2O_3, however TiO_2 is most studied and extensively used as photocatalyst due to its low cost, chemically stable nature, non-toxicity and easy availability. TiO_2 tubes have been utilized for the removal of organic water pollutants viz. azo dyes, congo red, toluene, dichlorophenol, trichlorobenzene, chlorinated ethane, phenol aromatic base pollutants etc. (Qu et al., 2013; Sadegh et al., 2017; Raliya et al., 2017; Liang et al., 2019; Bhatia et al., 2017). Moreover, the process of nanophotocatalysis may occur in two ways i.e. homogenously and heterogeneously. The heterogeneous method of nanophotocatalysis is widely employed due to its wider scope to decontaminate water. The photocatalyst works via two ways i.e. degradation and mineralization. The degradation of compounds using nanophotocatalyst implies to the decomposition of compound (pollutant) into its simpler intermediates whereas, the mineralization is the complete breakdown of compound into mineral form i.e. water, CO_2 and inorganic compounds. The extensive use and advantage of nanophotocatalysts in water treatment is limited by few factors viz. recovery of catalyst from mixture and toxicity. Although the researchers are continuously exploring new and diverse nanophotocatalyst which can be less or non-toxic to the nature. Moreover, one alternative has been suggested for the recovery of nanophotocatalyst from the mixture is to use magnetic nanophotocatalyst in which the magnetic waves can be applied to recollect the catalyst following completion of the reaction.

3.5.2 Nanomotors and micromotors

Nanomotors and micromotors are recently developed field of nanomaterial. The first nanomotor was synthesised in 2015, by researcher of university of California to deliver a cargo of nanoparticles to stomach. Nano and micromotors are devices which are designed in such a way that they respond to particular stimuli and response in the form of moments. These moments can be rotation, shuttling, delivery, rolling, contraction etc. These devices can work upon all the type of stimuli i.e. physical, chemical or biological. The major advantages of nanomotor are their high speed, specific control movement, high power, self-mix ability etc. (Review et al., 2020). Various applications of nanomotors and their mechanism has been described in Fig. 3.2.

3.5.3 Nanomembranes

Nanomembrane are the nanofibers aligned and coated on a membrane in a particular order which is then used for the remediation of undesirable pollutants. The membrane

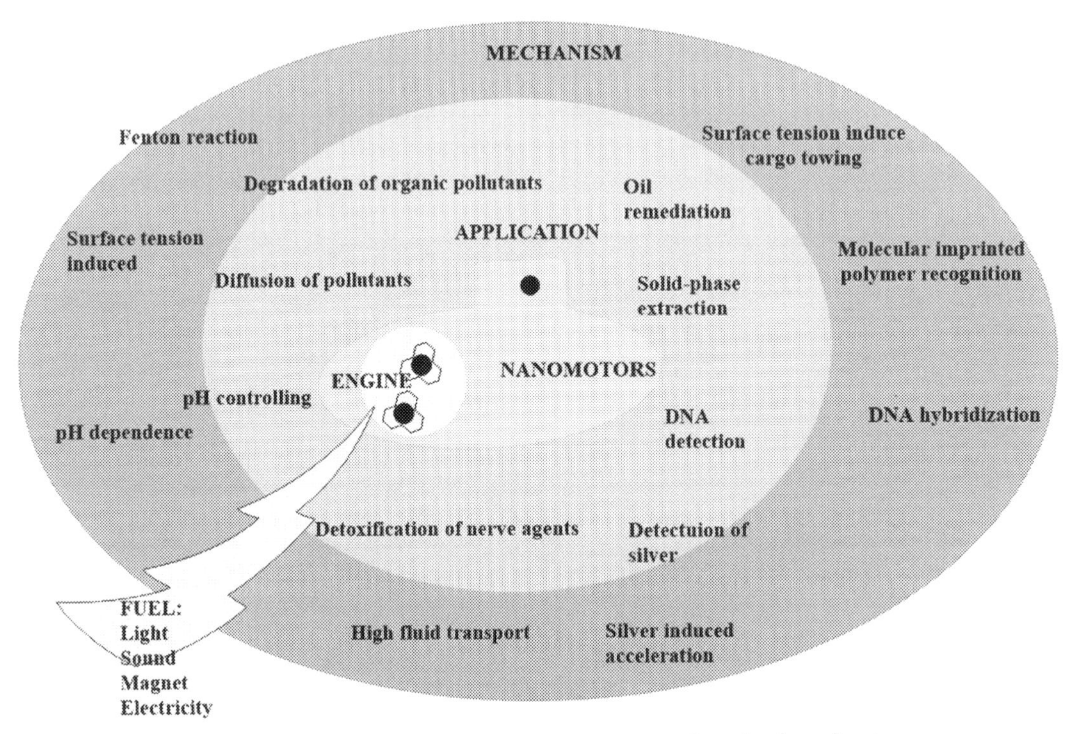

Fig. 3.2 *Applications of Nano-motors in waste water treatment and mechanism of action.*

used for the water treatment is a porous membrane which is used for the reverse osmosis, ultrafiltration, nanofilteration etc. This porous membrane is supported by another layer of membrane which is generally of carbon-based material (because CNT possess a property to reduce fouling, reduce chances of mechanical failure, anti-microbial in nature and produce biofilm) (Jaksic and Jaksic, 2020). Some antimicrobial materials are also doped into the membrane to enhance the efficiency of nanomembranes. There are various types of nanomembranes possessing advantages, uses and disadvantages.

3.5.3.1 Nanofiber membranes

Nanofiber membranes are used in filter cartridge for ultrafiltration and pre-filtration. The major advantage of nanofiber membrane lies in its excellent porosity and high permeate efficiency. However, the pores may get easily blocked which make the membrane of limited use (Peng et al., 2014).

3.5.3.2 Nanocomposite membrane

Nanocomposite membranes are fouling resistant, possess high hydrophilicity, exhibit notable mechanical and thermal properties in contrast to discharge high amount of nanoparticles in the surroundings (Yang et al., 2010).

3.5.3.3 Aquaporin-based membrane

These type of membranes are frequently used in desalination due to their high ion selectivity and better permeability. The only drawback of such membranes is less mechanical stability (Tang et al., 2015).

3.5.3.4 Self-assembling membranes

They are used for ultrafiltration as they possess homogenous nanopores. However, these membranes are difficult to use on industrial scale (Cornwell and Smith, 2015).

3.5.3.5 Nanofilteration membrane

The main application of nanofilteration membrane lies in its ability to remove colour, odour and hardness due to better selectivity and charge based repulsion method (Asatekin et al., 2006).

The main aim behind using nanomembranes lies in separation of toxic pollutants from water. Nanomembranes are so efficient that they are also used to check the water safety level (Bhat et al., 2018). Nanomembrane are better as compared to the standard methods because they independent on any charged ion for the exchange of pollutants in water like Na^+. However, the major drawbacks associated with nanomembrans are membrane stability and fouling. Nanomembranes are stable for a limited period of time and after that the stability decrease. The second issue is fouling of nanomembrane which arises after its one-time use making the membrane inefficient and expensive.

3.6 Conclusions

Water pollution is a key challenge that needs to be addressed by environmental scientists now-a-days. Pollutant detection and treatment are two critical aspects of waste water treatment. Conventional approaches for the same have been utilized even till now, however these have some limitations. Rapid industrialization and urbanization has led to the generation of novel pollutant species, whose detection could be done by specific biosensors. Biosensor based pollutant detection methods offer specificity and sensitivity for rapid and precise detection that could further ease the treatment. Intervention of nano-materials would further advance the detection methods. Designing of nanomaterials based biosensors for detection of specific pollutants would provide a sustainable way to enhance and develop proficient treatment processes for industrial effluents.

References

https://www.intechopen.com/books/biosensors-micro-and-nanoscaleapplications/immunosensors 2015.

https://www.intechopen.com/books/advances-in-microfluidic-technologies-for-energy-and-environmental-applications/application-of-microfluidics-in-biosensors 2020.

http://www.wepa-db.net/policies/law/laos/standards.htm 2021.

https://www.ncbi.nlm.nih.gov/books/NBK442375/ 2017.

https://www.who.int/news-room/fact-sheets/detail/drinking-water 2019.

Asatekin, A., Menniti, A., Kang, S., Elimelech, M., Morgenroth, E., Mayes, A.M., 2006. Antifouling nanofiltration membranes for membrane bioreactors from self-assembling graft copolymers. J. Memb. Sci. 285 (1–2), 81–89. http://doi.org/10.1016/j.memsci.2006.07.042.

Bahadir, E.B., Sezgintürk, M.K., 2015. Applications of commercial biosensors in clinical, food, environmental, and biothreat/biowarfare analyses. Anal. Biochem. 478 (March), 107–120. http://doi.org/10.1016/j.ab.2015.03.011.

Bhat, A.H., Rehman, W.U., Khan, I.U., et al., 2018. Nanocomposite Membrane for Environmental Remediation. Elsevier Ltd. http://doi.org/10.1016/B978-0-08-102262-7.00015-5.

Bhatia, D., Sharma, N.R., Singh, J., Kanwar, R.S., 2017. Biological methods for textile dye removal from wastewater: a review. Crit. Rev. Environ. Sci. Technol. 47 (19), 1836–1876. http://doi.org/10.1080/10643389.2017.1393263.

Brecht, A., 2005. Multianalyte bioanalytical devices: scientific potential and business requirements. Anal. Bioanal. Chem. 381 (5), 1025–1026. http://doi.org/10.1007/s00216-004-2912-7.

Chen, W., Liu, Q., Tian, S., Zhao, X., 2019. Exposed facet dependent stability of ZnO micro/nano crystals as a photocatalyst. Appl. Surf. Sci. 470 (August 2018), 807–816. http://doi.org/10.1016/j.apsusc.2018.11.206.

Cinti, S., Talarico, D., Palleschi, G., Moscone, D., Arduini, F., 2016. Novel reagentless paper-based screen-printed electrochemical sensor to detect phosphate. Anal. Chim. Acta 919, 78–84. http://doi.org/10.1016/j.aca.2016.03.011.

Claude, D., Houssemeddine, G., Andriy, B., Jean-Marc, C., 2007. Whole cell algal biosensors for urban waters monitoring. Novatech 7, 1507–1514.

Cornwell, D.J., Smith, D.K., 2015. Expanding the scope of gels - Combining polymers with low-molecular-weight gelators to yield modified self-assembling smart materials with high-tech applications. Mater Horizons 2 (3), 279–293. http://doi.org/10.1039/c4mh00245h.

Damborský, P., Švitel, J., Katrlík, J., 2016. Optical biosensors. Essays Biochem. 60 (1), 91–100. http://doi.org/10.1042/EBC20150010.

Dharshni, R., Dharshini, R., Sivarajasekar, N., 2019. Thermal Biosensors and Their Applications. Am. Int. J. Res. Sci. (December), 50. http://www.iasir.net.

Eggins, B.R., 2002. Chemical sensors and biosensors. Wiley, Chichester, United Kingdom.

Esfahani, M.R., Aktij, S.A., Dabaghian, Z., et al., 2019. Nanocomposite membranes for water separation and purification: fabrication, modification, and applications. Sep. Purif. Technol. 213, 465–499. http://doi.org/10.1016/j.seppur.2018.12.050.

Foley, J., Batstone, D., Keller, J., 2019. The R & D challenges of water recycling-technical and environmental horizons. In Advanced wastewater management centre.

Forzani, E.S., Zhang, H., Nagahara, L.A., Amlani, I., Tsui, R., Tao, N., 2004. A conducting polymer nanojunction sensor for glucose detection. Nano Lett. 4 (9), 1785–1788. http://doi.org/10.1021/nl049080l.

Gómez-Pastora, J., Dominguez, S., Bringas, E., Rivero, M.J., Ortiz, I., Dionysiou, D.D., 2017. Review and perspectives on the use of magnetic nanophotocatalysts (MNPCs) in water treatment. Chem. Eng. J. 310, 407–427. http://doi.org/10.1016/j.cej.2016.04.140.

Gruzina, T.G., Zadorozhnyaya, A.M., Gutnik, G.A., et al., 2007. A bacterial multisensor for determination of the contents of heavy metals in water. J. Water Chem. Technol. 29 (1), 50–53. http://doi.org/10.3103/S1063455X07010080.

Hoa, D.T., Srinivasa, R.S., Contractor, A.Q., Suresh Kumar, T.N., Punekar, N.S., Lal, R., 1992. Biosensor Based on Conducting Polymers. Anal. Chem. 64 (21), 2645–2646. http://doi.org/10.1021/ac00045a031.

Jaksic, Z., Jaksic, O., 2020. Biomimetic nanomembranes: an overview. Biomimetics, 5 (24) 1–45.

Jia, K., Eltzov, E., Toury, T., Marks, R.S., 2012. Ionescu RE. A lower limit of detection for atrazine was obtained using bioluminescent reporter bacteria via a lower incubation temperature. Ecotoxicol. Environ. Saf. 84, 221–226. http://doi.org/10.1016/j.ecoenv.2012.07.009.

Knecht, M.R., Sethi, M., 2009. Bio-inspired colorimetric detection of Hg2+ and Pb2+ heavy metal ions using Au nanoparticles. Anal. Bioanal. Chem. 394 (1), 33–46. http://doi.org/10.1007/s00216-008-2594-7.

Liana, D.D., Raguse, B., Justin Gooding, J., Chow, E., 2012. Recent advances in paper-based sensors. Sensors (Switzerland) 12 (9), 11505–11526. http://doi.org/10.3390/s120911505.

Liang, X., Cui, S., Li, H., Abdelhady, A., Wang, H., Zhou, H., 2019. Removal effect on stormwater runoff pollution of porous concrete treated with nanometer titanium dioxide. Transp Res Part D Transp Environ 73, 34–45. http://doi.org/10.1016/j.trd.2019.06.001.

Mao, S., Chang, J., Zhou, G., Chen, J., 2015. Nanomaterial-enabled Rapid Detection of Water Contaminants. Small 11 (40), 5336–5359. http://doi.org/10.1002/smll.201500831.

Mulchandani, A., Rogers, K.R., 1998. Enzyme and Microbial Biosensors Techniques and Protocols. Methods Biotechnol 6, 3–14.

Peng, F.Q., Ying, G.G., Yang, B., et al., 2014. Biotransformation of progesterone and norgestrel by two freshwater microalgae (Scenedesmus obliquus and Chlorella pyrenoidosa): transformation kinetics and products identification. Chemosphere, 581–588. http://doi.org/10.1016/j.chemosphere.2013.10.013.

Peyravi, M., Jahanshahi, M., Tourani, H., 2020. Analytical Methods of Water Pollutants Detection. doi: http://doi.org/10.1016/b978-0-12-818965-8.00006-8.

Pullano, S.A., Critello, C.D., Mahbub, I., et al., 2018. EGFET-based sensors for bioanalytical applications: a review. Sensors (Switzerland) 18 (11). http://doi.org/10.3390/s18114042.

Qu, X., Alvarez, P.J.J., Li, Q., 2013. Applications of nanotechnology in water and wastewater treatment. Water Res. 47 (12), 3931–3946. http://doi.org/10.1016/j.watres.2012.09.058.

Rackus, D.G., Shamsi, M.H., Wheeler, A.R., 2015. Electrochemistry, biosensors and microfluidics: a convergence of fields. Chem. Soc. Rev. 44 (15), 5320–5340. http://doi.org/10.1039/c4cs00369a.

Rajala, S., Siponkoski, T., Sarlin, E., et al., 2016. Cellulose Nanofibril Film as a Piezoelectric Sensor Material. ACS Appl. Mater. Interfaces 8 (24), 15607–15614. http://doi.org/10.1021/acsami.6b03597.

Raliya, R., Avery, C., Chakrabarti, S., Biswas, P., 2017. Photocatalytic degradation of methyl orange dye by pristine titanium dioxide, zinc oxide, and graphene oxide nanostructures and their composites under visible light irradiation. Appl. Nanosci. 7 (5), 253–259. http://doi.org/10.1007/s13204-017-0565-z.

Rattanarat, P., Dungchai, W., Cate, D., Volckens, J., Chailapakul, O., Henry, C.S., 2014. Multilayer paper-based device for colorimetric and electrochemical quantification of metals. Anal. Chem. 86 (7), 3555–3562. http://doi.org/10.1021/ac5000224.

Ravi, H.K., Simona, F., Hulliger, J., Cascella, M., 2012. Molecular origin of piezo- and pyroelectric properties in collagen investigated by molecular dynamics simulations. J. Phys. Chem. B 116 (6), 1901–1907. http://doi.org/10.1021/jp208436j.

Review, W.A., Yaqoob, A.A., Parveen, T., Umar, K., 2020. Role of Nanomaterials in the Treatment of. Water (Basel) 12, 495.
Rogers, K.R., 1995. Biosensors for environmental applications. Biosens. Bioelectron. 10 (6–7), 533–541. http://doi.org/10.1016/0956-5663(95)96929-S.
Sadegh, H., Ali, G.A.M., Gupta, V.K., et al., 2017. The role of nanomaterials as effective adsorbents and their applications in wastewater treatment. J. Nanostructure Chem. 7 (1), 1–14. http://doi.org/10.1007/s40097-017-0219-4.
Silva, L.M.C., Salgado, A.M., Coelho, M.A.Z., 2010. Agaricus bisporus as a source of tyrosinase for phenol detection for future biosensor development. Environ. Technol. 31 (6), 611–616. http://doi.org/10.1080/09593331003592238.
Tang, C., Wang, Z., Petrinić, I., Fane, A.G., 2015. Hélix-Nielsen C. Biomimetic aquaporin membranes coming of age. Desalination 368, 89–105. http://doi.org/10.1016/j.desal.2015.04.026.
Turner, A., Karube, I., Wilson, G.S., 1987. Biosensors: fundamentals and applications. Oxford university press.
Willner, M.R., Vikesland, P.J., 2018. Nanomaterial enabled sensors for environmental contaminants Prof Ueli Aebi, Prof Peter Gehr. J. Nanobiotechnology 16 (1), 1–16. http://doi.org/10.1186/s12951-018-0419-1.
Yang, Z., Peng, H., Wang, W., Liu, T., 2010. Crystallization behavior of poly(ε-caprolactone)/layered double hydroxide nanocomposites. J. Appl. Polym. Sci. 116 (5), 2658–2667. http://doi.org/10.1002/app.
Zaytseva, N., Miller, W., Goral, V., Hepburn, J., Fang, Y., 2011. Microfluidic resonant waveguide grating biosensor system for whole cell sensing. Appl. Phys. Lett. 98 (16). http://doi.org/10.1063/1.3582611.
Zhang, Z., Zeng, K., Liu, J., 2017. Immunochemical Detection of Emerging Organic Contaminants in Environmental Waters. Trends Anal. Chem. Published online. http://doi.org/10.1016/j.trac.2016.12.002.

CHAPTER 4

Removal of organic micro-pollutants by aerobic and anaerobic microorganism

Md. Lawshan Habib, M. Mehedi Hasan, Shovendu Biswas, Mohd. Tanzir Hossain, Md. Anwaruzzaman, Md. Kamruzzaman
Department of Applied Chemistry and Chemical Engineering, Bangabandhu Sheikh Mujibur Rahman Science and Technology University, Gopalganj, Bangladesh

4.1 Introduction

Micropollutants (MPs) can be defined as anthropogenic chemicals that occur in the (aquatic) environment well above a (potential) natural background level due to human activities but with concentrations remaining at trace levels (i.e. up to the microgram per liter range). Thus, MPs are defined by their anthropogenic origin and their occurrence at low concentrations. Thousands of chemicals fall into this category (Schwarzenbach et al., 2006). MPs can consist of purely synthetic chemicals, such as strongly halogenated molecules (e.g. fluorinated surfactants), or of natural compounds such as antibiotics (e.g. penicillines) or estrogens. MPs may originate from a wide range of sources (e.g. agriculture, households, traffic networks or industries) and enter water bodies through diverse entry paths. Sources of OMPs are: (1) industrial wastewater, (2) runoff from agriculture, livestock, and aquaculture; (3) landfill leachates; and (4) domestic and hospital effluents (Barbosa et al., 2016). The OMPs comprise a wide range of natural and synthetic organic compounds, which include pharmaceuticals, detergents, steroid hormones, industrial chemicals, pesticides, and many others (Wanda et al., 2018). Global OMPs production, between 1930 and 2000, has increased from 1 million to 400 million tons per each year (Gavrilescu et al., 2015). In the European Union (EU) more than 100,000 compounds have been registered of which 30,000 to 70,000 of them are used daily. Most of them end up to the aquatic environments. Natural waters receive about 300 million tons of synthetic compounds annually from industrial and consumer products effluents (Talib and Randhir, 2017).

Environmental pollution by organic micropollutants (OMPs) is nowadays of great concern, especially, when it affects the aquatic environment. For many years, quantification of water pollution was restricted to monitoring biochemical oxygen demand (BOD), chemical oxygen demand (COD), nitrates, phosphates and total suspended solids. Paralleling the bio/analytical progresses, the focus on macropollutants related to extensive industrial and agricultural activities is being enlarged to micropollutants belonging to diverse classes of chemicals which are detected in trace amounts (C. G. Daughton and Ternes, 1999). The most micropollutants are not included into

Biodegradation and Detoxification of Micropollutants in Industrial Wastewater
DOI: https://doi.org/10.1016/B978-0-323-88507-2.00003-8

routine monitoring for surface and drinking water yet (Wanda et al., 2017; Wanda et al., 2017), therefore, compared to other anthropogenic contaminants, the OMPs have largely been outside the scope of monitoring and worldwide regulations. Also there is not enough data on their levels, occurrence, and fate in environment (Wanda et al., 2017). Today wastewater effluents reuse for agricultural applications and land amendment is one of the main challenges among scientists, policy makers, and stakeholders (Cerreta et al., 2020).

Insufficient removal of OMPs by conventional wastewater treatment processes and lack of precautions and monitoring actions for OMPs have been caused that many of these compounds act as a threat for human and wildlife in the aquatic environments. The occurrence of OMPs at concentration between few nanogram per liter and several microgram per liter (Wanda et al., 2017) in the aquatic environment has been commonly associated with a number of negative effects such as chronic and acute toxicity, endocrine disrupting effects, and antibiotic resistance of microorganisms (Luo et al., 2014). Although the effects of micropollutants in aquatic environments are not very well known yet, there are clear indications that they have long-term impacts on ecosystem. Reasons for this are: (1) their potential to accumulate into aquatic organisms and human bodies (bioaccumulation), (2) their toxicity, and (3) their resistance to degradation in the environment (persistency). Regulations on their emission and discharge are thus vital for improving the aquatic environment and surface water quality (Chau et al., 2018).

Available water sources globally are limited because of several factors, such as domestic, agricultural and industrial uses, drought, global climate change, the increase of population density, and the continuous extraction of water from groundwater resources. Therefore for reuse of water, OMPs removal during water and wastewater treatment processes is necessary (Besha et al., 2017).

In this chapter, removal strategies of organic micro-pollutants by aerobic and anaerobic microorganism will be discussed extensively.

4.2 Organic micropollutants

Mostly organic micropollutants (OMPs) are found in wastewater and treated water and are of synthetic chemicals, remain at a concentration in the range of microorganism per liter, doesn't matter if it's too low. Since 1945, chemicals have been increased from 9 to 450 million tons due to its tremendous global production. Through the publish of 'Silent Spring' by Rachel Carson in 1962, the world had observed and investigated high and considerable detail on the harmful effects of these chemicals especially persistent organic pollutants. The effluents of many developed countries contain 5 out of 7 metabolites, 28 out of 35 pharmaceuticals with excessive concentration and sometime more than 1.7 μg/L (Larsen et al., 2004). Among them additives, excipients, EDS and pharmaceutical ingredients are most. The developing countries are going through a worse situation i.e., the concentrations and number of chemicals are exceedingly in danger. The major reason behind this may be due to the counter-compound used for neutralization for the other compounds, causes the concentration of micropollutants in environment (García-Galán et al., 2010). The far-reaching scientific field includes that the smart management and strategies must be increased and performed (Brack et al., 2015). However important rules and legislation should be developed for the permissible exposure of these chemicals and also be cautious about the fates and danger for the chemicals.

4.2.1 Sources

OMPs is an operational definition and they contain a group of compounds, which are covered by existing water quality regulations if they are in lower concentrations (i.e., nanogram per litter to microgram per litter) but are they have the potential risks to environmental ecosystems (Farré et al., 2008). The group assumes more than 22 types, which are found in the aquatic environment (Geissen et al., 2015). The active pharmaceutical compounds, additives, preservatives, detergents, surfactants, flame retardants, plasticizers, personal care products (PCPs), pesticides, industrial chemicals, disinfection by-products (DBPs), perfluorinated compounds (PFCs), endocrine-disrupting chemicals (EDCs), and their transformation products are the prominent classes (Ojajuni et al., 2015) (Table 4.1).

4.2.2 Types of organic micropollutants

OMPs to enter the environment in a different way; in general, we can distinguish between point sources and diffuse sources. The sources which are single locations are clearly distinguishable from other sources, but diffuse sources are rather intangible and are individualized for broad geographical scales (Lapworth et al., 2012). Examples for this are on the one side industrial effluents, hospital effluents, wastewater, and sewage treatment plants, waste disposal sites, septic tanks, and on the other side storm-water/urban runoffs, agricultural runoffs due to the application of sewage sludge, leakages from sewer systems (Lapworth et al., 2012). And micropollutants are classified depending on their source and occurrence OMPs, as is mentioned below.

4.2.2.1 Human and animal associated OMPs

Most OMPs are derivatives and pharmaceuticals and they are noted as drugs of human and animal, antibiotics, endocrine disruptive chemicals (EDCs), and (PCPs). The above chemicals are planned to govern the metabolic reactions into animals and human bodies. OMPs from pharmaceuticals have been directly found out in dirty water, deposited from the crap of humans and animals. Unused drugs produce active pharmaceutical compounds and waste from hospitals also find their waterways. Only some of these chemicals from water can be removed by Sewage treatment plants through a basic filtration process and metabolic way, while others pass through integral. It is affected by the

Table 4.1 Summary of major organic micropollutants (OMPs), their classes, mode of entry and example (Ellis, 2006; Farré et al., 2008; Verlicchi et al., 2010).

OMPs	Class	Entry	Example
Drugs of veterinary and human pharmaceuticals	Antibiotics	Discharges from hospital, accidental spills and wastes of farmland	Cefazolin, amoxicillin, chlortetracycline, ciprofloxacin, erythromycin, ciprofloxacin, lincomycin, doxycycline, ofloxacin, sulfamethoxazole, oxytetracycline, norfloxacin, tetracycline, penicillin, trimethoprim.
	Antidiabetics		Glibenclamide
	Analgesics		Codeine, dipyrone, diclofenac, acetaminophen, acetylsalicylic acid, indomethacin, fenoprofen, ibuprofen, fluoxetine, ketoprofen, paracetamol, mefenamic acid, propyphenazone, naproxen, salicylic acid.
	Psychiatric drugs		Carbamazepine, gabapentin, diazepam, phenytoin, primidone, salbutamol.
	X-ray contrast agents		Iopromide, diatrizoate, iopamodole.
	Cardiovascular drugs (β-blockers)		Atenolol, propranolol, metoprolol, solatolol, sotalol, timolol.
	Drugs of abuse		Cocaine, tetrahydrocannabinol, amphetamine.
	Veterinary drugs		Flunixin.
Endocrine disruptive chemicals (EDCs)	Hormones and steroids		Estradiol, estriol, estrone, androstenedione, diethylstilbestrol, ethinylestradiol, testosterone, progesterone.
Personal care products (PCPs)	Antiseptics	Disposal of industrial waste and shower waste	Triclosan, chlorophane
	Fragrances and synthetic musk		Galaxolide, musk ketone, nitro polycyclic, and macrocyclic musk; phthalates, tonalide.
	Stimulants		Caffeine
	Antihypertensives		Diltiazem
	UV Filters		Benzophenone, methyl benzylidene camphor.
	Insect repellents		N, N-diethyltoluamide.

OMPs	Class	Entry	Example
Agriculture	Pesticide	Agriculture waste	
	Herbicide		Terbuthylazine, MCPA, diuron, monocrop
Additives	Gasoline	Disposal of exhausted engine oil and mobile exhaust	Diallyl ether, methyl t-butyl ether
	Industrial	Municipal waste and food	Chelating agent (EDTA), aromatic sulfonates
Fluorinated compounds (PFCs) surfactants, detergents	Perfluorooctanoic acid Perfluoro octane sulfonate	Laundries, households, industries and agricultural pesticides, diluents dispersants	Alkylphenol ethoxylates, alkylphenols, alkylphenol carboxylates
Flame retardants		Industries and household items (furniture, electronics, appliances, baby products)	Polybrominated diphenyl ethers, tetrabromobisphenol A, hexabromocyclododecane, C10-C13 chloroalkane
New classes	Nanomaterials	Research industries	1,4-dioxane
	Genetic adaptation and mutations	Horizontal gene transfer in microorganisms	Tet (W), tet (O)

adsorption potential of the chemicals as a function of hydrophobic and/or electrostatic properties (Fent et al., 2006). These chemicals are made kinetically inert; so, most OMPs are allowed to pass through the organism's body which is not modified or in the conjugate's form. Finally, the wastewater contains those constituents and used as irrigation water, rivers, canals, and treatment plants in which the effects depend upon their chemical composition and characteristics (C. G. Daughton and Ternes, 1999). Such as, pharmaceuticals that are in acidic nature (being ionic at neutral pH) have a little adsorption probability to wastewater sludge more than basic pharmaceuticals (Ternes et al., 2004; Urase et al., 2005; Ã and Kikuta, 2005). Therefore, acidic pollutants have more possibility to stay in the water and thus are capable of moving through the surface waters. Basic pharmaceuticals may be removed from the water stream as adhered compounds they can easily adsorb onto the sludge sediments (Golet et al., 2002; Ternes et al., 2002; Ternes et al., 2004). For PCPs and EDCs there also develops a similar result. The high-adsorption efficiency (70–80 percent) of estrogens is a good example of fates of EDC in the wastewater treatment facilities where they exist in μg/L (Auriol et al., 2006; Johnson et al., 2007), and antidepressants found to be partitioned into sludge (Kinney et al., 2006; Anjum et al., 2017). Microbial and photodegradation processes are transferred by the adsorption-dependent result, a variety of OMPs (Klöpffer and Wagner, 2007; Saleem, 2016). Chemical entities with new physicochemical properties have resulted in that transformation, and hence they have fates and affects acute from the initial or primary compounds. Such as, sometimes metabolites present more than the initial chemicals in treated wastewater. This could be due to the fact that the resulting compound is metabolically more persistent (or pseudo persistent) compared to the parent compound after structural modifications (Boxall et al., 2004; C. G. Daughton, 2004; Escher et al., 2020). It is a good example that the formation of higher concentrations or high amounts of effluents in the domestic WWTP (Weigel et al., 2004). Likewise, another study revealed higher levels of desmethylcitalopram compared to its parent compound, citalopram, in effluent wastewater (Vasskog et al., 2008). However, it has not always the similarities as like OMPs which are capable of re-transformed into parent compounds. For instance, in the case of antimicrobials, the concentration of a metabolite of sulfamethoxazole (N4-Acetysulfamethoxazole) was significantly reduced in the effluent of the municipal wastewater (Göbel et al., 2007), though there is no change in the metabolite concentration now and before the treatment (Farré et al., 2008).

4.2.2.2 Agriculture associated OMPs

The environment contains a large number of persistent pesticides but only a little amount is degradable through physicochemical and biological processes and many of them still remain through the thin soil. particles and organic matter evaporate into the air, and/or leaches into the groundwater. Runoff water from agricultural lands carries these residues and their metabolites to lakes, rivers, and canals (Stuart et al., 2012). These residues, containing some of discontinued use, have been found out at trace amount in surface and groundwater the developing and developed countries. Although these pesticides, had been banned in the world for decades, experts have reported metabolites of DDT, heptachlor, and atrazine in groundwater (C. Sinclair et al., 2010). According to the authors, the surface drinking water contains 58 pesticide metabolites and it is imagined that all of these are from the parent compound usage, formation rates in soil, persistence, and mobility, and pesticide activity along with other parameters. But the drinking water containing persistent pesticides and other OMPs present at very low concentrations is highly demanding as well enough. It is also a great challenge for the measurement of transformation products. But their parent compounds are not so toxic as the transformation compounds (C. J. Sinclair and Boxall, 2003) and thus produce a potential risk to the biotic ecosystem. Sometimes, these metabolites show their higher degree more than the parent compounds in groundwater (Kolpin et al., 2004; Lapworth and Gooddy, 2006). A recent example is one of the most widely used herbicides glyphosates. The degradation of glyphosate by microbial process produces aminomethylphosphonic acid (AMPA) which has chronic accumulation and it may cause minor to moderate toxic effects in aquatic organisms and ecosystems. Not only glyphosate but also AMPA have highly water-soluble properties (Kolpin et al., 2006) and noted that the detection of AMPA six times more frequent than glyphosate in polluted surface water.

4.2.2.3 Industry-associated OMPs

Another group of OMPs includes chemicals used in or derived from industrial applications such as carpeting, upholstery, apparel, food paper wrappings, firefighting foams. The chemicals from these sources are declared as the maximum amount of pollutants ingredients, for example, adipates and phthalates, chlorinated solvents, fuel oxygenates, methyl tert-butyl ether, plasticizers/resins, and bisphenols (Moran et al., 2005; Moran et al., 2007; Verliefde et al., 2007) and depending on transformation, the pollutants are changed to metabolites as toxic, such as benzotriazole intermediates, and perfluorinated compounds (PFCs). These compounds are of major concern due to their hydrophilic properties with a resistance to degradation in nature but reports on toxicity for these chemicals are limited. A stabilizer used with 1,1,1-trichloroethane for instance, 1,4-dioxane has been found to readily leach to groundwater which is not bound to soil ingredients (Abe, 1999). At the same time, perfluorinated compounds (PFCs) such as perfluorooctane sulfonate (PFOS) and perfluorooctanoic acid are of notable concern because they have a property of weak carcinogenic (Jahan et al., 2008; Ju et al., 2008; Strynar and Lindstrom, 2008). Moreover, if a certain amount of PFCs (e.g., PFOS) are present at levels which is above the detection level, may occur endocrine disruptions. In accordance with a global survey, PFOS is observed not only in nonexposed but also in exposed populations, which makes an increase and the importance is considered as a maximum level pollutant. In this case, a noncompetitive enzyme-linked immunosorbent assay (ELISA) has been invented by which the investigation estrogenic activities of selected PFCs are measured (Liu et al., 2007). However, while many of the PFCs have been produced for five decades, their presence in the environment has only recently been reported along with their toxicity, persistence, and bio accumulative nature. As with other pollutants of the same group, PFCs also lives in water lives but no degradation occurs and thus it is found at far places like the deep ocean. Some PFCs with a relatively volatile nature are subject to transformation, leading to the formation of persistent sulfonate and carboxylic acid forms (Farré et al., 2008). A major amount of OMPs in the environment comes from the direct discharges from the pharmaceutical and pesticide from the industries.

4.2.3 Physicochemical characteristics of OMPs

Most Pharmaceuticals for example are polar, biologically active with relatively high hydrophilicity though physicochemical characteristics of OMPs are widely varied but and happening in order to get absorbed in the human body and to avoid degradation before they have a curing effect (C. C. Daughton, 2001). This phenomenon is related with the low amount or low concentration at which they are produced and their tinny size of molecular weights make their dismissal surely hard. The parameter which is used most in characterizing the size of molecules is molecular weight (MW). However, researchers have found that sometimes MW can't show a direct degree of the size and shape of the molecules (Kanani et al., 2010; Mehta and Zydney, 2005). This is very emergent as a size dismissal mechanism and also been thoughtful in solute membrane interaction. Using geometric Eqs., we can also evaluate the wide and length of molecules (Kanani et al., 2010) and the molecular size can easily be identified. The distance between the two atoms at a maximum distance is characterization as molecular length, while the molecular depth and width are measured by projecting the molecule perpendicularly to the plane of the long axis. Through the molecular volume, the transport characteristics of molecules can easily be evaluated. To indicate the hydrophobicity or hydrophilicity of organic compounds octanal – a water partition coefficient is usually used. The equilibrium concentration of a compound between octanol and water is also measured by this coefficient i.e., expressed as

$$\log \mathrm{Kow} = \log \left(\mathrm{Co} / \mathrm{Cw} \right)$$

Where, the concentration of the compound in the octanol phase and the unionized compound in the water phase is Co and Cw respectively.

An important indicator to determine the effect of a compound on its passage through a membrane or compound is its solubility. In the water a compound's solubility shows its affinity to water; so, highly water-soluble compounds have generally more liking to stay in the aqueous solution, not adsorbing the membrane surface of the compounds.

4.3 Effect of organic micropollutants on the environments

Continuous input of these non-biodegradable, persistent and bioaccumulative micropollutants in aquatic system raised several environmental concerns around the world (2-Schwarzenbach et al., 2006) as well as human health issues such as carcinogenicity, nervous system degradation, gastric troubles, dermal pathologies, etc. (Ahmed et al., 2017; Gwenzi et al., 2018). Though it was presumed that micropollutants with higher concentrations might be more harmful for organisms, later it was found that some of the micropollutants showed higher toxicity in extremely low concentrations (Oldenkamp et al., 2018). Until recently, versatilely used anthropogenic chemicals were one of the major sources of organic micropollutants. Despite of low concentrations, these OMPs were found to poison fish and other aquatic species, which resulted in environmental imbalance (Schwarzenbach et al., 2006). For example, one of the widely used biocide triclosan, has been frequently found in ng/L to μg/L in different water sources, was found to be highly toxic to green algae (Tatarazako et al., 2004), this may end up in fish later on in human beings by bioaccumulation process. Besides high toxicity, long term presence of these OMPs without any significant biodegradation is another reason of their toxicity (Shao et al., 2019). Different pharmaceutical residues such as carbamazepine and diclofenac were found to act as persistent or pseudo-persistent compounds which sustain a multigenerational exposure for the resident organisms (Shao et al., 2019).

Besides harming the individual organisms, these OMPs also effects the higher levels of biological organizations. For instance, OPMs in treated wastewater affected sex ratios of gammarid amphipods at population levels (Peschke et al., 2014). Moving on to higher level such as community levels, studies confirmed that presence of OMPs in wastewater decreased the fraction of sensitive species communities in the aquatic system. (Bunzel et al., 2013; Stamm et al., 2016). In addition, at the ecosystem level, pharmaceutical OMPs induced the behavioral changes in fish. which leads to their more intense predation behavior on prey communities in aquatic food webs (Heynen et al., 2016). Though the application of different micropollutant producing chemicals in agriculture field, pharmaceuticals, and different manufacturing plants are strictly monitor nowadays in developed countries, the circumstances are worse in developing countries (Gautam and Anbumani, 2020). In these third world countries, due to large population number and their listless approach towards environment as well as nonideal managemental system, concentrations of OMPs are exceedingly high. In addition, because of staying behind the technological advancement and apathy nature of environmental monitoring bureaus, the water treatment plants of these countries are inefficient in removing these

OMPs. From the above discussion it is confirmed that, if the negligence towards the control release of OMPs persists, environment for the upcoming generation will not be suitable for living.

4.4 Different methods for the removal of organic micropollutants

4.4.1 Nonbiological method for remediation

Advanced oxidation, activated carbon adsorption, membrane ultrafiltration, hydrostatic exclusions, and electrostatic exclusions are the most conventional modern nonbiological techniques for remediation of OMPs (Ojajuni et al., 2015). Among those the advanced oxidation is an aqueous phase oxidation procedure that can remove a wide range of OMPs primarily by mineralization organic micropollutants in the Environment: Potential of ecotoxicity and methods through the production of highly active species or converting them to less harmful products (Tawabini, 2014). Solar visible or near-ultraviolet irradiation, ozonization, Fenton's reaction, ultrasound, and wet air oxidation basis photocatalysis process which is the main methodology (Ikehata et al., 2008). Nowadays, newly invented procedures in oxidation are considered as ionizing radiation, pulsed plasma, microwaves, and Fenton's reagent. The technique has limitations when the effluents contain higher loads of organic and/or inorganic matter although the technique has gained tremendous popularity in the removal of OMPs. As it is hydrophobic in nature the pollutants (compounds with Kow> 2.0) favor their absorbance on the activated surface to reduce the problem of dealing with an effluent containing high organic load activated carbon adsorption (Jones et al., 2005). However, the surface area of activated carbon, i.e., granular and powdered promotes the adsorption capacity. Membrane-assisted processes also have abundant potential in removing a wide range of OMPs (Snyder et al., 2007; Schäfer et al., 2011; Ojajuni et al., 2015). Membranes play a role as a physical barrier that doesn't allow pollutants of dimensions more than the membrane's porosity but it allows water to penetrate through it (Kimura et al., 2003). Recently, experts have fulfilled several membrane processes to remove the OMPs which can be noted as microfiltration, ultrafiltration, nanofiltration, membrane adsorption, size exclusion, reverse osmosis, electrostatic exclusion, electrodialysis reversal, membrane bioreactors, and combinations of membranes in series (Kim, 2011; Ojajuni et al., 2015). For the improvement of the selection of the removal systems, lessening of secondary effluent, lowering of operating cost, and the possibility of waste recovery the research is still running, and hence, the existing methods more promising and potential alternatives compared to the currently employed methods.

4.4.2 Biological methods of remediation

Biological removal of OMPs considered a promising substitute due to minimized capital investment and eco-friendly nature as many of the nonbiological processes are expensive and may result in the generation of new waste products. Generally, many of the biological treatment procedures initially depend on microorganisms containing appropriate metabolic capabilities. The efficiency has always remained a challenge to environmental biologists but although biological degradation pathways have been reported to have a great influence on a wide range of contaminants. Many polar compounds are discharged that is not reduced in concentration because the conventional bioremediation process removes only a part of OMPs from the environment (Koh et al., 2008). To decrease OMPs load such as phytoremediation, plant-bacteria partnership system, constructed wetlands, ligninolytic enzymes application (e.g., laccase), and biofiltration different steps in the design and principle of the biological system have been adopted (J. A. Majeau et al., 2010; Drewes, 2010; Verlicchi et al., 2010).

4.4.3 Enzymatic degradation of OMPs

Direct application of ligninolytic enzymes can enhance the degradation of OMPs (e.g., laccases) not facilitating microbial concentrations as mentioned earlier (J. Majeau et al., 2010). The direct application of these enzymes is (1) it can control the microorganism evolvement with the help of a number of factors and there is also the presence of inhibitors which may affect the catabolic genes expression, and (2) it is not mandatory to be an effective enzyme for a particular OMP for the particular respect. In such situations, this direct application could bring desired results than bioaugmentation. As a result, the application of such enzymes can polymerize, oxidize, or transform a variety of phenolic compounds into less toxic compounds. These enzymes include phenols, EDCs, pesticides, dyes, and polycyclic aromatic hydrocarbons and the substrate ranges for such enzymes are quite diverse and have been widely using different types of fungi and microorganisms. Fungi produce laccase and this production depends on some parameters and these are a type of species, the procedure of cultivation, aeration, and agitation. But the crucial factors are the glucose and nitrogen sources, concerning their concentration and the ratio, and it is also mandatory to keep the balance in nature and concentration of the inducer. The culture medium can be controlled synthetically, naturally, or semi synthetically, such as solid lignocellulosic waste produced in an artificial liquid medium (J. A. Majeau et al., 2010). Basidiomycetes species have outstanding lignin-degrading capacity and so they have extensively studied organisms for laccase production. Some species have been declared to produce laccase carefully, as it can simplify purification procedures for industrial purposes. Some other laccase-producing organisms include Pleurites stratus (Hou et al., 2004) Marasmius Spermophilus (Farnet et al., 2000), Ganoderma adspersum (Songulashvili et al., 2006), Pycnoporus cinnabarinus, and Pycnoporus sanguineus (Eggert et al., 1996; Pointing et al., 2000). However, the presence of other ligninolytic enzymes is necessary to remove one or many OMPs in wastewater to improve the decontamination efficiency. The successful action of laccase for OMPs removal can be gained by the application of (1) enzyme not purified, (2) enzyme with purified immobilized, (3) culture broths derived enzyme and (4) bioremediation which is reactor-based with immobilized or free cells. Previous studies have generally focused on the oxidative capacity of purified laccase on a variety of contaminants; nevertheless, further efforts have to be made on the optimization of real wastewater treatment. At last, some chemists have declared that some phenolic compounds have the reverse property of the growth of white-rot fungi and thus affect overall yield (Buswell and Eriksson, 1994). However, appropriate selection of fungal growth stages and dilutions may decrease the desired adverse effects (Ryan et al., 2005) added small degree of stripped gas liquor from a coal gasification plant to Trametes pubescens culture so that it can induce laccase activity for removal of phenol in the effluent. Furthermore, exploiting the appropriate flow rate of wastewater for removal was shown to be particularly crucial in experiments with dye-laden effluent (Romero et al., 2006) and it led to laccase deactivation or inefficient system operation with higher or lower rates, respectively. Although they are enormous potential, it is still the state of knowledge of parameters regulating the production of such enzymes in microorganisms in their infancy (J.A. Majeau et al., 2010). At last, several phenolic compounds have been found to generate unwanted by-products by enzymatic degradation, and these are more toxic than the parent compounds. Therefore, enzymatic degradation shall be handled with great care. Before implying it at a wider scale risk assessment of OMPs and their by-products is necessary. although the availability of information is rare in terms of research is still being with high potential. For this, lignocellulosic waste has become a potential source of growing laccase-producing microbial cultures (Lorenzo et al., 2002; Howard et al., 2003) however, further research is necessary to unravel their potential with regard to the economic production of these enzymes.

4.4.4 Biofiltration of OMPs

During biofiltration, the removal of contaminants occurs due to the action of mixed cultures of microorganisms. Nowadays, it has been assumed that various physiological processes such as oxygen concentration, nutrients availability, temperature, sludge retention time (SRT), and hydraulic retention time (HRT) of the pollutant can increase the degradation of micropollutants with the help of the capacity of microorganisms (Sadef et al., 2015). Moreover, OMPs degradation reduces due to the lack of oxygen, and a higher concentration of inorganic nitrogen. By contrast, degradation or removal in a trickling filter cannot be sufficient for compounds for low solids retention time (SRT) and hydraulic retention time (HRT), e.g., estrogens (Koh et al. 2008). Another study informed that the highest removal may be achieved at plants with HRT >30 days and SRT >45 days (Servos et al., 2005).

4.5 Aerobic and anaerobic microorganism for the removal of micropollutants

Though both aerobic and anaerobic microbes have been utilized for the wastewater treatment, there are some significantly different outcomes has been reported because of their differentiations. While aerobic microbes require oxygen rich environment for surviving, anaerobic microbes flourish in oxygen free surrounding. Based on this characteristic of these microbes they can be used in different areas to treat wastewater. For example, as presence of air is compulsory for aerobic microbes so they can be used in open places like pond, effluent streams etc. On the contrary, anaerobic microbes can be used in dumped or closed areas where shortage of oxygen exists. However, in case of aerobic microbes based biological treatments constant and continuous flow of oxygen is required this results in an additional cost of using agitator or aerator which is not necessary for anaerobic process. Apart from the cost, level of contaminants also influences the selection of type of microbes such as, aerobic microbes perform well in low BOD/COD streams and removal of nitrogen and phosphorus, where anaerobic microbes have performed well high organic pollutant contaminated wastes.

4.5.1 Microbiological aspects of micropollutant degradation

Assimilations of organic micropollutants can be occurred in different routes for different organic pollutants (Ahmed et al., 2017; Tran et al., 2013). For example, in case of phthalates, suggested biodegradation, under aerobic condition in presences of *Rhodococcus pyridinivorans* at optimum 7.04 pH, 30.4 C temperature, is the breaking of long chain ester to short chain esters followed by several enzymatic reactions as depicted in Fig. 4.1A. Successive enzymetic reactions after the conversion of long chain ester to monoesters, results in the production of acetyl-CoA that enters the TCA cycles (Zhao et al., 2018).

While phthalates follow the degradation from long to small esters, some other aromatic pollutants follow ring opening reactions on ortho- or meta-position which leads to the formation of acetyl-CoA derivatives. As illustrated in Fig. 4.1B, it can be seen that pollutants such as naphthalene, benzene, phenanthrene, pyrene is converted to hydroxy rich substance by hydroxylating dioxygenase enzyme. Later, this hydroxyl rich substance, converted to different carboxylic and aldehyde based compounds by the ring structure opening on ortho- and meta-position in aid of enzymatic reaction of intradiol dioxygenase or extradiol dioxygenase. After that, this pathway leads to the formation of acetyl-CoA pyruvate and acetyl-CoA succinate by several successive chain reactions which are intermediates of Krebs cycles (Sahoo et al., 2013). Besides above mentioned microbes, different other microbes have also been successfully used for micropollutant removal as tabulated in Table 4.2.

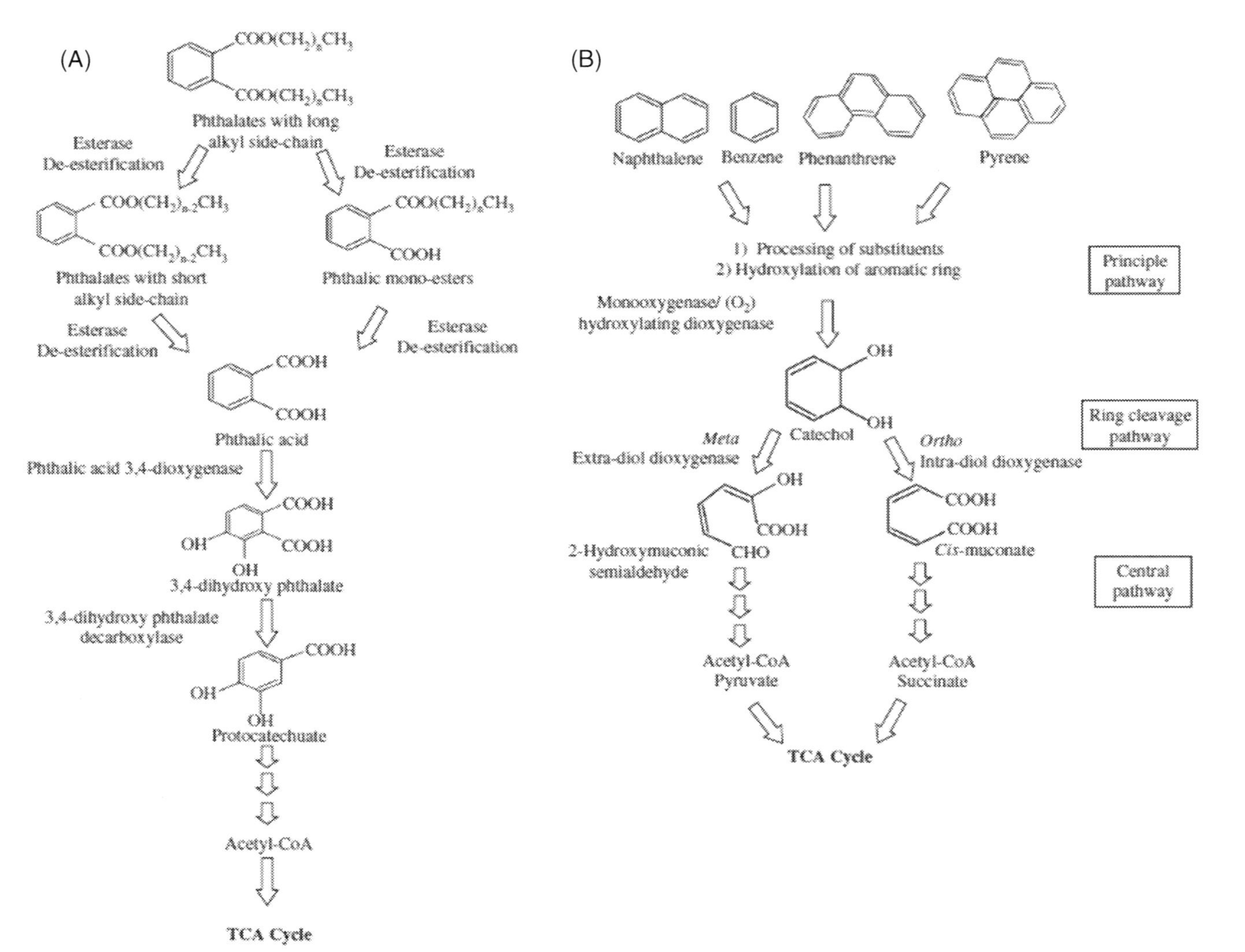

Fig. 4.1 Major metabolic pathway of bacterial degradation of micropollutants (Kanaujiya et al., 2019).

Table 4.2 Microorganisms capable of organic micropollutant degradation (Kanaujiya et al., 2019).

Microorganisms	Micropollutants	Concentration	Degradation efficiency (%)	Reference
Bacterial strains				
Gordonia sp. Dop5	DnOP	750 mg/L	100	(Sarkar et al., 2013)
Acinetobacter sp. HS-B1 Arthrobacter sp. C21	DBP	300 mg/L	100	(Wen et al., 2014)
Bacillus thuringiensis	Ibuprofen	25 mg/L	46.56	(Marchlewicz et al., 2016)
	Naproxen	6 mg/L	100	
Rhodococcus sp. JX-2	17β-Estradiol	30 mg/L	94	(Liu et al., 2016)
Pseudomonas putida	Estrone	5 mg/L	100	(Chen et al., 2017)
	Estradiol		94.86	
	17α-Ethinylestradiol		94.90	
	Estriol		94.56	
	Bisphenol A		96.56	
Ochrobactrum sp.	Erythromycin A	100 mg/L	97	(Zhang W, Qiu L, Gong A, n.d.)
Acinetobacter sp.	Sulfamethoxazole	240 mg/L	100	(Wang & Wang, 2018)
Rhodococcus sp. BCH2 *Bacillus* sp. PDK1 *Bacillus* sp. PDK2	Atrazine	1000 mg/L	> 90	(Gao et al., 2018)
Ochrobactrum sp. Jdc-41	DBP	500–1000 mg/L	99	(Feng et al., 2018)
Klebsiella sp.	Diclofenac	70 mg/L	> 90	(Stylianou et al., 2018)

4.5.2 Factors effecting the feasibility of microorganism selection

To maximize the removal efficiency of OMPs in aerobic and anaerobic and also in any other biological system, it is essential to understand the factors driving and limiting the biotransformation pathways, bearing in mind that enzymes are ultimately responsible for the biotransformation of OMPs if it occurs via either metabolism or co-metabolism.

4.5.2.1 Structure of compound

As biotransformation of compounds are directly depended on the compatibility of active sites of the compound with enzymes presents in the reactor, hence chemical structure of OMPs plays an important role in feasibility of the process. In addition, physicochemical properties of the compounds also depend on the chemical structure of the compounds. For example, presence of hydroxyl and methyl group as a functional group in the OMPs increases the probability of exhibiting hydrophilic and hydrophobic nature of these compounds, respectively. These physicochemical properties govern the fate of OMPs in biological treatments (Lorena Gonzalez-Gil et al., 2020).

Though some reactions like reductive dehalogenation and cleavage of ether bonds occurs for some OMPs during the anerobic conditions, it is still, for the most of the OMPs, unestablished which compound will follow which reaction mechanism. One of the main reasons of this problem is the presence of multiple electron-accepting and donating groups in the same compounds. While in some studies it has been found that electron donating groups like hydroxyl and amine governs the fate of OMPs in positive direction during biological treatment, presence of electron withdrawing groups exhibits inverse impact on the biotransformation (Wijekoon et al., 2015). However, outlier of this generalization has also found in case of some OMPs. For instance, in case of OMP atenolol, the electron withdrawing group (–$CONH_2$) undergoes the hydrolyzation process and electron donating groups (–OH, –NHR) remain unaltered (Lorena Gonzalez-Gil et al., 2019).

4.5.2.2 Operating condition

There is no clear correlative relationship has been not found till now among the variable parameters like temperature, organic loading rate, hydraulic retention time (HRT), solid retention time (SRT) and feed composition of anaerobic reactor. Psychrophilic (<20 °C), mesophilic (30–37 °C) and thermophilic (55 °C) are the three usually performed thermal conditions for anaerobic digestion processes (Lorena Gonzalez-Gil et al., 2020). Like normal reactions it can be expected that higher temperature favor higher reaction rate for digestion, however, it is still unsettling in case of OMPs anaerobic digestions. Type of OMPs influences this phenomenon of uncertainty. For example, biotransformation of OMPs like aromatic hydrocarbon, nonylphenol and its ethoxylates are promoted by thermophilic process (Trably and Patureau, 2006). On the other hand, digestion of pharmaceutical OMPs was not significantly governed by the change in thermal conditions (L. Gonzalez-Gil et al., 2016). Likewise, role of organic loading rate (OLR) and HRT on the efficacy of the digestion system are also perplexed. Both positive and negative influence of OLR is observed for the digestions for different OMPs. However, in case of HRT, efficiency of the digestion system has not been reported negative. It has been reported that increase of HRT in some reactors like upflow anaerobic sludge blanket and anaerobic sequencing batch reactors exhibited positive impact on the removal of some selective OMPs, while HRT does not play any significant role in case of sewage sludge anaerobic digester (Yang et al., 2016; Zhou et al., 2017). Apart from above discussed parameters, feeding composition could also affect the biological treatment process of the OMPs. It has been reported that addition of certain cofactors and electron acceptors materials as well as composition variation in the feed enhanced the digestion process by varying the cometabolic pathways (Lorena Gonzalez-Gil et al., 2020). Another highly important parameter is the pH of the entire process. pH of the system influences many major parameters such as protonation, hydrophilicity, hydrophobicity, acidity of the system. All of these properties have been found to impact the removal efficiency of the system. For example, 90 percent removal of ibuprofen was possible to achieve by lowering the pH to 6. Similar high removal of ketoprofen obtained when the pH dropped down below 5 (Urase et al., 2005). However, though pH played an important role in regulating the removal efficiency of the micropollutants, this has been least analyzed. Hence, more study is necessary to completely understand and finding the optimum conditions for the OMP removal (Cirja et al., 2008).

4.5.3 Application of aerobic and anaerobic system for micropollutants removal

One of the most convenient and cheap organic micro pollutant removal technique is activated sludge process. This process has been utilized for both domestic and industrial wastewater. In general, effluent or wastewater is mixed with microbial culture in an aeration tank where oxidation of carbon rich organic compound occurs (Eckenfelder and Cleary, 2014). Different studies over the years showed that ASP is capable to remove pollutants like diclofenac, phthalate from wastewater in significant amounts (Eckenfelder and Cleary, 2014). Though this process shows promising results but over the time the standards of the discharge effluent from ASP is not consistent. Moreover, this process requires a large area for the setup. However, addition of tertiary steps after the ASP system can elevate the quality of the discharge effluent (Kanaujiya et al., 2019).

Recently membrane bioreactors have gained wide attention in the treatment pf organic pollutants of industrial effluents. Earlier version of MBR were designed with a side stream system which required an extra pump to circulate the effluents for better performances. However, this additional side pump increased the costing of the process. Later on, to make MBR more affordable and efficient submerged MBR has been introduced. In this new design biomass are immobilized on the surface of membranes and membrane modules are kept inside the reactor. Effluent feeds pass through the membranes and concurrently treated by removing the micropollutants. Micropollutants are removed from the effluents by different mechanisms such as physical retention, air stripping, and biodegradation. However, recently photocatalytic MBR is also gaining attention for the removal of OMPs. Along pH and temperature, both HTR and STR greatly influence the performance of MBRs (Eckenfelder and Cleary, 2014).

Like submerged MBR, biomass is also immobilized in case of trickling filters. In here, wastewater with micropollutants percolate through the support and are degraded by immobilized biomass. However, long operation time and low efficiency limits its application in wastewater treatment (Eckenfelder and Cleary, 2014).

Apart from above mentioned process there are several other methods have also been reported for the removal of organic micropollutants. For example, moving bed biofilm reactor, where biofilms are grown on small supported biomass. This process is highly efficient and affordable compared to other process which makes this process more acceptable nowadays. Significant removal (above 71 percent at least) of iohexol, ibuprofen, acetaminophen, salicylic acid, estrogen, bisphenol-A and 21 other pollutants have been reported by this process (Casas et al., 2015; Luo et al., 2015). In addition, by using activated sludge along with membrane biofilm bioreactor, 72.8 percent of diclofenac was removed from wastewater (Abtahi et al., 2018). Beside membrane biofilm bioreactor, two-phase portioning bioreactors is another advance process which has been used for the degradation of toxic organic pollutants. In here, two different phases are found: organic pollutant content organic phase and microbial culture contain aqueous phase. By aid of high-speed agitation droplets of organic phase is dispersed in the aqueous phase, which accelerate the removal process of the pollutants. This process is found highly successful in case of the removal of hydrophobic pollutants like hydrocarbon, phenols, pyrene, styrene etc. by degradation (Kanaujiya et al., 2019).

Anaerobic digestion is used to stabilize primary and secondary sewage sludge in anaerobic digesters (ADs). However, it can also be applied to treat mainstream wastewater through up flow anaerobic sludge blanket (AnSB) reactors and anaerobic membrane bioreactors (AnMBRs). During the operation unlike aerobic processes, anaerobic microbes produce by products which can be used for the power generation purpose by degrading organic carbon into biogas (Harb et al., 2016). Nevertheless, like aerobic process some pollutants are completely removed from the streams by anerobic microbes while some pollutants remain unaltered. For example, biotransformation of micropollutants like galaxolide, naproxen, nonylphenol, diclofenac and bisphenol, by enzyme acetate kinase under anaerobic condition was reported by Gonzalez-Gil et al. (L. Gonzalez-Gil et al., 2016). Though significant research has been done on the aerobic microbe or process based biological treatments, literature regarding anaerobic microbes-based process is still limited.

4.6 Limitations and future prospects

Complete removal of micropollutants by employing biological treatment systems is a challenging task. The goal of designing advanced technologies is multi-directional: firstly, to improve the solubility of micropollutants for its enhanced bioavailability, as most of these compounds are less soluble in water at high concentration. Furthermore, there is a need to understand and elucidate a biodegradation mechanism of micropollutants for a better application of such processes. Such information will not only help in improving the process efficiency but will also enhance the development and use of genetically engineered strains for removal of micropollutants. Metabolic engineering and genetic engineering would further enhance the treatment efficiency of micropollutants in wastewater (Mardani et al., 2017). The effect of different bioreactor operation strategies and process conditions, such as HRT, SRT, pollutant load, and F/M ratio, needs to be examined in great detail in order to optimize the treatment process for achieving maximum performance. Application of nanomaterial to enhance the treatment efficiency of biological treatment systems is another upcoming research area that needs attention.

4.7 Conclusions

Micropollutants are among those emerging pollutants released from various industrial, agricultural, and domestic activities, which pose a prominent threat to the environment and human health. The presence of such micropollutants causes pollution of groundwater and surface water, thus leaving water unsafe and toxic for human consumption. Conventional physicochemical treatment technologies are inefficient in treating micropollutants containing wastewater or involve high cost, require high input, or generate a large amount of toxic sludge. Thus, the recent focus in this area is towards the biological treatment system. Different biological processes, particularly those employing aerobic and anaerobic microorganism, are proving to be highly efficient. In order to establish their application potential and improve their treatment efficiency, hybrid systems combining both biological and physicochemical methods seem to be promising.

References

Ã, T.U., Kikuta, T., 2005. Separate estimation of adsorption and degradation of pharmaceutical substances and estrogens in the activated sludge process. 39, 1289–1300. https://doi.org/10.1016/j.watres.2005.01.015.

Abe, A., 1999. Distribution of 1,4-dioxane in relation to possible sources in the water environment. Sci. Total Environ. 227 (1), 41–47. https://doi.org/10.1016/S0048-9697(99)00003-0.

Abtahi, S.M., Petermann, M., Juppeau Flambard, A., Beaufort, S., Terrisse, F., Trotouin, T., et al., 2018. Micropollutants removal in tertiary moving bed biofilm reactors (MBBRs): contribution of the biofilm and suspended biomass. Sci. Total Environ. 643, 1464–1480. https://doi.org/10.1016/j.scitotenv.2018.06.303.

Ahmed, M.B., Zhou, J.L., Ngo, H.H., Guo, W., Thomaidis, N.S., Xu, J., 2017. Progress in the biological and chemical treatment technologies for emerging contaminant removal from wastewater: a critical review. J. Hazard. Mater. 323, 274–298. https://doi.org/10.1016/j.jhazmat.2016.04.045.

Anjum, N.A., Gill, S.S., Tuteja, N., 2017. Enhancing cleanup of environmental pollutants. In Enhancing Cleanup of Environmental Pollutants 1 (May). https://doi.org/10.1007/978-3-319-55426-6.

Auriol, M., Filali-Meknassi, Y., Tyagi, R.D., Adams, C.D., Surampalli, R.Y., 2006. Endocrine disrupting compounds removal from wastewater, a new challenge. Process Biochem. 41 (3), 525–539. https://doi.org/10.1016/j.procbio.2005.09.017.

Barbosa, M.O., Moreira, N.F.F., Ribeiro, A.R., Pereira, M.F.R., Silva, A.M.T., 2016. Occurrence and removal of organic micropollutants: an overview of the watch list of EU Decision 2015/495. Water Res. 94, 257–279.

Besha, A.T., Gebreyohannes, A.Y., Tufa, R.A., Bekele, D.N., Curcio, E., Giorno, L., 2017. Removal of emerging micropollutants by activated sludge process and membrane bioreactors and the effects of micropollutants on membrane fouling: a review. J. Environ. Chem. Eng. 5 (3), 2395–2414.

Boxall, A.B.A., Fogg, L.A., Blackwell, P.A., Kay, P., Pemberton, E.J., Croxford, A., 2004. Veterinary medicines in the environment. Rev. Environ. Contam. Toxicol. 180, 1–91. https://doi.org/10.1007/0-387-21729-0_1.

Brack, W., Altenburger, R., Schüürmann, G., Krauss, M., López Herráez, D., van Gils, J., et al., 2015. The SOLUTIONS project: challenges and responses for present and future emerging pollutants in land and water resources management. Sci. Total Environ. 503–504, 22–31. https://doi.org/10.1016/j.scitotenv.2014.05.143.

Bunzel, K., Kattwinkel, M., Liess, M., 2013. Effects of organic pollutants from wastewater treatment plants on aquatic invertebrate communities. Water Res. 47 (2), 597–606. https://doi.org/10.1016/j.watres.2012.10.031.

Buswell, J.A., Eriksson, K.L., 1994. Effect of lignin-related phenols and their methylated derivatives on the growth of eight white-rot fungi. 10, 169–174.

Casas, M.E., Chhetri, R.K., Ooi, G., Hansen, K.M.S., Litty, K., Christensson, M., et al., 2015. Biodegradation of pharmaceuticals in hospital wastewater by staged Moving Bed Biofilm Reactors (MBBR). Water Res. 83, 293–302. https://doi.org/10.1016/j.watres.2015.06.042.

Cerreta, G., Roccamante, M.A., Plaza-Bolaños, P., Oller, I., Aguera, A., Malato, S., et al., 2020. Advanced treatment of urban wastewater by UV-C/free chlorine process: micro-pollutants removal and effect of UV-C radiation on trihalomethanes formation. Water Res. 169, 115220.

Chau, H.T.C., Kadokami, K., Duong, H.T., Kong, L., Nguyen, T.T., Nguyen, T.Q., et al., 2018. Occurrence of 1153 organic micropollutants in the aquatic environment of Vietnam. Environ. Sci. Pollut. Res. 25 (8), 7147–7156.

Cirja, M., Ivashechkin, P., Schäffer, A., Corvini, P.F.X., 2008. Factors affecting the removal of organic micropollutants from wastewater in conventional treatment plants (CTP) and membrane bioreactors (MBR). Rev. Environ. Sci. Bio/Technol. 7 (1), 61–78. https://doi.org/10.1007/s11157-007-9121-8.

Daughton, C.C., 2001. Pharmaceuticals and personal care products in the environment: overarching issues and overview. ACS Symp. Ser. 791, 2–38. https://doi.org/10.1021/bk-2001-0791.ch001.

Daughton, C.G., 2004. Non-regulated water contaminants : emerging research. 24, 711–732. https://doi.org/10.1016/j.eiar.2004.06.003.

Daughton, C.G., Ternes, T.A., 1999. Pharmaceuticals and Personal Care Products in the Environment: agents of Subtle Change? Environ. Health Perspect. 107, 907. https://doi.org/10.2307/3434573.

Drewes, J.E., 2010. The role of organic matter in the removal of emerging trace organic chemicals during managed aquifer recharge. Water Res. 44 (2), 449–460. https://doi.org/10.1016/j.watres.2009.08.027.

Eckenfelder, W.W., William, W., Cleary, J.G., Joseph, G., 2014. Activated sludge technologies for treating industrial wastewaters: design and troubleshooting. DEStech Publications, Inc. http://www.destechpub.com/links/catalogs/bookstore/environmental-technology-6/activated-sludge-technologies-for-treating-industrial-wastewaters/.

Eggert, C., Temp, U., Dean, J.F.D., Eriksson, K.L., 1996. A fungal metabolite mediates degradation of non-phenolic lignin structures and synthetic lignin by laccase. FEBS Letters 391, 144–148.

Escher, B.I., Stapleton, H.M., Schymanski, E.L., 2020. Our Changing Environment 392 (January), 388–392.

Farnet, A.M., Criquet, S., Tagger, S., Gil, G., Le Petit, J., 2000. Purification, partial characterization, and reactivity with aromatic compounds of two laccases from Marasmius quercophilus strain 17. Can. J. Microbiol. 46 (3), 189–194. https://doi.org/10.1139/w99-138.

Farré, M., Pérez, S., Kantiani, L., Barceló, D., 2008. Fate and toxicity of emerging pollutants, their metabolites and transformation products in the aquatic environment. TrAC - Trends in Analytical Chemistry 27 (11), 991–1007. https://doi.org/10.1016/j.trac.2008.09.010.

Fent, K., Weston, A.A., Caminada, D., 2006. Ecotoxicology of human pharmaceuticals. 76, 122–159. https://doi.org/10.1016/j.aquatox.2005.09.009.

García-Galán, M.J., Garrido, T., Fraile, J., Ginebreda, A., Díaz-Cruz, M.S., Barceló, D., 2010. Simultaneous occurrence of nitrates and sulfonamide antibiotics in two ground water bodies of Catalonia (Spain). J. Hydrol. 383 (1–2), 93–101. https://doi.org/10.1016/j.jhydrol.2009.06.042.

Gautam, K., Anbumani, S., 2020. Chapter 19 - Ecotoxicological effects of organic micro-pollutants on the environment (S. Varjani, A. Pandey, R.D. Tyagi, H. H. Ngo, C. B. T.-C. D. in B. and B. Larroche (Eds.), pp. 481–501). Elsevier. https://doi.org/10.1016/B978-0-12-819594-9.00019-X.

Gavrilescu, M., Demnerová, K., Aamand, J., Agathos, S., Fava, F., 2015. Emerging pollutants in the environment: present and future challenges in biomonitoring, ecological risks and bioremediation. New Biotechnol. 32 (1), 147–156.

Geissen, V., Mol, H., Klumpp, E., Umlauf, G., Nadal, M., van der Ploeg, M., et al., 2015. Emerging pollutants in the environment: a challenge for water resource management. Int. Soil Water Conserv. Res. 3 (1), 57–65. https://doi.org/10.1016/j.iswcr.2015.03.002.

Göbel, A., Mcardell, C.S., Joss, A., Siegrist, H., Giger, W., 2007. Fate of sulfonamides, macrolides, and trimethoprim in different wastewater treatment technologies. Sci. Total Environ. 372, 361–371. https://doi.org/10.1016/j.scitotenv.2006.07.039.

Golet, E.M., Strehler, A., Alder, A.C., Giger, W., 2002. Determination of Fluoroquinolone Antibacterial Agents in Sewage Sludge and Sludge-Treated Soil Using Accelerated Solvent Extraction Followed by Solid-Phase Extraction. Anal. Chem. 74 (21), 5455–5462.

Gonzalez-Gil, L., Papa, M., Feretti, D., Ceretti, E., Mazzoleni, G., Steimberg, N., et al., 2016. Is anaerobic digestion effective for the removal of organic micropollutants and biological activities from sewage sludge? Water Res. 102, 211–220. https://doi.org/10.1016/j.watres.2016.06.025.

Gonzalez-Gil, L., Carballa, M., Lema, J.M, 2020. Removal of organic micro-pollutants by anaerobic microbes and enzymes, Current Developments in Biotechnology and Bioengineering. Elsevier, pp. 397–426.

Gonzalez-Gil, L., Krah, D., Ghattas, A.-K., Carballa, M., Wick, A., Helmholz, L., et al., 2019. Biotransformation of organic micropollutants by anaerobic sludge enzymes. Water Res. 152, 202–214. https://doi.org/10.1016/j.watres.2018.12.064.

Gwenzi, W., Mangori, L., Danha, C., Chaukura, N., Dunjana, N., Sanganyado, E., 2018. Sources, behaviour, and environmental and human health risks of high-technology rare earth elements as emerging contaminants. Sci. Total Environ. 636, 299–313. https://doi.org/10.1016/j.scitotenv.2018.04.235.

Harb, M., Wei, C.H., Wang, N., Amy, G., Hong, P.Y., 2016. Organic micropollutants in aerobic and anaerobic membrane bioreactors: changes in microbial communities and gene expression. Bioresour. Technol. 218, 882–891. https://doi.org/10.1016/j.biortech.2016.07.036.

Heynen, M., Fick, J., Jonsson, M., Klaminder, J., Brodin, T., 2016. Effect of bioconcentration and trophic transfer on realized exposure to oxazepam in 2 predators, the dragonfly larvae (Aeshna grandis) and the Eurasian perch (Perca fluviatilis). Environ. Toxicol. Chem. 35 (4), 930–937. https://doi.org/10.1002/etc.3368.

Hou, H., Zhou, J., Wang, J., Du, C., Yan, B., 2004. Enhancement of laccase production by Pleurotus ostreatus and its use for the decolorization of anthraquinone dye. Process Biochem. 39 (11), 1415–1419. https://doi.org/10.1016/S0032-9592(03)00267-X.

Howard, R.L., Abotsi, E., Howard, S., 2003. Lignocellulose biotechnology: issues of bioconversion and enzyme production. Afr. J. Biotechnol. 2 (12), 602–619.

Ikehata, K., El-din, M.G., Snyder, S.A., Ikehata, K., El-din, M.G., Snyder, S.A., 2008. Ozone: science and Engineering Ozonation and Advanced Oxidation Treatment of Emerging Organic Pollutants in Water and Wastewater. Ozone: Sci. Eng. 9512, 20–26. https://doi.org/10.1080/01919510701728970.

Jahan, K., Ordóñez, R., Ramachandran, R., Balzer, S., Stern, M., 2008. Modeling Biodegradation of Nonylphenol. Water, Air, & Soil Pollut: Focus. 8 (3), 395–404. https://doi.org/10.1007/s11267-007-9148-4.

Johnson, A.C., Keller, V., Williams, R.J., Young, A., 2007. A practical demonstration in modelling diclofenac and propranolol river water concentrations using a GIS hydrology model in a rural UK catchment. Environ. Pollut. 146 (1), 155–165. https://doi.org/10.1016/j.envpol.2006.05.037.

Jones, O.A., Lester, J.N., Voulvoulis, N., 2005. Pharmaceuticals: a threat to drinking water? Trends Biotechnol. 23 (4), 163–167. https://doi.org/10.1016/j.tibtech.2005.02.001.

Ju, X., Jin, Y., Sasaki, K., Saito, N., 2008. Perfluorinated Surfactants in Surface, Subsurface Water and Microlayer from Dalian Coastal Waters in China. Environ. Sci. Technol. 42 (10), 3538–3542.

Kanani, D.M., Fissell, W.H., Roy, S., Dubnisheva, A., Fleischman, A., Zydney, A.L., 2010. Permeability – selectivity analysis for ultrafiltration: effect of pore geometry. J. Memb. Sci. 349, 405–410. https://doi.org/10.1016/j.memsci.2009.12.003.

Kanaujiya, D.K., Paul, T., Sinharoy, A., Pakshirajan, K., 2019. Biological Treatment Processes for the Removal of Organic Micropollutants from Wastewater: a Review. Current Pollution Reports 5 (3), 112–128. https://doi.org/10.1007/s40726-019-00110-x.

Kim, D.H., 2011. A review of desalting process techniques and economic analysis of the recovery of salts from retentates. DES 270 (1–3), 1–8. https://doi.org/10.1016/j.desal.2010.12.041.

Kimura, K., Amy, G., Drewes, J.E., Heberer, T., Kim, T., Watanabe, Y., 2003. Rejection of organic micropollutants (disinfection by-products, endocrine disrupting compounds, and pharmaceutically active compounds) by NF/RO membranes. J. Memb. Sci. 227, 113–121. https://doi.org/10.1016/j.memsci.2003.09.005.

Kinney, C.A., Furlong, E.T., Zaugg, S.D., Burkhardt, M.R., Werner, S.L., Cahill, J.D., et al., 2006. Survey of organic wastewater contaminants in biosolids destined for land application. Environ. Sci. Technol. 40 (23), 7207–7215. https://doi.org/10.1021/es0603406.

Klöpffer, W., Wagner, B.O., 2007. Atmospheric Degradation of Organic SubstancesAtmospheric Degradation of Organic Substances: Data for Persistence and Long-range Transport Potential. Wiley. https://doi.org/10.1002/9783527611638.

Koh, Y.K.K., Chiu, T.Y., Boobis, A., Cartmell, E., Scrimshaw, M.D., Lester, J.N., 2008. Treatment and removal strategies for estrogens from wastewater. Environ. Technol. 29 (3), 245–267. https://doi.org/10.1080/09593330802099122.

Kolpin, D.W., Schnoebelen, D.J., Thurman, E.M., 2004. Degradates Provide Insight to Spatial and Temporal Trends of Herbicides in Ground Water. Ground Water 42 (4), 601–608. https://doi.org/10.1111/j.1745-6584.2004.tb02628.x.

Kolpin, D.W., Thurman, E.M., Lee, E.A., Meyer, M.T., Furlong, E.T., Glassmeyer, S.T., 2006. Urban contributions of glyphosate and its degradate AMPA to streams in the United States. Sci. Total Environ. 354, 191–197. https://doi.org/10.1016/j.scitotenv.2005.01.028.

Lapworth, D.J., Baran, N., Stuart, M.E., Ward, R.S., 2012. Emerging organic contaminants in groundwater: a review of sources, fate and occurrence. Environ. Pollut. 163, 287–303. https://doi.org/10.1016/j.envpol.2011.12.034.

Lapworth, D.J., Gooddy, D.C., 2006. Source and persistence of pesticides in a semi-confined chalk aquifer of southeast England. Environ. Pollut. 144, 1031–1044. https://doi.org/10.1016/j.envpol.2005.12.055.

Larsen, T.A., Lienert, J., Joss, A., Siegrist, H., 2004. How to avoid pharmaceuticals in the aquatic environment. In: Larsen Tove, A (Ed.), J. Biotechnol. 113, 295–304. https://doi.org/10.1016/j.jbiotec.2004.03.033.

Liu, C., Du, Y., Zhou, B., 2007. Evaluation of estrogenic activities and mechanism of action of perfluorinated chemicals determined by vitellogenin induction in primary cultured tilapia hepatocytes. Aquat Toxicol. 85, 267–277. https://doi.org/10.1016/j.aquatox.2007.09.009.

Lorenzo, M., Moldes, D., Rodríguez Couto, S., Sanromán, A., 2002. Improving laccase production by employing different lignocellulosic wastes in submerged cultures of Trametes versicolor. Bioresour. Technol. 82 (2), 109–113. https://doi.org/10.1016/S0960-8524(01)00176-6.

Luo, Y., Guo, W., Ngo, H.H., Nghiem, L.D., Hai, F.I., Zhang, J., et al., 2014. A review on the occurrence of micropollutants in the aquatic environment and their fate and removal during wastewater treatment. Sci. Total Environ. 473, 619–641.

Luo, Y., Jiang, Q., Ngo, H.H., Nghiem, L.D., Hai, F.I., Price, W.E., et al., 2015. Evaluation of micropollutant removal and fouling reduction in a hybrid moving bed biofilm reactor–membrane bioreactor system. Bioresour. Technol. 191, 355–359. https://doi.org/10.1016/j.biortech.2015.05.073.

Majeau, J.A., Brar, S.K., Tyagi, R.D., 2010. Laccases for removal of recalcitrant and emerging pollutants. Bioresour. Technol. 101 (7), 2331–2350. https://doi.org/10.1016/j.biortech.2009.10.087.

Mardani, G., Mahvi, A.H., Hashemzadeh-Chaleshtori, M., Naseri, S., Dehghani, M.H., Ghasemi-Dehkordi, P., 2017. Application of genetically engineered dioxygenase producing Pseudomonas putida on decomposition of oil from spiked soil. Jundishapur J. Nat. Pharm. Prod. 12 (3 (Supp)).

Mehta, A., Zydney, A.L., 2005. Permeability and selectivity analysis for ultrafiltration membranes. J. Memb. Sci. 249, 245–249. https://doi.org/10.1016/j.memsci.2004.09.040.

Moran, M.J., Zogorski, J.S., Squillace, P.J., 2005. MTBE and Gasoline Hydrocarbons in Ground Water of the United States. Ground Water. 43 (4), 615–627.

Moran, M.J., Zogorski, J.S., Squillace, P.J., 2007. Chlorinated Solvents in Groundwater of the United States. Environ. Sci. Technol. 41 (1), 74–81.

Ojajuni, O., Saroj, D., Cavalli, G., 2015. Removal of organic micropollutants using membrane-assisted processes: a review of recent progress. In Environmental Technology Reviews. Taylor and Francis Ltd. 4(1), pp. 17–37. https://doi.org/10.1080/21622515.2015.1036788.

Oldenkamp, R., Hoeks, S., Čengić, M., Barbarossa, V., Burns, E.E., Boxall, A.B.A., et al., 2018. A High-Resolution Spatial Model to Predict Exposure to Pharmaceuticals in European Surface Waters: ePiE. Environ. Sci. Technol. 52 (21), 12494–12503. https://doi.org/10.1021/acs.est.8b03862.

Peschke, K., Geburzi, J., Köhler, H.-R., Wurm, K., Triebskorn, R., 2014. Invertebrates as indicators for chemical stress in sewage-influenced stream systems: toxic and endocrine effects in gammarids and reactions at the community level in two tributaries of Lake Constance, Schussen and Argen. Ecotoxicol. Environ. Saf. 106, 115–125. https://doi.org/10.1016/j.ecoenv.2014.04.011.

Pointing, S.B., Road, P., Avenue, T.C., 2000. Decolorization of azo and triphenylmethane dyes by Pycnoporus sanguineus producing laccase as the sole phenoloxidase. Fig. 1, 317–318.

Romero, S., Blánquez, P., Caminal, G., Font, X., Sarrà, M., Gabarrell, X., et al., 2006. Different approaches to improving the textile dye degradation capacity of Trametes versicolor. Biochem. Eng. J. 31 (1), 42–47. https://doi.org/10.1016/j.bej.2006.05.018.

Ryan, D.R., Leukes, W.D., Burton, S.G., 2005. Fungal Bioremediation of Phenolic Wastewaters in an Airlift Reactor. Biotechnol Prog. 1068–1074.

Sadef, Y., Poulsen, T.G., Bester, K., 2015. Impact of compost process conditions on organic micro pollutant degradation during full scale composting. Waste Manage. (Oxford) 40, 31–37. https://doi.org/10.1016/j.wasman.2015.03.003.

Sahoo, N.K., Ramesh, A., Pakshirajan, K., 2013. Bacterial degradation of aromatic xenobiotic compounds: an overview on metabolic pathways and molecular approaches. In: Satyanarayana, P.A.T, Johari, B.N. (Eds.), Microorganisms in Environmental Management: Microbes and Environment. Springer, pp. 201–220. https://doi.org/10.1007/978-94-007-2229-3_10.

Saleem, H., 2016. Plant - Bacteria Partnership: phytoremediation of Hydrocarbons Contaminated Soil and Expression of Catabolic Genes Bioremediation: use of living organisms toward degradation Phytoremediation: use of plants to decontaminate pollutants. Bull. Environ. Stud. 1 (1), 18–28.

Schäfer, A.I., Akanyeti, I., Semião, A.J.C., 2011. Micropollutant Sorption to Membrane Polymers: a Review of Mechanisms for Estrogens. Adv Colloid Interface Sci. 164(1-2), 100-117. https://doi.org/10.1016/j.cis.2010.09.006.

Schwarzenbach, R.P., Escher, B.I., Fenner, K., Hofstetter, T.B., Johnson, C.A., von Gunten, U., et al., 2006. The Challenge of Micropollutants in Aquatic Systems. Science 313 (5790), 1072. LP –1077. https://doi.org/10.1126/science.1127291.

Servos, M.R., Bennie, D.T., Burnison, B.K., Jurkovic, A., Mcinnis, R., 2005. Distribution of estrogens, 17β-estradiol and estrone, in Canadian municipal wastewater treatment plants. Sci. Total Environ. 336, 155–170. https://doi.org/10.1016/j.scitotenv.2004.05.025.

Shao, Y., Chen, Z., Hollert, H., Zhou, S., Deutschmann, B., Seiler, T.-.B., 2019. Toxicity of 10 organic micropollutants and their mixture: implications for aquatic risk assessment. Sci. Total Environ. 666, 1273–1282. https://doi.org/10.1016/j.scitotenv.2019.02.047.

Sinclair, C.J., Boxall, A.B.A., 2003. Assessing the Ecotoxicity of Pesticide Transformation Products. Environ. Sci. Technol. 37 (20), 4617–4625.

Sinclair, C., Beinum, W.V., Adams, C., Bevan, R., Levy, L., Parsons, S., et al., 2010. A Desk Study on Pesticide Metabolites, Degradation and Reaction Products to Inform the Inspectorate's Position on Monitoring Requirements. Food and Environment Research Agency (February).

Snyder, S.A., Adham, S., Redding, A.M., Cannon, F.S., Decarolis, J., Oppenheimer, J., et al., 2007. Role of membranes and activated carbon in the removal of endocrine disruptors and pharmaceuticals. Desalination 202, 156–181. https://doi.org/10.1016/j.desal.2005.12.052.

Songulashvili, G., Elisashvili, V., Wasser, S., Nevo, E., Hadar, Y., 2006. Laccase and manganese peroxidase activities of Phellinus robustus and Ganoderma adspersum grown on food industry wastes in submerged fermentation. Biotechnol. Lett. 28 (18), 1425–1429. https://doi.org/10.1007/s10529-006-9109-4.

Stamm, C., Räsänen, K., Burdon, F.J., Altermatt, F., Jokela, J., Joss, A., et al., 2016. Chapter Four - Unravelling the Impacts of Micropollutants in Aquatic Ecosystems: interdisciplinary Studies at the Interface of Large-Scale Ecology. In: Dumbrell, A.J., Kordas, R.L., Woodward, E.R. (Eds.), Large-Scale Ecology: Model Systems to Global Perspectives. Academic Press, 55, pp. 183–223. https://doi.org/10.1016/bs.aecr.2016.07.002.

Strynar, M.J., Lindstrom, A.B., 2008. Perfluorinated Compounds in House Dust from Ohio and North Carolina, USA. Environ. Sci. Technol. 42 (10), 3751–3756.

Stuart, M., Lapworth, D., Crane, E., Hart, A., 2012. Review of risk from potential emerging contaminants in UK groundwater. In Sci. Total Environ. 416, 1–21. https://doi.org/10.1016/j.scitotenv.2011.11.072.

Talib, A., Randhir, T.O., 2017. Climate change and land use impacts on hydrologic processes of watershed systems. J. Water Clim. Change 8 (3), 363–374.

Tatarazako, N., Ishibashi, H., Teshima, K., Kishi, K., Arizono, K., 2004. Effects of triclosan on various aquatic organisms. Environ. Sci.: Int. J. Environ. Physiol. Toxicol. 11 (2), 133–140.

Tawabini, B.S., 2014. Simultaneous Removal of MTBE and Benzene from Contaminated Groundwater Using Ultraviolet-Based Ozone and Hydrogen Peroxide. Int. J. Photoenergy 2014, 1–7. https://doi.org/10.1155/2014/452356.

Ternes, T.A., Andersen, H., Gilberg, D., Bonerz, M., Flo, D.-., 2002. Determination of Estrogens in Sludge and Sediments by Liquid Extraction and GC /MS / MS 74 (14), 3498–3504.

Ternes, T.A., Joss, A., Siegrist, H., 2004. Scrutinizing pharmaceuticals and personal care products in wastewater treatment. In Environ. Sci. Technol. 38(20). https://doi.org/10.1021/es040639t.

Trably, E., Patureau, D., 2006. Successful Treatment of Low PAH-Contaminated Sewage Sludge in Aerobic Bioreactors (7 pp) ★. Environ. Sci. Pollut. Res. 13 (3), 170–176. https://doi.org/10.1065/espr2005.06.263.

Tran, N.H., Urase, T., Ngo, H.H., Hu, J., Ong, S.L., 2013. Insight into metabolic and cometabolic activities of autotrophic and heterotrophic microorganisms in the biodegradation of emerging trace organic contaminants. Bioresour. Technol. 146, 721–731. https://doi.org/10.1016/j.biortech.2013.07.083.

Urase, T., Kagawa, C., Kikuta, T., 2005. Factors affecting removal of pharmaceutical substances and estrogens in membrane separation bioreactors. Desalination 178 (1–3 SPEC. ISS.), 107–113. https://doi.org/10.1016/j.desal.2004.11.031.

Vasskog, T., Anderssen, T., Pedersen-bjergaard, S., Kallenborn, R., Jensen, E., 2008. Occurrence of selective serotonin reuptake inhibitors in sewage and receiving waters at Spitsbergen and in Norway. J. Chromatogr A 1185, 194–205. https://doi.org/10.1016/j.chroma.2008.01.063.

Verlicchi, P., Galletti, A., Petrovic, M., Barceló, D., 2010. Hospital effluents as a source of emerging pollutants : an overview of micropollutants and sustainable treatment options. J. Hydrol. 389 (3–4), 416–428. https://doi.org/10.1016/j.jhydrol.2010.06.005.

Verliefde, A., Cornelissen, E., Amy, G., Bruggen, B.V.D., Dijk, H.V., 2007. Priority organic micropollutants in water sources in Flanders and the Netherlands and assessment of removal possibilities with nanofiltration. Environ. Pollut. 146, 281–289. https://doi.org/10.1016/j.envpol.2006.01.051.

Wanda, E.M.M., Nyoni, H., Mamba, B.B., Msagati, T.A.M., 2017. Occurrence of emerging micropollutants in water systems in Gauteng, Mpumalanga, and North West Provinces, South Africa. Int. J. Environ. Res. Public Health 14 (1), 79.

Wanda, E.M.M., Nyoni, H., Mamba, B.B., Msagati, T.A.M., 2018. Application of silica and germanium dioxide nanoparticles/polyethersulfone blend membranes for removal of emerging micropollutants from water. Physics and Chemistry of the Earth, Parts A/B/C, 108, 28–47.

Weigel, S., Berger, U., Jensen, E., Kallenborn, R., Thoresen, H., Heinrich, H., 2004. Determination of selected pharmaceuticals and caffeine in sewage and seawater from Tromsø/Norway with emphasis on ibuprofen and its metabolites. Chemosphere 56, 583–592. https://doi.org/10.1016/j.chemosphere.2004.04.015.

Wijekoon, K.C., McDonald, J.A., Khan, S.J., Hai, F.I., Price, W.E., Nghiem, L.D., 2015. Development of a predictive framework to assess the removal of trace organic chemicals by anaerobic membrane bioreactor. Bioresour. Technol. 189, 391–398. https://doi.org/10.1016/j.biortech.2015.04.034.

Yang, S., Hai, F.I., Price, W.E., McDonald, J., Khan, S.J., Nghiem, L.D., 2016. Occurrence of trace organic contaminants in wastewater sludge and their removals by anaerobic digestion. Bioresour. Technol. 210, 153–159. https://doi.org/10.1016/j.biortech.2015.12.080.

Zhao, H.-.M., Hu, R.-.W., Chen, X.-.X., Chen, X.-.B., Lü, H., Li, Y.-.W., et al., 2018. Biodegradation pathway of di-(2-ethylhexyl) phthalate by a novel Rhodococcus pyridinivorans XB and its bioaugmentation for remediation of DEHP contaminated soil. Sci. Total Environ. 640–641, 1121–1131. https://doi.org/10.1016/j.scitotenv.2018.05.334.

Zhou, H., Zhang, Z., Wang, M., Hu, T., Wang, Z., 2017. Enhancement with physicochemical and biological treatments in the removal of pharmaceutically active compounds during sewage sludge anaerobic digestion processes. Chem. Eng. J. 316, 361–369. https://doi.org/10.1016/j.cej.2017.01.104.

CHAPTER 5

Emerging dye contaminants of industrial origin and their enzyme-assisted biodegradation

Sougata Ghosh[a], Bishwarup Sarkar[b]
[a]Department of Microbiology, School of Science, RK University, Rajkot, Gujarat, India
[b]Department of Microbiology and Biotechnology Centre, The Maharaja Sayajirao University of Baroda, Vadodara, Gujarat, India

5.1 Introduction

Continuous rise in population have impacted almost all spheres of life, the notable being the clean water. Several ecosystems are damaged or altered due to toxicity of the pollutants released in the industrial effluents. Dyestuffs or the chemicals that impart colour to the final products are used in various industries that are associated with pharmaceuticals, textiles, cosmetics, plastics, food and paper (Saratale et al., 2011). These industries also consume large volumes of water during the dying or colouring process and in turn generates huge amount of coloured effluents. On an average 7×10^5 tnes of dyestuffs is produced globally that is comprised of 100,000 commercially available dyes (Supaka et al., 2004). Environmental threat is posed by the release of 5–50 percent of the reactive dyes that enters the waste water due to inability to bind with the fibres during textile processing (Rai et al., 2005). Broadly, dyes are classified as anthraquinone, azo, heterocyclic, polymeric, and triphenylmethane dyes depending upon the chemical structure of the chromophoric group. Among the aforementioned dyes, diverse types of azo and triphenylmethane dyes are most predominant in the effluents of textile industries (Yang et al., 2009).

The recalcitrant nature of the synthetic dyes is attributed to the complex chemical backbone that makes it more difficult for removal (Lin et al., 2010). Indiscriminate release of the dye contaminated effluents in the water bodies leads to reduced light penetration, thereby affecting the aquatic flora. This reduces the photosynthetic ability of the aquatic plants and thereby adversely impacts dissolved oxygen concentration and water quality. Moreover, the hazardous dyes result in acute toxicity to the aquatic flora and fauna which is associated with the impairment of the physiological and metabolic activities (Luikham et al., 2018). Further, such toxic dyes also alter the total organic carbon (TOC) and chemical oxygen demand (COD) degrading the overall quality of the water. Several synthetic dyes are not only highly toxic but also carcinogenic and

Biodegradation and Detoxification of Micropollutants in Industrial Wastewater
DOI: https://doi.org/10.1016/B978-0-323-88507-2.00005-1

mutagenic in nature. Hence on entering the food chain, they pose potential health hazard to humans. Dye contaminated water also adversely affect crops and natural vegetation which are responsible for providing habitat to the wildlife and prevention of soil erosion. Contamination of the irrigational water with dyes may cause reduction in seed germination, flowering, growth, and overall yield of the crop plants (Karmakar et al., 2020; Ghosh and Webster, 2020).

Conventional methods of dye treatment include both physical and chemical processes. Advanced oxidation processes that employ ultraviolet light, ozone, and hydrogen peroxide, are not only expensive, but also limited to small scale treatment (Hai et al., 2007; Kurade et al., 2012). Hereby, alternative strategy for dye removal involves biological methods which are more specific, effective, rapid, environmentally benign and less energy consuming. Biological treatment of dye removal involve bacteria, fungi, algae and plants and their enzymes that can either partially or completely degrade the dyes. Moreover, in this strategy, stable and non-toxic end products are formed (Bloch et al., 2021; Ghosh et al., 2021; Pandey et al., 2007). In view of the background, this chapter is voluminous detail of various enzymes involved in the process of dye bioremediation.

5.2 Enzymes for dye degradation

Microbial decolorization of hazardous dyes is either aerobic or anaerobic where various enzymes like azoreductases, laccases, peroxidises are used that are discussed in the following section.

5.2.1 Azoreductase

Microbes from diverse taxonomic groups like bacteria, actinomycetes, fungi and algae are able to cleave azo bond by an enzyme called azoreductase resulting in effective degradation of toxic dyes (Table 5.1). In this process colorless aromatic amines are formed which are further degraded using multi-step enzymatic metabolic process to simpler non toxic form (Das and Mishra, 2016).

Ramya et al. (2010) reported degradation of acid red 37 dye using *Acinetobacter radioresistens* isolated from textile dye contaminated soil. A high decolourisation percentage (87.2 ± 0.45%) was achieved with an optimal temperature ranging from 40 to 50 °C. Further, a maximum decolourisation of 90 percent was obtained in a pH range of 6–7. Azoreductase enzyme played a key role in the dye degradation producing 7, 8-diamino-3[(aminoxy)sulfonyl] naphthalene-1-ol and 1-{3-amino-5-[(aminoxy)sulfonyl] phenyl} ethanol as major degradation products. It was speculated that a hydrazo intermediate was formed that was converted to aromatic amines. One of the possible mechanisms proposed was cleavage of azo bond through a two-step NADH dependent reduction reaction resulting in degradation of the acid red dye.

Table 5.1 Bacterial enzymes for dye degradation.

Bacteria	Enzyme	Dye	Removal (%)	References
Acinetobacter radioresistans	Azoreductase	Acid red 37	90	Ramya et al., 2010
Alcaligenes aquatilis	–	Synazol red 6HBN	82	Ajaz et al., 2019
Bacillus lentus BI377	Azoreductase	Reactive Red 141	99	Oturkar et al., 2013
Bacillus subtilis ORB7106	AzoR1	Methyl Red	89	Leelakriangsak and Borisut, 2012
Chromobacterium violaceum	CV_RS09840	Amaranth dye	–	Verma et al., 2019
Kocuria rosea	Azoreductase	Methyl Orange	100 percent	Parshetti et al., 2010
Pseudomonas aeruginosa NGKCTS	–	Reactive Red BS	91	Sheth and Dave, 2009
Rhodococcus opacus 1CP	AzoRo	Methyl Red	89	Qi et al., 2016
Streptomyces sp. S27	AzoRed2	Methyl Red	99	Dong et al., 2019
Bacillus subtilis	CotA-laccase	Sudan Orange G	50	Pereira et al., 2009
Bacillus vallismortis fmb-103	Laccase	Brilliant Green	95	Zhang et al., 2012
Stenotrophomonas maltophilia AAP56	SmLac	Reactive Black 5	99	Galai et al., 2014
Streptomyces cyaneus CECT 3335	Laccase	New coccine	92	Moya et al., 2010
Bacillus sp. VUS	Lignin peroxidase	Navy Blue 2GL	94	Dawkar et al., 2009
Pseudomonas sp. SUK1	Peroxidase	Methyl Orange	72	Kalyani et al., 2011
Raoultella ornithinolytica OKOH-1	RaoPrx	Congo Red	65	Falade et al., 2019
Serratia marcescens	Manganese peroxidase	Ranocid Fast Blue	90	Verma et al., 2003

In another study, Ajaz et al. (2019) reported that *Alcaligenes aquatilis* isolated from industrial effluent sample was able to degrade Synazol red 6HBN. Effective dye degrading potential was obtained in mineral salt medium (MSM) with supplementation of saw dust as carbon source and yeast extract as nitrogen source with an inoculum size of 6 percent and dye concentration of 10 mg/L. A maximum decolourisation percentage of 82 percent was obtained after incubation at 37 °C and pH 7 for a time period of 4 days under static conditions. Fig. 5.1 shows that the enzymatic degradation of dye led to formation of various end products, some of which included pentadecanal, palmitic acid, chlorobenzene, 2-acetyl-3-methylhexahydryopyrrolo[1,2-α] pyrazine-1,4-dione,

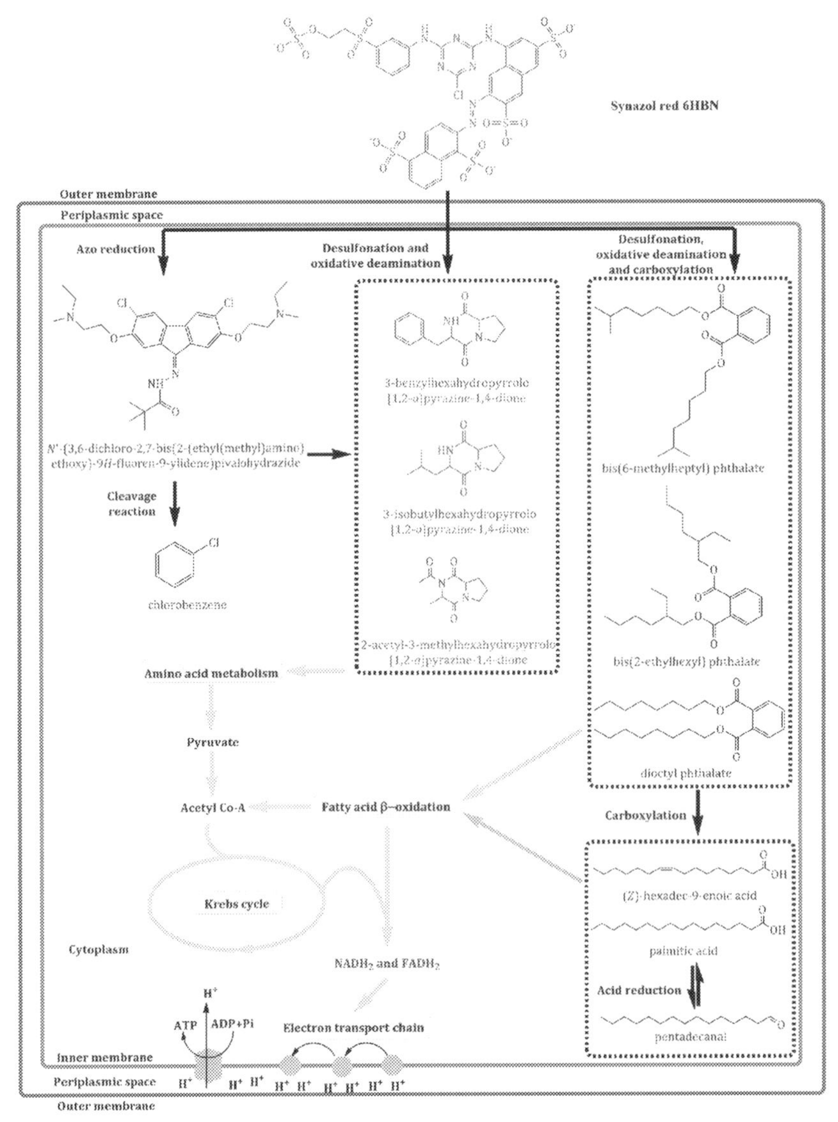

Fig. 5.1 ***Synazol red 6HBN dye enters into the cell (A. aquatilis 3c) by unknown mechanism.*** Upon entrance into the cell the dye is enzymatically processed into various end products. Most probably, azo group (*–N = N–*) is reduced, followed by cleavage reaction and different end products are produced. Secondly, desulfonation and oxidative deamination results in synthesis of pyrrolo[*1,2-a*]pyrazine-1,4-dione derivative which can be used as substrates in amino acid metabolism. The amino acid catabolism can synthesize pyruvate (3C compound) which can be converted into acetyl-CoA. The acetyl-CoA undergoes Krebs cycle to produce $NADH_2$ and $FADH_2$ (substrates of electron transport chain). Moreover, dye desulfonation, oxidative deamination and carboxylation lead to produce phthalate derivatives, which can be transformed into different fatty acids and aldehydes. The phthalate, fatty acids and aldehydes can directly/indirectly enter into fatty acid oxidation reactions (β-oxidation) to produce acetyl-CoA, $NADH_2$ and $FADH_2$. Reprinted from Ajaz, M., Rehman, A., Khan, Z., Nisar, MA, Hussain, S., 2019. Degradation of azo dyes by *Alcaligenes aquatilis* 3c and its potential use in the wastewater treatment. AMB Express, 9:64. (Open access)

bis (6 methylheptyl) phthalate, 3-isobutylhexahydropyrrolo[1,2-α] pyrazine-1,4-dione as well as [Z]-hexadec-9-enoic acid. It was proposed that these end-products can be utilised through various metabolic pathways. For example, several of the end-products formed by azo dye degradation such as pyrrolo[1,2-α] pyrazine-1,4-dione derivates and phthalate derivates can be used as a substrate in amino acid metabolism and fatty acid β-oxidation reactions, respectively, which can be further directed for production of reduced molecules (that is, NADH and $FADH_2$) that may be used for synthesis of ATP.

An alkaliphilic *Bacillus lentus* BI377 strain was reported by Oturkar et al. (2013) for its contribution in decolorization of Reactive red 141 dye with the help of azoreductase enzyme. Maximum degradation of 99.11 percent was observed after 6 h of incubation. Structural analysis revealed formation of two products- [2-amino-8- (2- (4- (6- (7-amino-3,6,8-trihydroxynaphthalene −1-ylamino) pyridine − 2-ylamino) phenylamino) pyrimidin-4 ylamino) naphthalene-1,3,6-triol] and 8-(4,6-dichloro-1,3,5-triazin-2-ylamino)−2-diazenylnaphthalene-1,3,6-triol after 2 h of incubation. After 4 h of incubation, the first product further is converted to benzene-1, 4-diamine and 2,8-diaminonaphthalene-1,3,6-triol. And finally, after 6 h, these products are transformed into simpler metabolites. The intermediates formed during degradation were less toxic as compared to untreated dye and complete decolorization may lead to mineralization.

Leelakriangsak and Borisut (2012) reported decolourisation of methyl red using *Bacillus subtilis* ORB7106, which is a deletion mutant of the *yodB* gene and thus constitutively expresses *azoR1* and produce increased levels of azoreductase enzyme. The efficacy of decolorization was increased up to 1.8-fold upon inducing the wild type parent *B. subtilis* strain JH642 with methyl red dye. Further, comparative analysis of decolourising activity of the mutant strain with its wild type parent revealed better efficiency of decolourisation, hence providing evidence that *azoR1* gene product may be an azoreductase which is capable of degrading azo dyes. 40–98 percent efficiency of removal of colour was obtained over a wide range of concentrations of methyl red (10–200 mg/L). Maximal colour removal of more than 89 percent was achieved at 45 °C and pH 7 in 48 h under anoxic conditions.

Similarly, an actinomycetes, *Streptomyces* sp. S27 was reported to produce a novel azoreductase enzyme- AzoRed2 as identified by Dong et al. (2019) which was efficiently able to reduce methyl red. AzoRed2 was observed to be a novel FMN-dependent azoreductase that was able to use either NADH or NADPH as electron donors. A linear increase in azoreductase activity was found in the temperature range of 25 °C to 55 °C with maximum activity at 55 °C. However, the azoreductase activity decreased rapidly beyond 55 °C. A higher activity was observed under relatively acidic conditions, which was proposed to be useful for effluent treatment. A good tolerance toward pH change was obtained with high activity between pH of 4.5 and 5.5, optimal pH being 5.0. Impressive tolerance against several organic solvents such as dimethyl sulfoxide (DMSO), isopropanol, ethanol, and methanol were observed which may be useful in degradation of azo dyes in industrial wastewater effluents containing considerable concentration of organic solvents.

Further, an efficient whole-cell biocatalyst containing an integrated enzyme system was constructed having high expression levels of two enzymes-azoreductase (AzoRed2) from *Streptomyces* sp. S27 and *Bacillus subtilis* glucose 1-dehyrogenase (BsGDH), which acted as an *in-situ* coenzyme regeneration system. This resulted in complete degradation of azo dye with more than 99 percent efficiency without any external addition of NAD^+, thus contributed toward efficient decolourisation of methyl red. Glucose was used as the hydrogen donor in this coenzyme regeneration reaction, in order to reduce NAD^+ to NADH. A 98.42 percent decolourization of methyl red was obtained at 10 mM of glucose concentration and the decolourization percentage of methyl reached up to 99 percent in 120 min. with a loading of over 5.0 mg/mL of this whole-cell biocatalyst, highlighting the importance of whole-cell biocatalysts in efficient degradation of azo dyes in wastewater without any external addition of NAD^+.

Parshetti et al. (2010) reported that *Kocuria rosea* completely decolorized (100 percent) methyl orange (50 mg/L) under static conditions. However, no decolorization was obtained at shaking condition suggesting depletion in oxygen content promoted decolorization by *K. rosea.* Optimal temperature was 30 °C which facilitated maximum decolorization. Optimal pH range was found to be in between 6.5–6.8. Up to 50 mg/L of the dye could be effectively decolourised while high concentration of dye showed reduced percentage decolorization even with longer incubation time of 120 h A 90 percent yeast extract showed maximum percent of decolorization as compared to other supplements of carbon and nitrogen source within 48 h, however, presence of peptone and NH_4Cl exhibited only 15 percent decolourizing ability, thus, indicating maximum decolorization of methyl orange was obtained in presence of yeast extract as nitrogen source in synthetic media. Minimal decolourization of 6 percent was obtained with glucose as a carbon source, while lactose, maltose and sucrose showed 43.33 percent, 33.33 percent and 23.7 percent decolorization, respectively. Less time was taken to decolourise repeated addition of methyl orange which could be useful when implemented on a large scale. Activities of two enzymes namely-azoreductase and NADH-DCIP reductase were significantly increased suggesting their involvement in decolorization of methyl orange. The degradation products formed were found to be less toxic when exposed to *A. vinelandii* and *P. aeruginosa* as compared to methyl orange. Further, detoxification of methyl orange was confirmed when percentage of seed germination and length of plumules and radicles of *T. aestivum* and *P. mungo* seeds were found to be significantly higher in degradation products and distilled water as compared to presence of methyl orange. A pathway of degradation of methyl orange was proposed involving symmetric cleavage of azo bond by azoreductase resulting in the formation of 4-amino sulfonic acid and N, N-dimethyl *p*-phenylenediamine.

Verma et al. (2019) recently reported the functional role of azoreductase (CV_RS09840) in efficient degradation of amaranth dye. The enzyme was isolated from *Chromobacterium violaceum* and subsequently cloned in *E. coli* BL21 (DE3) cells. This azoreductase enzyme remained active in a wider temperature and pH range with maximum activity between 30 °C to 37 °C and at pH 7.2, thus, highlighting its potential applications for decolorization and degradation of several industrial effluent dyes. Also, a low K_m value was obtained indicating high substrate affinity of azoreductase against these dyes. The activity of azoreductase was observed to be dependent upon NADH, while the native enzyme always remained bound with FAD, indicating that this azoreductase enzyme may be a flavo-enzyme. Amaranth dyes are commonly used in both domestic and industrial activities, thus, levels of toxicity were determined on fibroblast cell line (L929). 10 percent of cells were viable at a concentration of 1000 μg mL^{-1} of amaranth dye indicating its toxicity. However, after degradation of same concentration of dye, the percentage viability of cells increased to 70 percent, suggesting that the degraded products are non-toxic as compared to untreated dye. Hence, it can be inferred that azoreductase from *C. violaceum* can decolorize as well as detoxify azo dyes effectively.

Pseudomonas aeruginosa NGKCTS was isolated by Sheth and Dave (2009) from dye contaminated soil sample. The isolated strain was continuously grown at 31± 2 °C and at 7.4 ± 0.2 pH in nutrient broth containing peptone, meat extract and NaCl. During the initial phase of transfer in nutrient broth, the organism was able to grow actively and required 10 h of incubation for decolorization of 300 ppm of reactive red BS dye. After 5 successful transfers in broth, the culture adapted to the dye concentration and decolorized 91.1 ± 3.0 percent of 300 ppm of reactive red BS dye in only 6 h under static conditions. Decolorization was not observed in shaking conditions, indicating that a facultative anaerobic condition may be favourable, which could be due to degradation by azoreductase enzyme. A decolourization rate of 47.53 mg L^{-1} h^{-1} was obtained when the optimum inoculum concentration was at 20 percent (v/v) with a dry biomass weight of 0.31 g L^{-1}. Hence, the removal rate of dye was found to be 153 mg dye g $cell^{-1}$ h^{-1}. Optimum pH range was found to be 6.0–7.5, with highest decolorization observed at pH 7.0. Increase in decolorization rate up to 55.45 mg L^{-1} h^{-1} was observed at 700 ppm dye concentration with 81.2 percent decolorization. Further increase in dye concentration to 800 ppm and above led to inhibition of decolorization. The decolorization rate was observed to decrease with increased time of incubation. This was speculated to occur because of utilisation of actively growing culture which do not have any lag phase and thus, decolorization could start just after inoculation and the decrease in decolorization rate with increasing amounts of time could be due to less amount of dye available in the broth. Continuous addition of dye in a fed batch manner led to gradual decrease in decolorization time from 5.5 h to 2 h up to the sixth addition which was suggested to happen because of formation of increased biomass in the system. Subsequent addition of dye gradually increased the time of decolorization up to 6 h which could be due to accumulation of metabolites. Efficient decolorization was observed in absence of glucose and in presence of 20 g L^{-1} of peptone and 2.0 g L^{-1} of yeast extract in the medium.

A flavin-containing azoreductase (AzoRo) was identified by Qi et al. (2016). This enzyme was obtained from a contaminated soil isolate of *Rhodococcus opacus* 1CP. A maximum removal efficiency of 89 percent of methyl red was achieved after 28 h of incubation in presence of *R. opacus* 1CP. An *azoRo*-gene was functionally expressed in *E. coli* BL21 using pAzoRo_P01 as the expression plasmid. However, it was observed that AzoRo was not suitable for *in vivo* biotransformation and is useful only as an *in vitro* biocatalyst. Purified AzoRo enzyme from the cell-free extracts was bright yellow in colour, thus, suggesting presence of flavin in the enzyme complex. The molecular weight of the enzyme was found to be 25 kDa. Highest activity of purified AzoRowas observed at an optimal pH of 4.0 and at 53 °C. Cu^{2+} and Mn^{2+} increased the activity while Mg^{2+}, Ca^{2+}, Zn^{2+}, Fe^{3+} and Fe^{2+} slightly inhibited the performance of AzoRo. Analysis of degraded products led to identification of mainly two metabolites- dimethyl-p-phenylenediamin (DMPD) and 2,2′-azino-bis (3-ethylbenzothiazoline-6-sulfonic acid) (ABTS) suggesting that AzoRo is a true azoreductase. The suggested degradation pathway suggested for this enzyme involve the ping-pong mechanism where the flavin cofactor is first reduced by NAD(P)H and then these reduction equivalents are transferred to the azo bond forming a hydrazo intermediate. The reaction is then repeated to generate enough electrons to completely reduce the azo dyes into amines.

5.2.2 Laccase

Another enzyme laccase, mostly found in fungi, also play a critical role in dye degradation as seen in Table 5.2. Hadibarata et al. (2012) reported laccase production from a newly isolated *Armillaria* sp. strain F022 that was able to degrade an anthraquinone dye-Remazol Brilliant Blue R (RBBR) very efficiently. The white-rot fungus was isolated from a decayed wood sample in a tropical rainforest and was maintained on malt extract medium forming a white spore. More than 80 percent of 300 mg L^{-1} of RBBR dye was degraded by purified laccase that was isolated from this *Armillaria* sp. Maximum decolouration was obtained within 94 h at an optimum pH value of 4.0 and optimal temperature of 40 °C. The apparent K_m and V_{max} values of laccase enzyme with ABTS as a substrate was reported to be 325 ± 76 μM and 95 ± 11 μmol min^{-1}, respectively.

Telke et al. (2010) reported an intracellular laccase production in the filamentous fungus *Aspergillus ochraceus* NCIM-1146. The laccase was able to efficiently decolorise 56 percent of methyl orange dye without any redox mediators. Maximal amount of laccase was produced by *A. ochraceus* NCIM-1146 when it was allowed to grown in potato dextrose broth for 96 h. Further purification by anion exchange and size

Table 5.2 Fungal enzymes for dye degradation.

Fungi	Enzyme	Dye	Removal (%)	References
Armillaria sp. strain F022	Laccase	Remazol Brilliant Blue R	80	Hadibarata et al., 2012
Aspergillus ochraceus NCIM-1146	Laccase	Methyl Orange	56	Telke et al., 2010
Ganoderma sp. En3	Laccase	Bromophenol Blue	98	Zhuo et al., 2011
Lentinus polychrous Lév	Laccase	Rhodamine B	90	Suwannawong et al., 2010
Peroneutypa scoparia	Laccase	Acid Red 97	75	Pandi et al., 2019
Trametes pubescens	Tplac	Congo Red	80	Si et al., 2013
Irpex lateus F17	Manganese peroxidase	Malachite Green	96	Yang et al., 2016
Penicillium ochrochloron MTCC 517	Lignin peroxidase	Cotton Blue	93	Shedbalkar et al., 2008

exclusion chromatography gave a yield of 1.66 mg of pure intracellular laccase with 22-fold purification and a total laccase activity of 15 percent. The purified laccase obtained from *A. ochraceus* was found to have a molecular weight of 68 kDa with a single protein band on SDS-PAGE and was observed to have a type-2 binuclear and type-1 copper sites. Hence, the purified enzyme appeared blue in colour in presence of copper. Using ABTS as a substrate, the enzyme remained stable in the pH range of 5.0–7.0. The optimum temperature and pH of purified *A. ochraceus* laccase was found to be 4.0 and 60 °C, respectively. The K_{cat} value of purified laccase was obtained as 111/*sec*. Several metal salts such as $HgCl_2$ and $FeSO_4$ completely inhibited enzyme activity. A 400 mM of NaCl completely inhibited oxidation of ABTS by purified laccase. Other salts such as $CaCl_2$, $MgCl_2$, $MnSO_4$, and $ZnSO_4$ significantly inhibited laccase activity as well. Laccase activity was also strongly inhibited by sodium azide, EDTA, L-cysteine and dithiothreitol. Compounds such as o-tolidine, ABTS, hydroquinone, guaiacol, pyrogallol, o-dianisidine and other methyl and methoxy group substituted phenolic and nonphenolic compounds were observed to be oxidised by purified laccase. However, laccase was able to oxidise veratryl alcohol in presence of appropriate electron transfer mediators such as ABTS and HBT. p-N, N′-dimethylamine phenyldiazine was observed to be the only metabolite that was formed after the degradation suggesting a symmetric cleavage of azo bond. It was proposed that this cleavage may lead to formation of two intermediates- p-N, N′-dimethylamine phenyldiazine and p-hydroxybenzene sulfonic acid.

Pereira et al. (2009) reported a recombinant CotA-laccase from *Bacillus subtilis* that was able to decolorize Sudan orange G (SOG) azo dye. Decolouration of 0.5 mM of SOG was obtained at an optimal pH value of 8.0. An optimal temperature of 75 °C was observed to be suitable for SOG biotransformation. 50 percent of SOG was transformed in 1 h with a rate of biotransformation of 23 ± 2 nmol min^{-1} mg^{-1} by the recombinant enzyme. Further, a two-fold increase in rate of biotransformation was observed upon addition of 10 μM of ABTS as a redox mediator. The calculated K_m, V_{max} and k_{cat} values were found to be 44 ± 7 M, 57 nmol min^{-1} mg $protein^{-1}$ and 0.05 s^{-1}, respectively. It was speculated that the deprotonated dye at its acidic hydroxyl group was more prone to oxidation by the enzyme, hence the optimal pH was found to be at 8.0. The proposed mechanism of action of CotA-laccase on degradation involve oxidation of SOG dye through a single-electron transfer resulting in formation of a phenoxyl radical that sequentially gets oxidised to a carbonium ion. Further, a nucleophilic attack on the N—C bond of phenolic carbonium was speculated to produce diazenylbenzene and 4-hydroxy-1,2-benzoquinone. Diazenylbenzene can then lose a nitrogen molecule to subsequent formation of benzene radical which can undergo hydrogen radical addition or participate in coupling reactions. A detoxification of 67 ± 6 percent was achieved after enzymatic treatment of SOG

Zhang et al. (2012) isolated *Bacillus vallismortis* fmb-103 from a textile industry waste disposal site that produced a spore laccase without any pigment formation. The enzyme was observed to have an optimal activity at a temperature of 85 °C, with highest stability at 70 °C retaining more than 90 percent of its original activity even after 10 h incubation. Similarly, the enzyme remained stable without any loss of activity at 20 °C even after incubation for 10 days. These results indicated that the enzyme have a good thermal stability and long-term stability. Oxidation of ABTS as a substrate was found to be optimal at a pH value of 4.4 and 50 percent of catalytic activity was retained at pH 6.4. Optimal pH values increased in presence of syringaldehyde suggesting that this particular spore laccase can remain stable at both acidic and alkaline pH. The enzyme was reported to completely lose its activity in presence of NaN_3 while inhibitors such as L-cysteine and dithiothreitol was also found to inhibit laccase activity significantly. The enzyme was able to decolorize 95 percent of brilliant green dye in presence of mediators such as ABTS and acetosyringone. Furthermore, similar efficiency in decolorization was observed during five repeated cycles of reaction of spore laccase with brilliant green dye in presence of acetosyringone at pH 6.0 highlighting excellent stability and reusability for repetitive operations which could be particularly useful in industrial applications.

Ganoderma sp. En3 was isolated from the forest of Tzu-chin Mountain, China by Zhuo et al. (2011) that was reported to secrete a laccase whose activity was significantly enhanced upon addition of Cu^{2+}. The enzyme was able to rapidly decolorize 98.3 percent of 50 mg/L of bromophenol blue within a short incubation time of 12 h, thus highlighting its potentiality for decolorizing industrial effluents. However, an increase in decolorization efficiency was not observed upon subsequent increase in dye concentration. A significant reduction of 86.1 percent of total organic carbon (TOC) was observed after decolorization. Degradation of bromophenol blue led to detoxification as well, since the decolorization products were reported to have a lower phytotoxicity with respect to growth of root and shoot of *Triticum aestivum* and *Oryza sativa* seeds when compared with untreated dyes. Since crude enzyme is supposed to be more feasible and inexpensive for its application on a large scale, decolorization of culture supernatants of *Ganoderma* sp. En3 grown in GYP medium were observed. Decolourisation of 90.88 percent of bromophenol blue was reported within 12 h of incubation. Laccase activity was inhibited by addition of 30 mM kojic acid which led to subsequent decrease in decolorization efficiency of crude enzyme down to 13.89 percent indicating that decolorization of dye was positively linked to activity of laccase. Additionally, whole cultures of *Ganoderma* sp. En3 were able to effectively decolorize 91.3 percent and 90.9 percent of simulated dye effluent-I and II within 8 days, respectively. These simulated dye effluents mimic the dye effluent released by various industries. Thus, indicating that this white rot fungi was effectively able to tolerate as well as degrade and detoxify higher concentrations of complex reactive dye effluents. Furthermore, real indigo effluent containing indigo dyes, collected from Puqi Textile Dyeing Factory, China was effectively decolorized up to 91.38 percent by *Ganoderma* sp. En3 within 14 days. In order to utilise this laccase, full-length cDNA of *lac-En3-1* gene which encoded this enzyme was cloned in pPIC3.5K plasmid and transformed into *Pichia pastoris* GS115. Decolorization efficiency of *Pichi pastoris* GS115 transformants were analysed against bromophenol blue as synthetic dye, simulated dye effluents and real indigo effluents, respectively. Effective degradation of 80.27 percent of bromophenol blue, 81.54 percent simulated dye effluent-I and 83.72 percent of real indigo effluents were observed by the transformants, thus, highlighting the functionality of *lac-En3-1* gene encoding active laccase enzyme. Analysis of 5'-flanking sequence present upstream of the gene indicated the presence of putative cis-acting responsive elements in the promoter region suggesting regulation of *lac-En3-1* gene expression at a transcriptional level.

Decolorization of rhodamine B dye was reported by Suwannawong et al. (2010) using a partially purified laccase obtained from *Lentinus polychrous* Lév, which was collected from a mushroom farm and maintained on potato dextrose agar at a temperature of 4 °C. Laccase enzyme was partially purified giving a yield of 5.6 percent and a specific activity of 14.3 U mg^{-1}. The molecular mass was suggested to be around 45 kDa. The efficiency of decolorization o was increased significantly in presence of redox mediators such as 25 μM of ABTS. The optimum pH range for decolorization was found to lie in between 4.0–5.0 and the optimal temperature was at 50 °C. A maximum decolorization of 90 percent was observed within 52 h of incubation at 50 °C with a dye concentration of 20 μM.

Pandi et al. (2019) for the first-time reported an extracellular laccase production from *Peroneutypa scoparia* showing an activity of 12 U mL^{-1} and also within a very short time period of fermentation of less than 4 days. The enzyme was reported to maximally decolorize 75 percent of acid red 97 dye within 6 h of incubation. The enzyme activity was mesophilic in nature where it remained stable within a pH range of 5.0–8.0 with optimal enzymatic activity at 6.0. Also, stability was observed up to 50 °C with highest activity obtained at 40 °C. This activity was lost at 60 °C. Hence, due to the mesophilic optima of laccase, it has been proposed to have a potential for several industrial applications such as textile, food etc. In order to further improve the decolorization ability of laccase, enzyme mediators were added. Addition of a natural mediator- syringaldehyde (SA) increased decolorization of acid red 97 up to 87 percent. Synthetic mediator such as 1-hydroxybenzotriazole (HBT) showed similar results. The increase in decolorization efficiency of laccase upon introducing mediators such as SA could be due to the electron donor effect of methoxy substituents between the phenol and the enzyme as well as a decrease in redox potential caused by the presence of benzene ring in SA. The degradation product of acid red 97 was found to be naphthalene 1,2-dione and 3-(2-hydroxy-1-naphthylazo) benzenesulfonic acid. The mechanism of degradation was proposed to be composed of three phases. In the first phase, phenolic group of the dye is oxidized by laccase with a single electron transfer leading to formation of a phenoxy radical. The second phase is followed by oxidation of carbonium ion. A nucleophilic attack on phenolic ring carbon bearing the azo linkage by water takes place in the final phase which lead to the production of naphthalene 1,2-dione and 3-(2-hydroxy-1-naphthylazo) benzenesulfonic acid.

In another study, *Stenotrophomonas maltophilia* AAP56 was isolated from soil by Galai et al. (2014) and was reported to produce a laccase enzyme (SmLac) that showed a 99 percent decolorization rate of the diazoic dye reactive black 5 (RB5) in the presence of redox mediators such as 2,2-azino-bis (3-ethylbenzthiazoline-6-sulfonic acid (ABTS) within 1 h. The bacterium was grown in Luria-Bertani (LB) medium at 30 °C and 150 rpm. 400 μmol L^{-1} of $CuSO_4$ was supplemented in the medium to induce the activity of SmLac. SmLac activity was observed using 2,2′-azino-bis (3-ethylbenzthiaz oline-6-sulfonic acid) (ABTS) as its substrate for oxidation. The optimal pH and temperature for oxidation of ABTS was found to be 5.0 and 40 °C, respectively. SmLac was found to be specific and efficient towards oxidation of ABTS with a catalytic efficiency of 7.38×10^3 M^{-1} s^{-1}. The redox potential of laccase is one of the parameters that determine the catalytic activity of laccase. Without the presence of oxygen, the redox potential of native SmLac was found to be 638 mV. Further, an increase of SmLac redox potential was found upon addition of oxygen or copper ions, indicating their aid during reduction process. It was proposed that the interaction of O_2 with the catalytic centre of the enzyme followed by its reduction to form water could lead to increase of reduction current and in the case of copper addition, increase in current could be due to reduction of Cu (II) to Cu (I) in the catalytic centre of enzyme. Hence, SmLac could act as a reductase in the presence of O_2 and in the absence of an oxidative substrate. The optimal pH range for decolourisation was found to be in between 7.0 to 8.0 with 80 percent of removal of reactive black 5 within 15 min in presence of ABTS as redox mediator. The degradation products were proposed to form because of reduction of the azoic bridge with formation of a partial mineralized by-product.

Laccase produced by *Streptomyces cyaneus* CECT 3335 was reported by Moya et al. (2010) to degrade new coccine azo dye. The culture was routinely grown on Soy-Mannitol agar at 28 °C for 4–6 days until formation of spores are observed. The enzyme was purified from 14 days old culture having a yield of 60.41%, laccase activity of 0.2 U mL^{-1} and a purification factor of 8.57. A low decolouration rate of 18 percent was obtained against new coccine dye with this purified laccase after 3 h of incubation. Hence, 0.1 mM of acetosyringone was used as an effective oxidation mediator increasing the decolorization rate up to 92% in 4 hrs of incubation. Toxicity analysis revealed that decolouration of new coccine resulted in a 300% decrease in toxicity as compared to untreated dye as control. However, it was shown that disappearance of colour does not involve the detoxification of dyes. Further, analysis revealed that acetosyringone mediator was also degraded by the laccase enzyme which could affect cost-effectiveness of this laccase-acetosyringone mediator system for its use in industrial applications.

Si et al. (2013) reported an extracellular laccase (Tplac) obtained from *Trametes pubescens* with a relatively high specific activity of 18.542 U mg^{-1} from a 6-day culture. It was suggested that the laccase produced by this basidiomycete fungi is a monomeric protein in composition. A typical type-III binuclear copper center was found to be the nature of catalytic center of the enzyme. Interestingly, the enzyme was found to be highly stable over a broad range of pH of 4.5–10.0 with an optimum pH at 5.0, with an activity of 20.218 U mL^{-1} after 72 h, and maintaining 75% of its original activity at 25 °C. In comparison, the enzyme activity was highly unstable at a pH range of 1.0–4.0, thus, indicating Tplac exhibits alkali-resistant activity which could be useful for special applications. Optimum temperature of Tplac was observed at 50 °C, exhibiting a laccase activity of 20.744 U mL^{-1} after incubation for 2 h Another unique feature of Tplac was found to be its higher tolerance toward metal ions. Surprisingly, the enzyme activity was enhanced in presence of metal ions such as Cu^{2+}, Mn^{2+}, Na^{+}, Zn^{2+}, and Mg^{2+} with a maximum of 111.32% increase over addition of Cu^{2+} after incubation at 50 °C. 1.0 U mL^{-1} of Tplac was able to decolorize 80.53% of 50.0 mg L^{-1} of congo red dye after incubation period of 72 h Congo red dye degradation led to formation of metabolites that were identified as naphthalene amine, biphenyl amine, biphenyl, and naphthalene diazonium. The first step of the proposed pathway for the mechanism of degradation involves reduction of the $-N = N-$ bond, resulting in the formation of two reactive intermediates which are further degraded to form stable intermediates. Also, degradation using Tplac was thought to detoxify azo dyes as well since the degradation products of congo red were found to be less toxic as the germination percentage and lengths of the plumule and radicle of *P. mungo*, *S. vulgare*, and *T. aestivum* seeds were found to be greater when compared to untreated Congo Red as a control.

5.2.3 Peroxidase

In addition to phenoloxidases (laccases) and reductases, peroxidases from several microbes can effectively transform specific recalcitrant dye pollutants to less toxic intermediates that allow effective removal during the final treatment processes. Dawkar et al. (2009) reported decolorization and biodegradation of Navy Blue 2GL dye by a lignin peroxidase (LiP) enzyme isolated from *Bacillus* sp. VUS. 94 percent of effective decolorization of 50 mg L^{-1} of navy blue 2GL dye was observed only under static and anoxic conditions in 48 h using yeast extract medium. LiP enzyme was able to decolorize dye concentrations between the range of 50–350 mg L^{-1}. Substrates such as *n*-propanol and phenolic compounds which include L-DOPA, hydroxyquinone, ethanol and veratrole were oxidised by LiP. The time taken to oxidise *n*-propanol was much less as compared to other substrates. However, no activity was observed against xylidine and indole. A 1.33-fold increase in LiP activity was observed in presence of H_2O_2, with a K_m value of 0.046 mM indicating high affinity of enzyme towards H_2O_2. $CaCl_2$ was found to act as an inducer and decreased the time from 48 h taken by *Bacillus* sp. VUS to 18 h However, no significant change in LiP activity was observed in presence of $CaCl_2$. Based on the analysis of degradation products, a pathway had been proposed for biodegradation of navy blue 2GL which involve asymmetric cleavage of carbon of aromatic ring and nitrogen by LiP resulting in formation of intermediates which undergo demethylation reaction to produce a stable product- [2-(2-Bromo-4, 6-dinitro phenylazo)−4-methoxy-phenylamino]-methanol. This product is further degraded to produce 4-Amino-3-(2-bromo-4, 6-dinitro-phenylazo)-phenol and acetic acid 2-(2-acetoxy-ethylamino)-ethyl ester as the final metabolites. No germination inhibition was found in *S. bicolor* and *T. aestivum* when grown with metabolites formed after complete degradation, while 80 percent and 70 percent germination inhibition was observed in *S. bicolor* and *T. aestivum*, respectively when grown under navy blue 2GL as control, thus indicating a decrease in toxicity along with decolorization of dye.

A white rot fungus *Irpex lateus* F17 was reported by Yang et al. (2016) to produce a manganese peroxidase (MnP), which was able to decolorize and detoxify Malachite Green- a triphenylmethane dye as seen in Fig. 5.2. The crude enzyme showed an activity of 923.1 ± 29.5 UL^{-1} with K_m, V_{max} and k_{cat} values as 109.9 μmol L^{-1}, 152.8 μmol L^{-1} min^{-1} and 4.5 s^{-1}, respectively. Comparison between efficiency of decolorization of dye by purified and crude MnP was found to be 90.3 and 87 percent, respectively, indicating easy and low-cost preparation process of crude enzyme may be a good option for its application on a large scale. Highest decolorization was obtained at pH 3.5 after 1 h of incubation. Further, 96.4 percent decolorization was obtained at 40 °C indicating optimal thermal range of enzyme to be 35–50 °C. A significant decrease in dye decolorization was observed upon subsequent increase in concentration of malachite green. However, decolorization efficiency increased with an increasing concentration of H_2O_2 for malachite green concentrations greater than 300 mg L^{-1}. Optimum concentration

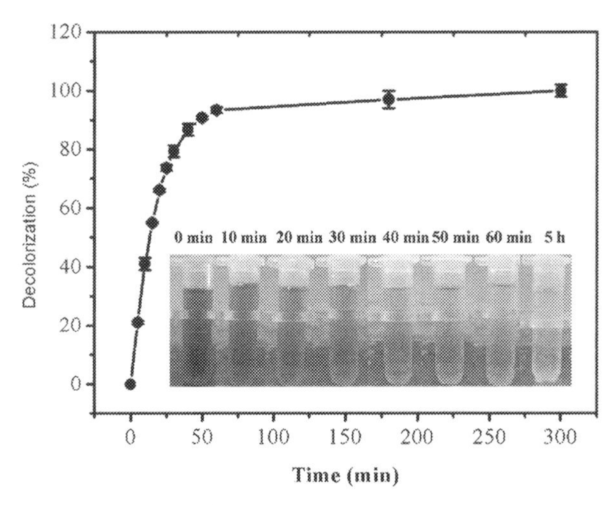

Fig. 5.2 ***MnP-catalyzed decolorization of MG under optimal conditions at various time periods and the change in colour of MG samples was shown in the inset.*** Error bars correspond to standard deviation of three samples. Reprinted with permission from Yang, X., Zheng, J., Lu, Y. and Jia, R., 2016. Degradation and detoxification of the triphenylmethane dye malachite green catalyzed by crude manganese peroxidase from *Irpex lacteus* F17. Environ. Sci. Pollut. Res. 23(10), 9585–9597. Copyright © Springer-Verlag Berlin Heidelberg 2016.

of concentration of Mn^{2+} and H_2O_2 was found to be 2.0 mM and 1.0 mM, respectively, where it showed maximum enzymatic activity. Further, under all previously stated optimal conditions, decolorization efficiency increased from 96 percent to 100 percent upon increasing the reaction time from 1 hr to 5 h Toxicity studies of degradation products compared with untreated dye against bacteria (*E. coli*, *B. subtilis*, *P. vulgaris* and *S. aureus*) and algae (*C. vulgaris* and *S. obliquus*) showed significant decrease in growth inhibition percentage indicating detoxification of malachite green along with decolorization by MnP. Two degradation pathways of malachite green by MnP were proposed. In the first pathway, initiation of reaction by oxidative cleavage of C—C bond present in the dye with the help of the Mn^{3+} generate by the enzyme. This may lead to formation of 4-(dimethylamino) benzophenone (DLBP) as an intermediate which can undergo successive N-demethylation reaction to form 4-(methylamino) benzophenone and 4-aminobenzophenone. DBLP may also be converted into 4-dimethylaminobenzaldehyde due to removal of benzene. Another pathway that was proposed follows N-demethylation and cleavage of the conjugated structure of malachite green.

A lignin peroxidase mediated decolorization of cotton blue dye was reported by Shedbalkar et al. (2008). The enzyme was produced by *Penicillium ochrochloron* MTCC 517. Optimal biodegradation was observed at pH 6.5 and at temperature 25 °C exhibiting 93% decolorization of 50 mg L^{-1} of cotton blue dye using 96 h grown fungal biomass. Analysis of metabolites extracted after decolorization indicated formation of sulphoxides and sulphonamides. Degradation pathway proposed for this peroxidase against triphenylmethane dyes such as Cotton Blue, involve asymmetric cleavage of dye by the enzyme. Role of lignin peroxidase was confirmed with presence of high activity of this enzyme in supernatant after addition of dye indicating appearance of extracellular activity of peroxidase in presence of cotton blue dye. Further phytotoxicity studies revealed significant increase in root and shoot growth of *T. aestivum* and *E. lens Linn* seeds in presence of metabolites after dye degradation as compared to untreated dye, thus, biodegradation of dye by *P. ochrochloron* peroxidase led to detoxification as well.

Pseudomonas sp. SUK1 was isolated by Kalyani et al. (2011) from waste disposal site of textile processing and dye manufacturing industries. This bacterial strain was reported to produce intracellular peroxidase that showed effective decolorization activity against methyl orange dye. The purified enzyme had an apparent molecular mass of 86 kDa with a 17.10-fold increase in peroxidase activity and a yield of 1.67%. The optimum pH and temperature of this purified peroxidase was found to be 3.0 and 40 °C, respectively. The enzyme activity was reduced drastically upon further increase in temperature above 40 °C. The enzyme was observed to be a heme-containing protein. Various substrates were observed to be oxidised by the purified peroxidase such as 3-(3, 4-dihydroxy phenyl) L-alanine (L-DOPA), catechol and dimethoxyphenol with maximum oxidation of *n*-propanol. The low K_m value of 0.061 mM was obtained indicating a high substrate affinity by the enzyme. V_{max} was observed to be 2.7×10^2 μM. 72% of methyl orange was efficiently decolorized within 12 h by this protease enzyme. Analysis of degradation products suggested that no sulfamides or sulfonamides were formed upon degradation of methyl orange dye. The mechanism of degradation of methyl orange that was proposed involves symmetric cleavage of the azo bond present in the dye by peroxidase enzyme leading to synthesis of 1,4-benzenediamine, N, N-dimethyl and an intermediate 4-aminobenzenesulfonic acid which further gets desulfonated to form aniline. Recently, Falade et al. (2019) reported a crude peroxidase (RaoPrx) obtained from *Raoultella ornithinolytica* OKOH-1 which had the ability to effectively degrade congo red dye. This lignolytic bacteria was a river isolate and was maintained at 4 °C on minimal salt medium agar slant supplemented with 1 g/L of kraft lignin. K_m and V_{max} of RaoPrx was found to be 3.8 mM L^{-1} and 4.65 μM mL^{-1} min^{-1}, respectively. The enzyme was observed to oxidise numerous substrates some of which includes ABTS, guaiacol, veratryl alcohol and pyrogallol with maximum peroxidase activity exhibited for pyrogallol. The optimum activity for RaoPrx was found to be at pH 6.0 with a stability over a short pH range of 5.0–7.0 after 120 min of incubation. Likewise, optimum temperature of RaoPrx was determined to be 50 °C and was stable at a relatively high temperature range of 50–70 °C for an incubation time of 120 min indicating high thermostability, one of the important traits for industrial applications. Significant improvement of enzymatic activity was observed upon addition of metal ions such as

Ag^{+}, Cu^{2+}, Zn^{2+} and Fe^{2+} in both high (10 mM L^{-1}) and low (1 mM L^{-1}) concentrations, indicating that may be RaoPrx is a metalloenzyme. However, activity of RaoPrx was inhibited in presence of Ca^{2+}, Mg^{2+}, Ba^{2+}, Al^{3+} and NaN_3. Complete inhibition of enzyme was observed at low concentrations of Co^{2+} and EDTA. This suggested that the enzyme may require certain cofactors for its catalytic function which are made unavailable due to presence of metal chelating agents such as EDTA. The molecular weight and isoelectric point of RaoPrx was estimated to be 17.587 kDa and 4.51, respectively. Moreover, the nucleotide sequence of RaoPrx was found to be 99% similar to that of Class B type of DyP-type peroxidase family found in *Raoultella ornithinolytica* B6. Also, a single nucleotide polymorphism (SNP) was identified revealing a genetic variation, thus making RaoPrx unique from previously reported peroxidase genes of *R. ornithinolytica*. RaoPrx exhibited efficient decolorisation of congo red with a maximum decolorization activity of 65.03% within 30 min of incubation suggesting affinity of RaoPrx towards azo and *ortho* positioned arene substituent dyes.

Serratia marcescens was isolated from decomposed neem hull waste by Verma et al. (2003) that was reported to efficiently decolorize ranocid fast blue (RFB) dye. The organism was maintained on nutrient medium supplemented with 0.1% yeast extract, 0.1% glucose and 50 mg $RFBL^{-1}$. Under high concentrations of nutrient medium, complete decolorization of PBB-HGR was observed in 5 days when kept under static conditions while it took 6 days under limiting concentrations of nutrient medium components. Both laccase and a manganese peroxidase (MnP) activity were detected in presence of dye. Activity of MnP was increased two-fold in presence of RFB indicating maybe azo dyes such as RFB is a substrate of MnP instead of laccase. Further, it was observed that dye decolorization was obtained only when culture supernatant was supplemented with 1 mM of Manganese and 1 mM of H_2O_2 suggesting primary involvement of MnP in dye decolorization. An optimum concentration 50 μM of Mn^{2+} and 25 μM of oxalic acid significantly increased decolorization. A decolorization efficiency of 90% was obtained by *S. marcescens* at pH 7.0 in 5 days when incubated with 100 mg L^{-1} of RFB dye. The optimum temperature for production of enzyme and dye decolorization was found to be 26 °C. Analysis of degraded products suggested formation of metanalic acid as one of the metabolites.

5.3 Immobilized enzymes

Daâssi et al. (2014) reported immobilization of a partially-purified acid fungal laccase from white-rot basidiomycete *Coriolopsis gallica* KJ412304 into calcium-alginate beads. The immobilized enzyme was able to decolorize the anthraquinone dye Remazol Brilliant Blue R (RBBR). *C. gallica* KJ412304 was a locally-isolated strain that was cultured in semi-solid-state fermentation conditions containing sawdust and minimal medium for hydration. $CuSO_4$ was supplemented in the medium to induce laccase enzyme production. The cell-free extracellular liquid was collected and used as enzyme source for enzyme quantification. Laccase was immobilized sodium alginate at a concentration of 2% (w/v) with the loading efficiency value being 88.1 ± 0.5%. Concentration of alginate played a key role for trapping the enzyme in Ca-alginate beads. 2% (w/v) $CaCl_2$ was used as a cross linking agent whose concentration affects the activity and the stability of immobilized enzyme. A ratio of 1:4 for the Enzyme: Alginate (E/A) concentrations was optimal giving a high immobilized yield of 93.3 ± 1.1% for immobilized laccase. Immobilized laccase showed better thermal stability after 90 min of incubation at 55 °C as compared to free laccase enzyme. Likewise, higher stability of immobilized enzyme was observed in acidic as well as alkaline pH range. After 20 days of storage, immobilized laccase retained 82.7 ± 0.9% of its activity while free enzyme showed only 22.4 ± 1.8% activity thus highlighting the fact that immobilization may also help in increasing the shelf-life of laccase. Immobilized laccase was able to decolorize 90.3% of RBBR dye after 90 min of treatment as compared to 74.6% decolorization by free laccase enzyme. In the control, Ca-alginate beads were able to remove about 12% of initial RBBR dye within 90 min. Only 12% of initial RBBR dye was decolorized by Ca-alginate beads as control within 90 min, thus, indicating that the predominant mechanism for RBBR removal involve biodegradation by laccase enzyme. Further, the reusability of the immobilized laccase for seven successive batches was investigated which showed that at the end of fourth cycle, more than 70% decolorization was observed. However, lower yield was obtained at the end of seven cycles.

Chhabra et al. (2015) reported decolorization of Basic Green 4 and Acid Red 27 by an immobilized laccase obtained from *Cyathus bulleri* using poly vinyl alcohol (PVA) beads cross-linked with either nitrate or boric acid. PVA-nitrate gave an immobilization efficiency of 90% while retaining 75% laccase activity after incubation for 108 h at 30 °C and at 100 rpm. Thus, a higher percentage of laccase was immobilized and the PVA beads were stable enough for continuous use in column reactors. However, the immobilized enzyme showed lower affinity towards substrate such as ABTS as compared to the free enzyme, which eventually led to decrease in catalytic efficiency. The reason behind this was proposed to be the low permeability of PVA beads which in turn allow less diffusion of dyes and oxygen required for laccase activity. Also, immobilization did not affect any change in resistance against common laccase inhibitors such as EDTA and chlorides. The storage stability was improved significantly where PVA-immobilized enzyme retained 80% of laccase activity after 4 months of storage at 4 °C while free enzyme completely lost its activity in less than 1 month. The optimum pH and temperature of immobilized enzyme remained same as that of free enzyme. 95% decolorization of a simulated effluent containing basic green 4 dye was observed using PVA-immobilized enzyme in batch mode even up to 20 cycles. Further, Acid Red 27 was decolorized by immobilized laccase in presence of hydroxybenzotriazole monohydrate

(HOBT) as a mediator and subsequent analysis led to identification of four major degradation products which were 4-((2-oxo-3,6-disulfo-2,3-dihydronaphthalen-1-yl) diazenyl) naphthalene-1-sulfonate, 4-diazenylnaphthalene-1-sulfonate, naphthalene-1-sulfonate and 3,4-dioxo-7-sulfo-2,3,4,4a-tetrahydronaphthalene-2-sulfonate. The pathway for decolorization of Acid Red 27 that was proposed involve laccase-mediated two electron oxidation of phenolic group of dye to a quinine which leads to formation of 4-((2-oxo-3,6-disulfo-2,3-dihydronaphthalen-1-yl) diazenyl) naphthalene-1-sulfonate which further oxidized resulting in splitting of dye to form the other 3 degradation products. A packed bed bioreactor based continuous decolorization for 5 days resulted in 90 percent decolorization.

Immobilized MnP obtained from *Ganoderma lucidum* was reported by Iqbal and Asgher (2013) to effectively decolorize textile effluents. MnP production by *G. lucidum* was carried out in solid state bio-processing using wheat straw under optimized growth conditions having 2% fructose as carbon source and 0.02% yeast extract as nitrogen source. A 2 mg mL^{-1} of purified MnP was then immobilized using sol-gel entrapment technique having an immobilization efficiency of 93.7%. 84.6% enzymatic activity of this sol-gel entrapped MnP was retained even after 10 successive cycles of $MNSO_4$ oxidation. Also, sol-gel entrapped MnP was found to remain more stable when stored at 25 °C for up to 75 days retaining 68% of its catalytic activity. Interestingly, 99.2% of crescent textile effluent sample was found to be decolourised by sol-gel immobilized MnP within 4 h of reaction time under continuous shaking conditions. Nitro-amines were observed to be the intermediates that were formed during effluent decolorization that were ultimately degraded by immobilized MnP leading to detoxification of effluent sample along with decolorization.

MnP obtained from *Phanerochaete chrysosporium* was immobilized by Bilal et al. (2016) onto glutaraldehyde activated chitosan beads by cross-linkage. Chitosan beads proved to be an excellent biocompatible surface as high efficiency of immobilization (84.8 ± 2.34%) was achieved. An optimum concentration of 2.5% (w/v) of chitosan was used for improved enzyme loading efficiency and was subsequently activated by 2.0% glutaraldehyde which was suggested to provide reinforcement to chitosan beads by cross-linkage within the matrix and also between chitosan support and enzyme molecules. Chitosan-immobilized MnP (CI-MnP) showed better stability as well as catalytic activity under both acidic and alkaline conditions as compared to free MnP, hence, suggesting the fact that immobilization may have broadened the pH stability range of enzyme compared to its free enzyme counterpart. Free MnP exhibited maximum activity at 35 °C, while CI-MnP exhibited maximum activity at a relatively high temperature of 60 °C. The stability of immobilized enzyme at higher temperature was thought to occur because of decreased molecular mobility when the enzyme binds to the support which may strengthen the structural rigidity of enzyme. Further, activity of free enzyme was observed to be reduced down to 2.2% after 10 h incubation at 60 °C while CI-MnP retained 38.4% of its original activity in same conditions. To check for the efficiency of decolorization, textile effluent samples were treated with free MnP and CI-MnP. Removal of 97.31% of colour was observed in 25% diluted wastewater sample in case of CI-MnP, while 67.04% decolorization was observed for free MnP indicating higher decolorization efficiency by CI-MnP. Chitosan beads without MnP was used as control and it showed only 20% colour reduction which can be due to absorption of dye on the beads, thus it was concluded that degradation was predominantly because of enzymatic activity of MnP only. Reduction in COD values was higher in CI-MnP as well when compared to free MnP. However, higher TOC removal was observed when effluent sample was treated with CI-MnP. Cytotoxicity of treated and untreated samples were assayed against *Allium cepa* and *Artemia salina* and it was observed that toxicity was significantly reduced in both free and CI-MnP treated samples. Mutagenicity tests suggested that the effluent sample may contain alkylating agents responsible for mutagenesis. Results of Ames assay revealed that textile effluents may possess mutagenic agents which are removed after MnP treatment, highlighting an attractive strategy to remove these mutagenic agents from wastewater. It was also observed that CI-MnP retained most of its activity up to 5 repeated cycles highlighting its reusability and recyclability which is important for practical implementation.

Ramírez-Montoya et al. (2015) reported decolorization of Acid Orange 7 (AO7) dye to a reasonable extent using immobilized laccase enzyme on a mesoporous carbon obtained from pecan shells. The immobilization of enzyme obtained from *Trametes versicolor* was carried out in batch systems under continuous agitation at 150 rpm and 25 °C leading to adsorption of the enzyme on the support. Phosphate buffer was used to dissolve laccase and this solution was mixed with 50 mg of carbon support. After 48 h, a maximum specific activity of 0.038 U mg^{-1} was obtained at pH 5.0. The activity of supported laccase remained constant over a pH range of 3.0–5.0, suggesting that maximum values of the enzyme activity are observed at acidic pH and immobilization may have increased the stability of laccase in terms of pH. However, the activity of immobilized enzyme was decreased as compared to free enzyme which had been proposed because of steric hindrance caused by the support, thus, limiting the enzyme flexibility and diffusion of either substrate or products. Immobilized laccase was found to be effective against degradation of acid dyes such as Acid Orange 7 with almost 90 percent decolorization within 6 h Further, it was observed that decolorization of dyes was contributed by adsorption by unloaded carbon support as well. Hence, both adsorption and degradation mechanisms were speculated to be involved in the decolorization of these acidic dyes. Degradation of Acid Orange 7 was proposed to involve two oxidative steps and a hydroxylation which may result in formation of 1,2-naphthoquinone and 4-diazenylbenzenesulfonate.

5.4 Conclusions and future prospects

Industrial effluent contaminated with dyestuff is not only detrimental to the environment but also hazardous to the health. Owing to stringent regulatory guidelines for reducing permissible limits of toxic dyes in water, it is highly recommended to develop rapid, efficient, economic and environmentally benign route for dye removal. The limitations of the existing physical and chemical dye removal technologies include high cost of operation and infrastructure, enormous production of sludge and low versatility. In view of the background there is a growing need to consider complementary and alternative strategies for removal and degradation of the reactive and toxic dyes. Microbial enzymes are excellent biocatalysts that lead to rapid and effective degradation of various dyes. Enzymes like azo-reductase, laccase, and peroxidase can cleave certain bonds that facilitate the formation of intermediate products that can be effectively metabolized and removed.

Enzymatic process for dye degradation can be made more effective by rational optimization of the reaction parameters like duration, temperature, pH, aeration, and enzyme concentration. Moreover, thorough characterization of the intermediate during enzymatic degradation should be carried out in order to assess the toxicity and environmental impact of the same. Enzyme immobilization is another attractive technology which not only ensure the stability of the enzymes but also reusability and recyclability for effective dye removal. Enzymes can be also functionalized on the magnetic nanoparticles which will ensure easy recovery after each batch of dye degradation. Advanced technology like genetic engineering can be employed to develop genetically improved microbes with higher dye degrading capability. Further scale-up of the laboratory processes to industrial and community level water treatment process would certainly help to ensure clean water. Hereby, microbial enzyme mediated dye removal strategy can revolutionize the water treatment process.

References

Ajaz, M., Rehman, A., Khan, Z., Nisar, M.A., Hussain, S., 2019. Degradation of azo dyes by *Alcaligenes aquatilis* 3c and its potential use in the wastewater treatment. AMB Express 9 (1), 64.

Bilal, M., Asgher, M., Iqbal, M., Hu, H., Zhang, X., 2016. Chitosan beads immobilized manganese peroxidase catalytic potential for detoxification and decolorization of textile effluent. Int. J. Biol. Macromol. 89, 181–189.

Bloch, K., Webster, T.J., Ghosh, S., 2021. Mycogenic synthesis of metallic nanostructures and their use in dye degradation. In: Dave, S., Das, J., Shah, M. (Eds.), Photocatalytic degradation of dyes: Current trends and future. Elsevier, pp. 509–525.

Chhabra, M., Mishra, S., Sreekrishnan, T.R., 2015. Immobilized laccase mediated dye decolorization and transformation pathway of azo dye acid red 27. J. Environ. Health Sci. Engineer. 13 (1), 38.

Daâssi, D., Rodríguez-Couto, S., Nasri, M., Mechichi, T., 2014. Biodegradation of textile dyes by immobilized laccase from *Coriolopsis gallica* into Ca-alginate beads. Int. Biodeterior. Biodegrad. 90, 71–78.

Das, A., Mishra, S., 2016. Decolorization of different textile azo dyes using an isolated bacterium *Enterococcus durans* GM13. Int. J. Curr. Microbiol. Appl. Sci. 5, 675–686.

Dawkar, V.V., Jadhav, U.U., Ghodake, G.S., Govindwar, S.P., 2009. Effect of inducers on the decolorization and biodegradation of textile azo dye Navy blue 2GL by *Bacillus* sp.VUS. Biodegradation. 20 (6), 777–787.

Dong, H., Guo, T., Zhang, W., Ying, H., Wang, P., Wang, Y., et al., 2019. Biochemical characterization of a novel azoreductase from *Streptomyces* sp.: application in eco-friendly decolorization of azo dye wastewater. Int. J. Biol. Macromol. 140, 1037–1046.

Falade, A.O., Mabinya, L.V., Okoh, A.I., Nwodo, U.U., 2019. Biochemical and molecular characterization of a novel dye-decolourizing peroxidase from *Raoultella ornithinolytica* OKOH-1. Int. J. Biol. Macromol. 121, 454–462.

Galai, S., Korri-Youssoufi, H., Marzouki, M.N., 2014. Characterization of yellow bacterial laccase SmLac/role of redox mediators in azo dye decolorization. J. Chem. Technol. Biotechnol. 89 (11), 1741–1750.

Ghosh, S., Bhagwat, T., Webster, T.J., 2021. Cyanobacteria mediated bioremediation of hazardous dyes. In: Shah, M.P. (Ed.), Prokaryotes: Microbial Remediation of Azo Dyes. CRC Press, Boca Raton (In Press).

Ghosh, S., Webster, T.J., 2020. Biologically Synthesized Nanoparticles for Dye Removal. In: Shah, M., Couto, S.R., Biswas, J.K. (Eds.), Removal of Emerging Contaminants from Wastewater through Bio-nanotechnology. Elsevier (In Press).

Hadibarata, T., Yusoff, A.R.M., Aris, A., Hidayat, T., Kristanti, R.A., 2012. Decolorization of azo, triphenylmethane and anthraquinone dyes by laccase of a newly isolated *Armillaria* sp. F022. Water Air Soil Pollut. 223 (3), 1045–1054.

Hai, F., Yamamoto, K., Fukushi, K., 2007. Hybrid treatment system for dye wastewater. Crit. Rev. Environ. Sci. Technol. 37, 315–377.

Iqbal, H.M.N., Asgher, M., 2013. Decolorization applicability of sol–gel matrix immobilized manganese peroxidase produced from an indigenous white rot fungal strain *Ganoderma lucidum*. BMC Biotechnol. 13 (1), 1–7.

Kalyani, D.C., Phugare, S.S., Shedbalkar, U.U., Jadhav, J.P., 2011. Purification and characterization of a bacterial peroxidase from the isolated strain *Pseudomonas* sp. SUK1 and its application for textile dye decolorization. Ann. Microbiol. 61 (3), 483–491.

Karmakar, S., Ghosh, S., Kumbhakar, P., 2020. High Photocatalytic Activity under Visible Light for Dye Degradation. In: Dave, S., Das, J., Shah, M.P. (Eds.), Photocatalytic Degradation of Dyes: Current Trends and Future. Elsevier (In Press).

Kurade, M.B., Waghmode, T.R., Kagalkar, A.N., Govindwar, S.P., 2012. Decolorization of textile industry effluent containing disperse dye Scarlet RR by a newly developed bacterial- yeast consortium BL-GG. Chem. Eng. J. 184, 33–41.

Leelakriangsak, M., Borisut, S., 2012. Characterization of the decolorizing activity of azo dyes by *Bacillus subtilis* azoreductase AzoR1. Songklanakarin J. Sci. Technol. 34 (5), 509–516.

Lin, J., Zhang, X., Li, Z., Lei, L., 2010. Biodegradation of Reactive Blue 13 in a two-stage anaerobic/aerobic fluidized beds system with a Pseudomonas sp. isolate. Bioresour. Technol. 101, 34–40.

Luikham, S., Malve, S., Gawali, P., Ghosh, S., 2018. A novel strategy towards agro-waste mediated dye biosorption for water treatment. World J. Pharm. Res. 7 (4), 197–208.

Moya, R., Hernández, M., García-Martín, A.B., Ball, A.S., Arias, M.E., 2010. Contributions to a better comprehension of redox-mediated decolouration and detoxification of azo dyes by a laccase produced by *Streptomyces cyaneus* CECT 3335. Bioresour. Technol. 101 (7), 2224–2229.

Oturkar, C.C., Patole, M.S., Gawai, K.R., Madamwar, D., 2013. Enzyme based cleavage strategy of *Bacillus lentus* BI377 in response to metabolism of azoic recalcitrant. Bioresour. Technol. 130, 360–365.

Pandey, A., Singh, P., Iyengar, L., 2007. Bacterial decolorization and degradation of azo dyes. Int. Biodeter. Biodegrad. 59, 73.

Pandi, A., Kuppuswami, G.M., Ramudu, K.N., Palanivel, S., 2019. A sustainable approach for degradation of leather dyes by a new fungal laccase. J. Clean. Prod. 211, 590–597.

Parshetti, G.K., Telke, A.A., Kalyani, D.C., Govindwar, S.P., 2010. Decolorization and detoxification of sulfonated azo dye methyl orange by *Kocuria rosea* MTCC 1532. J. Hazard. Mater. 176 (1–3), 503–509.

Pereira, L., Coelho, A.V., Viegas, C.A., dos Santos, M.M.C., Robalo, M.P., Martins, L.O., 2009. Enzymatic biotransformation of the azo dye Sudan Orange G with bacterial CotA-laccase. J. Biotechnol. 139 (1), 68–77.

Qi, J., Schlömann, M., Tischler, D., 2016. Biochemical characterization of an azoreductase from *Rhodococcus opacus* 1CP possessing methyl red degradation ability. J. Mol. Catal. B Enzym. 130, 9–17.

Rai, H., Bhattacharya, M., Singh, J., Bansal, T.K., Vats, P., Banerjee, U.C., 2005. Removal of dyes from the effluent of textile and dyestuff manufacturing industry: a review of emerging techniques with reference to biological treatment. Crit. Rev. Environ. Sci. Technol. 35, 219–238.

Ramírez-Montoya, L.A., Hernández-Montoya, V., Montes-Morán, M.A., Jáuregui-Rincón, J., Cervantes, F.J., 2015. Decolorization of dyes with different molecular properties using free and immobilized laccases from *Trametes versicolor*. J. Mol. Liq. 212, 30–37.

Ramya, M., Iyappan, S., Manju, A., Jiffe, J.S., 2010. Biodegradation and decolorization of acid red by *Acinetobacter radioresistens*. J. Bioremed. Biodegrad. 1, 105.

Saratale, R.G., Saratale, G.D., Chang, J.S., Govindwar, S.P., 2011. Bacterial decolorization and degradation of azo dyes: a review. J. Taiwan Inst. Chem. Eng. 42, 138–157.

Shedbalkar, U., Dhanve, R., Jadhav, J., 2008. Biodegradation of triphenylmethane dye cotton blue by *Penicillium ochrochloron* MTCC 517. J. Hazard. Mater. 157 (2–3), 472–479.

Sheth, N.T., Dave, S.R., 2009. Optimisation for enhanced decolourization and degradation of Reactive Red BS CI 111 by *Pseudomonas aeruginosa* NGKCTS. Biodegradation 20 (6), 827–836.

Si, J., Peng, F., Cui, B., 2013. Purification, biochemical characterization and dye decolorization capacity of an alkali-resistant and metal-tolerant laccase from *Trametes pubescens*. Bioresour. Technol. 128, 49–57.

Supaka, N.J., Somsak, D.K.D., Pierre, M.L.S., 2004. Microbial decolorization of reactive azo dyes in a sequential anaerobic–aerobic system. Chem. Eng. J. 99, 169–176.

Suwannawong, P., Khammuang, S., Sarnthima, R., 2010. Decolorization of rhodamine B and congo red by partial purified laccase from *Lentinus polychrous* Lév. J. Biochem. Tech. 2 (3), 182–186.

Telke, A.A., Kadam, A.A., Jagtap, S.S., Jadhav, J.P., Govindwar, S.P., 2010. Biochemical characterization and potential for textile dye degradation of blue laccase from *Aspergillus ochraceus* NCIM-1146. Biotechnol. Bioproc. E. 15 (4), 696–703.

Verma, K., Saha, G., Kundu, L.M., Dubey, V.K., 2019. Biochemical characterization of a stable azoreductase enzyme from *Chromobacterium violaceum*: application in industrial effluent dye degradation. Int. J. Biol. Macromol. 121, 1011–1018.

Verma, P., Madamwar, D., 2003. Decolourization of synthetic dyes by a newly isolated strain of *Serratia marcescens*. World. J. Microbiol. Biotechnol. 19 (6), 615–618.

Yang, X., Zheng, J., Lu, Y., Jia, R., 2016. Degradation and detoxification of the triphenylmethane dye malachite green catalyzed by crude manganese peroxidase from *Irpex lacteus* F17. Environ. Sci. Pollut. Res. 23 (10), 9585–9597.

Yang, X.Q., Zhao, X.X., Liu, C.Y., Zheng, Y., Qian, S.J., 2009. Decolorization of azo, triphenylmethane and anthraquinone dyes by a newly isolated *Trametes* sp. SQ01 and its laccase. Process Biochem. 44, 1185–1189.

Zhang, C., Diao, H., Lu, F., Bie, X., Wang, Y., Lu, Z., 2012. Degradation of triphenylmethane dyes using a temperature and pH stable spore laccase from a novel strain of *Bacillus vallismortis*. Bioresour. Technol. 126, 80–86.

Zhuo, R., Ma, L., Fan, F., Gong, Y., Wan, X., Jiang, M., et al., 2011. Decolorization of different dyes by a newly isolated white-rot fungi strain *Ganoderma* sp. En3 and cloning and functional analysis of its laccase gene. J. Hazard. Mater. 192 (2), 855–873.

CHAPTER 6

An overview on the application of constructed wetlands for the treatment of metallic wastewater

Shweta Singh, Christy K Benny, Saswati Chakraborty
Department of Civil Engineering, Indian Institute of Technology, Guwahati, Assam, India

6.1 Introduction

Heavy metals are elements that exist in the earth's crust, often defined as metals with relatively higher density (more than 5.0 g cm^{-3}) and extremely toxic even at lower concentrations. Heavy metal toxicity due to the elevated concentration of metals in water bodies has been a grave peril to the environment and biodiversity. Many heavy metals (such as iron and zinc) form an essential part of a human diet but become very poisonous when exposed to higher concentrations. The primary culprit of heavy metal discharge into the aquatic environment is from diversified industrial processes such as petroleum refining, electroplating, mining, mineral tanning (use of chromium salt) in tanneries, battery industry, landfill leachates, steel and fertilizer production. Often, heavy metal contaminated industrial wastewater is directly disposed into the nearby freshwater bodies and ocean without realizing the degree of ecological damage. Heavy metals can quickly enter the food chain from polluted streams due to their high solubility in water. Over time, large amounts of heavy metals get bio-accumulated and bio-magnified at a higher hierarchical level of the food chain (Kurniawan et al., 2006). Therefore, removing heavy metals from metallic effluent below the permissible discharge limits before its disposal is indispensable. Latest findings revealed that developing regions (Africa, Asia, and South America) have higher concentration of heavy metals in aquatic environment as well as greater number of heavy metals exceeded the threshold concentration values of the WHO and USEPA standards than compared to developed regions (Europe and North America) (Zhou et al., 2020).

Several treatment technologies have been developed and adopted to combat heavy metal pollution depending on its suitability and complexity of the metallic wastewater. The most conventional physicochemical treatment include chemical precipitation, coagulation-flocculation, flotation, adsorption, ion exchange and membrane filtration (Kurniawan et al., 2006). Lately, many advanced technologies such as electrodialysis (Babilas and Dydo, 2018), photocatalysis (Chen et al., 2019) and electrochemical

Biodegradation and Detoxification of Micropollutants in Industrial Wastewater
DOI: https://doi.org/10.1016/B978-0-323-88507-2.00004-X

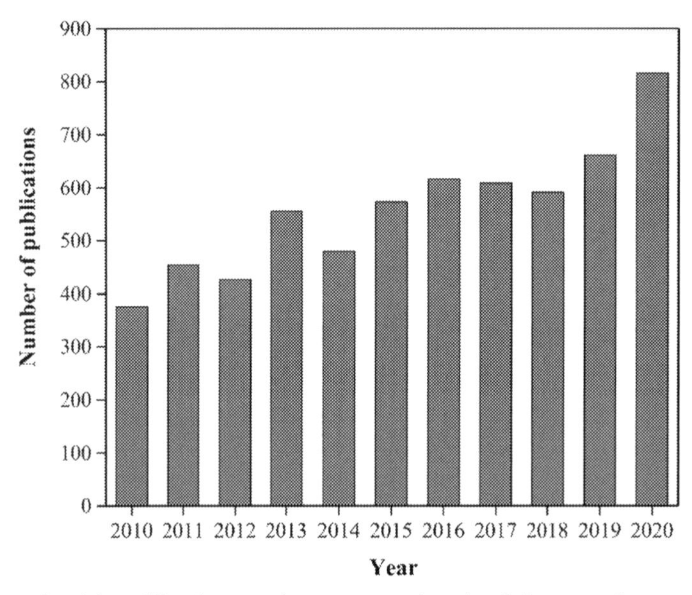

Fig. 6.1 ***Number of article publications on the constructed wetlands for water/ wastewater treatment from 2010–2020.*** (Source: Scopus by Elsevier).

precipitation (Tran et al., 2017) have gained considerable popularity due to its advantages over conventional techniques.

Constructed wetlands (CWs) are artificial wetlands that are engineered to provide biological approach for metallic wastewater. CWs can potentially remove trace amount of metal concentrations, which even conventional techniques fail to remove and thus CWs offer a long-term treatment outlook. CWs employ a multifaceted network of biotic and abiotic routes (or mechanisms) for metal attenuation. The underlying metal removal mechanisms in CW involve physical (settling, sedimentation and filtration), chemical (sorption, oxidation, hydrolysis, precipitation and co-precipitation reactions) and biological (plant uptake and bacterial metabolism) processes (Sheoran and Sheoran, 2006). The application of CW treatment has tremendously increased for different industrial wastewater containing challenging recalcitrants. From the Scopus statistics retrieved in January 2021 using TITLE-ABS-KEY (constructed AND wetlands) AND (LIMIT-TO (DOCTYPE, "ar")), as shown in Fig. 6.1, the number of research articles published on constructed wetlands for water/wastewater treatment has immensely amplified over the past decade from 2010 to 2020. Therefore, most likely that in the coming years, the application of CWs would increase many folds for water and wastewater treatment.

6.2 Sources of metal pollution and its environmental implications

Table 6.1 depicts the various origin of heavy metal release and its associated level of contamination.

Table 6.1 Major origins of heavy metal release and its associated level of contamination.

Source of metal enrichment	Major metal ions	Concentration (mg L^{-1})	References
Natural sources			
Hot Springs	Au Ag As Hg	0.01–85 0.2–500 22–20,713 0.1–2000	Steinmüller (2001), Weissberg et al. (1979)
Volcanic eruptions	Pb Zn Cu Cr As Hg	2.5–37.5 7.5–111.0 5.0–93.0 111–133 6.0–20, 713 0.1–282.0	Steinmüller (2001), Cimino and Ziino (1983)
Weathering from undisturbed mineral deposits	Cu Zn Pb Cd	0.01–68.0 0.02–16.0 0.02–0.34 0.26	Runnells et al. (1992)
Industrial effluents			
Mine effluents	Fe Al Zn Cu Cd Cr Mn	4.8–12,600 36.4–772 7.1–1030 0.15–5,55,500 0.09–9.1 0.19–1.3 1.6–10.2	Campbell et al. (2020), Pandey et al. (2007)
Oil refineries	Cd Pb Ni Co V	0.78–0.88 1.92–2.18 49.6–50.1 8.08–8.28 10.9–11.3	Aghel et al. (2020)
Electroplating effluents	Cr Cu Zn Ni Fe Pb	34–123.5 0.92–57.81 24–239 28–190 2.60–8.55 0–2.50	Kumar et al. (2011), Wei et al. (2013)
Pulp and paper mills	Fe Cu Mn Ni Zn Cd	1.43–67.5 0.21–2.15 0.16–11.0 0.05–3.30 0.39–13.90 0.01–0.255	Arivoli et al. (2015), Chandra et al. (2017)

(*continued*)

Table 6.1 (Cont'd)

Source of metal enrichment	Major metal ions	Concentration (mg L^{-1})	References
Fertilizers manufacturing	Fe	0.6–295	Anaga and Abu (1996), Fernández-González et al. (2020)
	Zn	1.30–20	
	Ni	3.5	
	Pb	1.9	
	Cu	4.6–198.36	
	Al	0.68–177	
	Cr	1.27–1.9	
Landfill leachates	Fe	200–1000	Carvajal-Flórez and Cardona-Gallo (2019), Kjeldsen et al. (2002)
	Cd	0.01–0.4	
	Cu	0.01–10.0	
	Cr	0.02–1.5	
	Pb	0.02–5.0	
	Ni	0.05–13.0	
	Zn	0.03–1000	
Metallurgical industries (steel making, copper smelting etc.)	As	689–1979	Basha et al. (2008), Balladares et al. (2018)
	Cd	76.1–162	
	Cu	164–1325	
	Fe	88–167	
	Ni	12–26	
	Pb	4.6–46	
	Zn	456–2832	
Battery industries	Cr	0.14–3	Ribeiro et al. (2018), Kumar et al. (2015)
	Mn	0.5–2	
	Fe	255–480	
	Zn	45–78	
	Pb	0.22–33	
	As	0.03–0.09	
Textile mills, dyes and paints	Cu	4.02–20.4	Kakoi et al. (2017), Patel et al. (2015)
	Cd	0.20–0.69	
	Zn	0.78–7.85	
	Fe	4.20–10.24	
	Cr	1.38–3.48	
	Mn	0.38–7.53	
	Ni	0.19–1.21	
	Pb	0.17–16.74	
Leather tanneries	Cr	8.30–5500	Benhadji et al. (2011), Mustapha et al. (2019)
	Cu	0.1–1.92	
	Pb	1.21–1.70	
	Zn	0.50–2.15	
	Fe	3.12–7.20	

6.2.1 Natural origins

The release of the heavy metals from natural sources includes geothermal activities (such as hot springs, volcanic eruptions, seismic activities, geysers and fumaroles), weathering from mineralized zones, degassing from oceans and other water bodies, and animal or human excretions. Volcanic eruptions are major causes of emitting substantial metal quantities into the atmosphere and oceans (Garrett, 2000). It has been estimated that the eruption of Mount Pinatubo in the Philippines released 10^4 M tonnes of rock to the surface that contained large quantities of lead (100,000 tonnes) and mercury (800 tonnes) (Garrett, 2000). Some of the rock-forming minerals like Olivine, Hornblende, Augite, Biotite, Apatite, Anorthite and Andesine are highly prone to weathering, which subsequently release various trace heavy metals (Co, Cu, Ga, Mn, Ni, Li, Mo, V and Zn) into the environment (Runnells et al., 1992). There are numerous transport pathways of heavy metals being transported to groundwater and surface water. The solubilized metals could either be sorbed unto the mineral phases, bound unto the organic matter or taken up by the plants, which increases the metal ion mobility in the different spheres of the environment.

6.2.2 Industrial (or anthropogenic) origins

The discharge of untreated industrial effluents without scrutinizing the various pollutants present and their impact on the well-being of living beings is often practiced. Several industrial sources directly or indirectly supplement the metal enrichment of water bodies. Some of the major metal polluting industries are mentioned in Table 6.1.

The generation of mine effluents (or rock drainage/acid mine drainage, AMD) from coal mines and polymetallic mines is a major environmental menace. The weathering of exposed pyrite (FeS_2) and sulfide-bearing mineral rock produces acidic metal-laden sulfate-rich wastewater. The catastrophic effect of heavy metals on Kocacay River from waste rock dump (WRD) of abandoned Balya mine (lead/zinc) in Turkey is reported, where high amounts of arsenic mobilized from mining upstream to wells and resulted in many cancer cases among the residents of Balya town (Aykol et al., 2003). The acidic, metallic waters continue to flow from active and abandoned mines and pollute streams and rivers downstream, which can have toxic effects on biota.

Wastewater from oil refineries or petroleum refineries is made up of different pollutants like oil, grease, phenol, sulfide, ammonia, suspended solids, cyanide and heavy metals like Cr, Fe, Ni, Cu, Mo, Se, V and Zn. The electroplating process has undoubtedly emerged as an important source of metal pollution with the rise in the industries such as automotive, manufacturing and electronics. The major pollutants of concern associated with the electroplating industry are spent solutions containing heavy metals and other noxious chemicals. Most commonly used metals for plating include Ni, Cd, Cu, Zn, Ag, Cr and Pb.

Pulp and paper mills are described to be the most water-intensive and polluting industries, generating large amount of effluents during the process of pulp making and paper finishing, which mainly depends on washing and bleaching of the pulp, wood, and kraft. The chief inorganic metallic pollutant found in paper mill effluent are Fe, Mn, Cd and Cr. Hanchette et al. (2018) revealed the significant correlation of ovarian cancer incidence with water pollution arising from the pulp and paper industry in the US.

Landfill leachates are potentially obnoxious fluid discharges from landfill sites, which may pollute the groundwater and surface water around the areas under uncontrolled circumstances. Leather tanneries are known for severe water pollution due to toxic hexavalent chromium from the chrome tanning process, which involves dichromate, sodium chromate, and chromium sulfate. About 50–70% chromium is not absorbed during the commercial chrome tanning process and results in excessive surplus and ecological imbalance (Rao et al., 1997).

Metallurgical industries (steel manufacturing, aluminium and copper industry) produce most of their waste during the smelting and beneficiation process. Other miscellaneous metal polluting industries include battery industries, textile mills, dyes and paints, which are reported to generate effluents rich in metals like Cu, Cd, Fe, Cr, Mn, Ni, Zn and Pb (Kakoi et al., 2017; Ribeiro et al., 2018). The use of chrome-, lead- and cadmium-containing colored formulator pigments in automotive and refinish markets is now restricted owing to the environmental concerns and rise in ongoing demands for zinc-phosphate-based pigments.

6.3 Environmental impacts

Fig. 6.2 illustrates the heavy metal contamination pathways of the aquatic environment and its deleterious impact on the environment and living beings. Heavy metals often leach and move along the waterways affecting soil and finally depositing in the aquifer, resulting in groundwater pollution. The toxicity impact of metals on marine life is very evident across the globe. A new study by Gao et al. (2021) revealed metal enrichment (Cd, Cu, Hg, Pb and Zn) in different marine organisms (0.001–29 $\mu g\ g^{-1}$ in fish, 0.002–20 $\mu g\ g^{-1}$ in mollusc and 0.001–28 $\mu g\ g^{-1}$ in crustacean) and metal concentrations in organisms and water (or sediment) established a positive correlation. Heavy metal toxicity at high concentrations occurs through ligand interactions or through the displacement of essential metal ions via (a) impairment of cell membrane, (b) shift of enzyme specificity, (c) disruption of cellular functions and (d) mutilation of DNA structure (Bruins et al., 2000).

Plant-metal uptake, translocation, sequestration and storage in plants pose a serious danger to the growth of the plants by inducing oxidative and genotoxic stress, and in many cases, plants (especially agriculture crops) themselves become the major source of metal pollution that eventually pose a risk to human health. Heavy metal accumulation

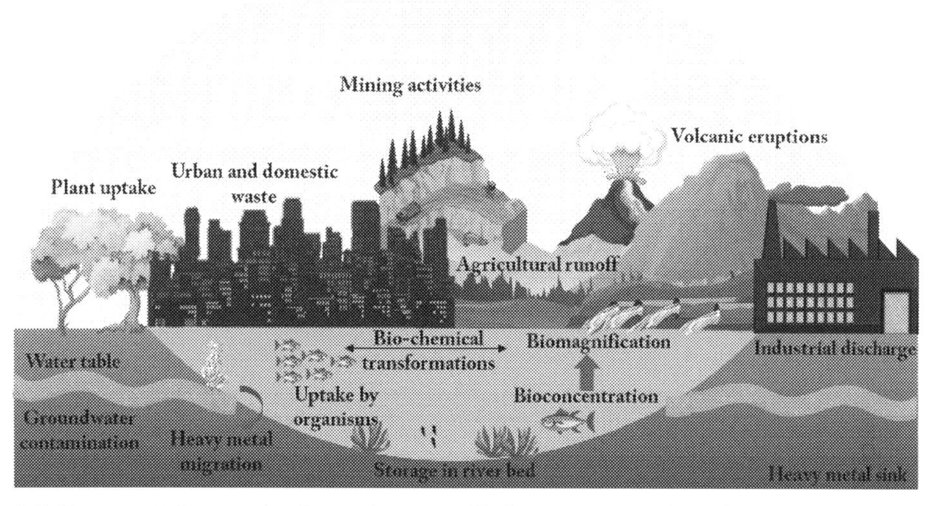

Fig. 6.2 ***Heavy metal contamination pathways and its impact on aquatic environment.***

in plants beyond the permissible limits inhibits electron transport, reduces CO_2 fixation, and causes chloroplast disorganization depending upon the type of plant species, metal element and metal bioavailability in soil. Metal toxicity may also induce reduced plant growth through amplifying the generation of free radicals and reactive oxygen species. Some heavy metals are carcinogenic in nature like As, Pb, Hg, Ni, Cd and Cr. These metals primarily target proteins involved in signaling or cellular regulatory processes such as apoptosis, cell cycle regulation, cell growth, DNA repair and methylation. Additionally, heavy metals also target the nervous system by causing inhibition of neurotransmitter and neuron damage leading to neurotoxicity (Engwa et al., 2019).

6.4 Treatment of metal-laden industrial wastewater

Table 6.2 entails the various treatment technologies for metal removal with its advantages and disadvantages.

6.4.1 Physico-chemical treatment

Chemical precipitation is the most widely adopted technology because it is effective, economical and straightforward, where metal ions are precipitated out of the solution as insoluble hydroxides by controlling the pH (9–11). Similar to chemical precipitation, the coagulation-flocculation technique employs the principle of destabilization of colloidal particles using pH adjustment (most effective range between 11.0–11.5) (Kurniawan et al., 2006) and the addition of different coagulants (e.g., ferric/alum salts)

Table 6.2 Comparative summary of various treatment techniques available for heavy metal removal.

Treatment technology	Principle mechanism	Advantages	Disadvantages	References
Physico-chemical treatment				
Chemical precipitation	Applied chemicals (lime, caustic soda) and heavy metal ions react to form insoluble precipitates in alkaline pH	• Simple and inexpensive process • Convenient and safe operations	• Slow precipitation, poor settling and aggregation of precipitates • Large chemical requirement • Excessive sludge generation which needs further treatment	Kurniawan et al. (2006)
Coagulation-flocculation	Destabilization of colloidal particles into bulky floccules by addition of coagulants (aluminium sulfate, ferric chloride)	• Cost effective • Dewatering qualities	• High operational cost • Generation of toxic sludge • Incomplete removal or treatment of heavy metals	Kurniawan et al. (2006)
Adsorption	Pollutant bound to the solid surface by physical or chemical interactions	• Cheap and inexpensive materials • Less sludge production • High removal efficiency	• Requires adsorbent synthesis • Pre-treatment is required • Adsorbents are metal specific • Desorption	Kurniawan et al. (2006)
Ion-exchange	Exchanged of metal ion for a similarly charged ion attached to an immobile solid resin	• No sludge problems • Convenient for fieldwork • Easy to use and less time consuming	• Requires prior removal of suspended solids • Sensitive to pH change • Removes only limited metals • High capital/operational cost	Kurniawan et al. (2006)

Treatment technology	Principle mechanism	Advantages	Disadvantages	References
Flotation	Heavy metals escape from wastewater by attaching to the injected bubbles and removal of concentrated hydrophobic particles eventually	• Removal of smaller particles • Short hydraulic retention time • Less sludge generation	• High operational and maintenance cost • Requires further treatment	Deliyanni et al. (2017)
Membrane filtration	Separation principle on the basis of particle size, pH, concentration and pressure	• High removal efficacy • Less space requirement	• Expensive • Membrane fouling • Complex process	Abdullah et al. (2019)
Advanced oxidation processes (AOPs)	Destruction of metal-binding complexes by hydroxyl and other radicals to liberate free metal ions	• No electricity requirement • Simultaneous metal recovery	• Large amount of acid requirement increases operation cost • Large dosage of catalyst agents • Rusting	Du et al. (2020)
Electrochemical treatment (electrodialysis, electrocoagulation, electrodeposition)	Plating-out of metal ions on a cathode surface and recovery of metals in the elemental metal state.	• Efficient for metal removal • Less chemical requirement	• Substantial initial investment • High electrical supply • Membrane clogging	Babilas and Dydo (2018), Tran et al. (2017)
Biological treatment				
Sulfidogenic bioreactors	Metal precipitation aided by SRBs through sulfate reduction	• Selective recovery of heavy metals • Predictable and readily controllable performance • High metal removal efficiency	• Substantial construction, operation and maintenance cost • Sensitive towards toxic substances	Kiran et al. (2017)

(continued)

Table 6.2 (cont'd)

Treatment technology	Principle mechanism	Advantages	Disadvantages	References
Bioelectrochemical systems (BES)	Oxidation of organic matter by electroactive microbes at the anode is coupled to reduction of metal ions at cathode	• Possible metal recovery • Simultaneous energy production	• High operation and maintenance cost • High metal concentration may cause toxicity	Kaushik and Singh (2020)
Bioremediation using algae	Electronegative charge on the algal cell wall allows strong bonding with metal ions via intracellular ionuptake (active uptake) or biosorption (passive uptake)	• Low cost cultivation • Extraction of valuable by-products (bioethanol and biodiesel) • High metal uptake • Metal selectivity	• Posioning of living cells • Limited sorption capacity • Sensitive to pH, light intensity and level of dissolved nitrates • Scale-up issues	Zeraatkar et al. (2016)
Bioremediation using fungi	Intracellular (or extracellular) accumulation, complexation, biosorption to cell wall and sequestration	• More metal tolerant • High biomass yield and high uptake rate • Possibility of metal recovery	• Sensitive to low pH • Toxicity to living cells • Disposal of unwanted by-product (mycelia) • Scale-up issues	Ghosh et al. (2016)
Phytoremediation	Metal-ion interaction via phytoextraction, phytodegradation, rhizofiltration, phytoaccumulation, phytovolatilization and phytostabilization	• Solar driven • Relatively inexpensive • Recyclable metal-rich plant residue • Eliminates secondary air or waterborne wastes	• May contaminate the food chain • Need space and proper care • Results in non-edible plant products • Volatilization of compounds may transform a groundwater pollution to an air pollution	Muthusaravanan et al. (2018)

to form insoluble metal salt precipitates. In many cases, to ensure complete elimination of heavy metals from wastewater, coagulation-flocculation technique is often coupled with sedimentation and filtration process. Flotation is a gravity separation method, which is primarily applied to separate out solids from the aqueous solution using the rising action of bubble attachment (Deliyanni et al., 2017).

Lately, due to the ease in operation of membrane filtration, it has been extensively employed worldwide to separate a variety of pollutants based on the size of the solid required to be captured (Abdullah et al., 2019). The membrane filtration technique segregates the particle on the basis of size, pH, concentration and pressure applied. Membranes are fabricated with a definite porous matter that plays a significant role in removing heavy metals from the wastewater. Depending on the particle size that needs to be removed, ultrafiltration (0.002–0.1 μm), nanofiltration (0.001 μm) and reverse osmosis (0.0001 μm) are practiced for heavy metal removal. Ion exchange is a reversible process where the exchange of ions occurs between solid and liquid phases through an ion-exchange resin (solid insoluble natural or synthetic resin), and more suitable for the pretreated secondary effluents (Kurniawan et al., 2006). The sludge generation in the ion exchange process is low as compared to the coagulation process. Adsorption is a mass transfer phenomenon where pollutant is bound to the surface of a solid phase (sorbent material such as activated carbon, bio-adsorbents, carbon nanotubes and low-cost adsorbents) from the liquid phase. Currently, adsorption technology for selective removal of heavy metal is mainly explored on newly synthesized low-cost adsorbents depending upon their surface area and adsorption capacity.

6.4.2 Biological treatment

The bioremediation process uses living organisms such as bacteria, fungi, algae, etc., to absorb and degrade contaminants. Bioremediation can be understood through bioleaching, immobilization, chelation, methylation, and bioprecipitation. Sulfidogenic bioreactors are a very promising technique for treating metallic wastewater through microbial sulfate reduction and biogenic metal-sulfide precipitation by sulfate-reducing bacteria (SRB). However, presently, sulfidogenic bioreactors are largely researched at the lab scale and need more exploration at full scale application to realize the potential of these systems in metal removal and recovery (Kiran et al., 2017). Numerous designs of sulfidogenic reactors are being designed as anaerobic filter reactor (AFRs), upflow anaerobic sludge blanket reactor (UASBR), fluidized bed reactor (FBR), anaerobic baffled reactor (ABR), continuous stirred tank reactor (CSTR), anaerobic fixed-film reactors, anaerobic packed bed reactors, column reactors, rotating biological contact or systems and infiltration beds. The efficacy of these sulfidogenic bioreactors and the nature of metal precipitate formed is mainly governed by pH, for the sustenance of SRBs, which depends on metal toxicity, retention time and sulfide toxicity.

In recent years, research attention has inclined more towards sustainable treatment technologies such as application of bioelectrochemical systems (BES) for metal removal with concurrent energy generation and metal recovery. It involves the biodegradation (or oxidation) of organics present in wastewater by electroactive microorganisms via extracellular electron transfer at the anode, combined with the transformation (or reduction) of metal ions to less toxic form at the cathode. It may also be associated with other biological processes of biotransformation and bioaccumulation. Major factors which determine the efficacy of these systems are the oxidation/reduction potential of the metal, initial pH, metal concentration, source of carbon as substrate in the anode, biofilm composition and external resistance (Kaushik and Singh, 2020).

Biological treatment utilizing different microorganisms (algae and fungi) as biosorbents involves various mechanisms such as exchange of ions, complexation, and electrostatic interaction (Zeraatkar et al., 2016). Various cellular pathways (active and passive) such as intracellular uptake (only in live cells), biosorption (in both live and dead cells), extracellular accumulation/precipitation, crystallization, biosorption to the surface of cell wall, metal transformation and sequestration are suggested to take part in heavy metal removal (Ghosh et al., 2016). Most of the studies focus chiefly on utilization of non-living biomass due to its high reusability and recyclability, chemically resistant, elimination of toxicity risk and low cost as it does not require any addition of growth nutrients.

Phytoremediation relies on various processes such as phytoextraction, rhizofiltration, rhizodegradation, phytoaccumulation, phytovolatilization and phytostabilization (Muthusaravanan et al., 2018). Aquatic macrophytes are reported to accumulate metals very high up to 10^5 times greater than the associated amount present in wastewater. However, proper disposal of metal-rich plant biomass is required to restrict the metal re-release and its entry into the food chain.

6.5 Constructed wetlands for heavy metal removal

Traditionally, CWs are broadly distinguished into two groups on the basis of water depth with respect to the media bed. These are surface flow (SF) or free-water surface (FWS) and sub-surface flow (SSF) wetlands. In SF CWs, the water surface is above the media and thus completely flooded. In contrast, in SSF CWs, the water level is maintained below the media layer and forms interaction with different microorganisms. SSF CWs can be further categorized depending on the flow path as horizontal flow (HF) or vertical flow (VSF). Hybrid systems incorporate both SF and SSF CWs. CWs are cost-effective, technically feasible, require low operation and maintenance, offer aesthetic enhancement, environmental friendly, low volume of sludge generation, high long-term performance, ability to remove metal and sulfates and metal precipitates formed are stable and thus provide lasting retention. However, factors such as large area requirement, media clogging, less-predictable performance, uncertainty about long-term fate and stability of the accumulated deposits within CW often hinder the large-scale application.

CWs are an effective artificial sink for heavy metals. Most of the earliest records on metals removal using CWs are from systems employed for treating mine wastewater (Burris et al., 1984). One such example of AMD treatment is presented by Wildeman et al. (1990) at Idaho Springs-Central City, Colorado, using a different mixture of organic substrates, where fresh mushroom compost exhibited consistently good performance with respect to Cu, Fe and Zn removal. Table 6.3 presents comprehensive literature on the application of CWs for various metallic wastewater.

6.5.1 Metal removal mechanisms

Major metal removal mechanisms in CWs are illustrated in Fig. 6.3. Each wetland component (such as media, plant and microorganisms) functions specifically in many ways to eliminate metals. Depending on the wastewater characteristics, wetland configuration, loading rate, type of media and plant species, comparable performance efficiencies have been reported in the literature. The entire process of metal attenuation is inter-dependent and co-existing, and therefore it is very complex to decipher the level of individual involvement of different wetland mechanisms.

6.5.1.1 Filtration and sedimentation

Filtration and sedimentation play a major part in retaining metals once metal form precipitates or in the colloidal state. Settling and interception of metal particulates primarily occurs in sand/gravel media (in the case of HF and VF wetlands), where metals get trapped and strongly bound with other incoming organic and suspended matter, which affects the hydraulic conductivity and wetland performance over time. In the case of FWS or FTW, low inflow velocity accompanied by plant litter presence promotes settling and interception of metal solids (Kadlec and Wallace, 2008). Dense vegetation increases the retention times and thus enhances the sedimentation process. Particulates denser than water can be readily settled out under relatively low-velocity flow conditions, whereas for lighter particles, floc-formation is essential for sedimentation. The formation of floc is improved by increasing pH, suspended solids concentration, algal densities and ionic strength of wastewater (Matagi et al., 1998). Flocs tend to settle rapidly and may result in adsorption of other suspended particles and heavy metals (Sheoran and Sheoran, 2006).

Liu et al. (2018) investigated the filtration functions of wetland from the upper delta plain region to its shallow sea wetland, northeastern China and revealed a significant high buildup of particulate metals within the upper delta plain wetlands ecosystem and thus revealed the its buffering mechanism to coastal marine.

Table 6.3 Examples of the application of CWs for various metallic wastewater.

Type of wastewater	Type of CW	Plant species	Substrates used	Metal concentration (in mg L^{-1})	Removal (%)	Reference
Simulated surface run-off	HSSF (lab-scale)	*Typha latifolia*, *Phragmites australis*, *Schoenoplectus lacustris* and *Iris pseudacorus*	Cocopeat and gravel media	Cu (1–10), Pb (1–10) and Zn (1–10)	Cu (81.7–91.8), Pb (75.8–95.3) and Zn (82.8–90.4)	Mungur et al. (1997)
Simulated polluted river water	VSSF (microcosmic)	*Iris sibirica*	Composite filling (ceramsite, blast furnace-granulated slag, soil and sawdust at 3:3:2:1 ratio)	Cd (1–6)	Cd (89.5–93.8)	Gao et al. (2015)
Primary-treated domestic wastewater spiked with metals	HSSF (lab-scale)	*Typha latifolia*	Graded gravel (diameter size 2.4–4.8 mm)	Zn (10), Pb (2.5), Cd (0.5) and Cu (5.0)	Zn (> 99), Pb (>99), Cd (>99) and Cu (>99)	Lim et al. (2003)
Pulp and paper industry effluent	VSSF (pilot-scale)	*Erianthus arundinaceus*	Red soil aggregates and clay soil aggregates	Fe (1.56), Mn (0.21), Cu (0.246), Ni (0.06), Zn (0.39) and Cd (0.0128)	Fe (74), Mn (60), Cu (80), Ni (71), Zn (70) and Cd (70)	Arivoli et al. (2015)
Municipal landfill leachate	Series of ten FWS (full-scale)	*Lemna* sp., *Phragmites australis*, *Scirpus palla*, *Typha latifolia* and *Carex* sp.	–	Fe (146.69), Mn (8.89), Zn (8.31), Pb (1.15), Cr (3.60), Ni (0.44), Cu (2.34) and Cd (0.19)	Fe (99.1), Mn (97.6), Zn (98.4), Pb (80), Cr (96.7), Ni (77.8), Cu (96.5) and Cd (97.7)	Wojciechowska and Waara (2011)

Type of wastewater	Type of CW	Plant species	Substrates used	Metal concentration (in mg L^{-1})	Removal (%)	Reference
Synthetic mine drainage	VSSF (lab-scale)	*Phragmites australis*	Pumice stones, limestone and loamy soil	Cu (20), Zn (30), Cd (0.30), Mn (0.50), Pb (2.50) and Fe (4.50)	Cu (99.63), Zn (20.7), Cd (40) and Pb (90)	Blesson et al. (2021)
Simulated metallic wastewater	VSSF (pilot-scale)	*Iris pseudoacorus*	Gravel, vermiculite and soil	Cu (0.35), Pb (0.28) and Cd (0.60)	Cu (65.7), Pb (64.3) and Cd (81.7)	Liu et al. (2020)
Simulated AMD	HSSF (lab-scale)	*Typha latifolia*	Gravel, bamboo chips and cow manure media	Fe (100), Al (25), Mn (6.0), Co (1.0), Ni (1.0) and Cr (1.0)	Fe (91.6), Al (59.7), Mn (no net removal), Co (93.7), Ni (97.8) and Cr (99.7)	Singh and Chakraborty (2020)
Simulated river water impacted by AMD	HSSF (lab-scale)	*Phragmites australis*	Limestone and zeolite	Fe (49–57), Pb (0.88–0.9), Zn (10–12) and As (2.1–3.7)	Fe (>96), Pb (> 94), Zn (>40) and As (>96)	Lizama-Allende et al. (2021)
Oil refinery wastewater	VSSF (pilot-scale)	*Phragmites karka*	Sand and organic compost	Fe (3–11), Cu (5–9) and Zn (5–11)	Fe (48), Cu (56) and Zn (61)	Aslam et al. (2007)
Synthetic electroplating wastewater	Up-flow (UF) VSFCW	*Phragmites australis*	Gravel–peat mixture	Cu (1–5), Ni (1–5), Pb (1) and Zn (1–5)	Cu (97.4), Ni (98.1), Pb (66) and Zn (93.4)	Sochacki et al. (2014)
Simulated pretreated tannery wastewater	HSSF (pilot-scale)	*Phragmites* sp.	Medium gravel (5–10 mm diameter) and fine sand	Cr (0–10)	Cr (99.9–100)	Ramírez et al. (2019)
Textile wastewater	VSSF (pilot-scale)	*Brachiaria mutica*	Coarse gravel, fine gravel and river sand	Cr (9.7), Fe (14.3), Ni (7.6) and Cd (0.88)	Cr (97), Fe (89), Ni (88) and Cd (72)	Hussain et al. (2018)

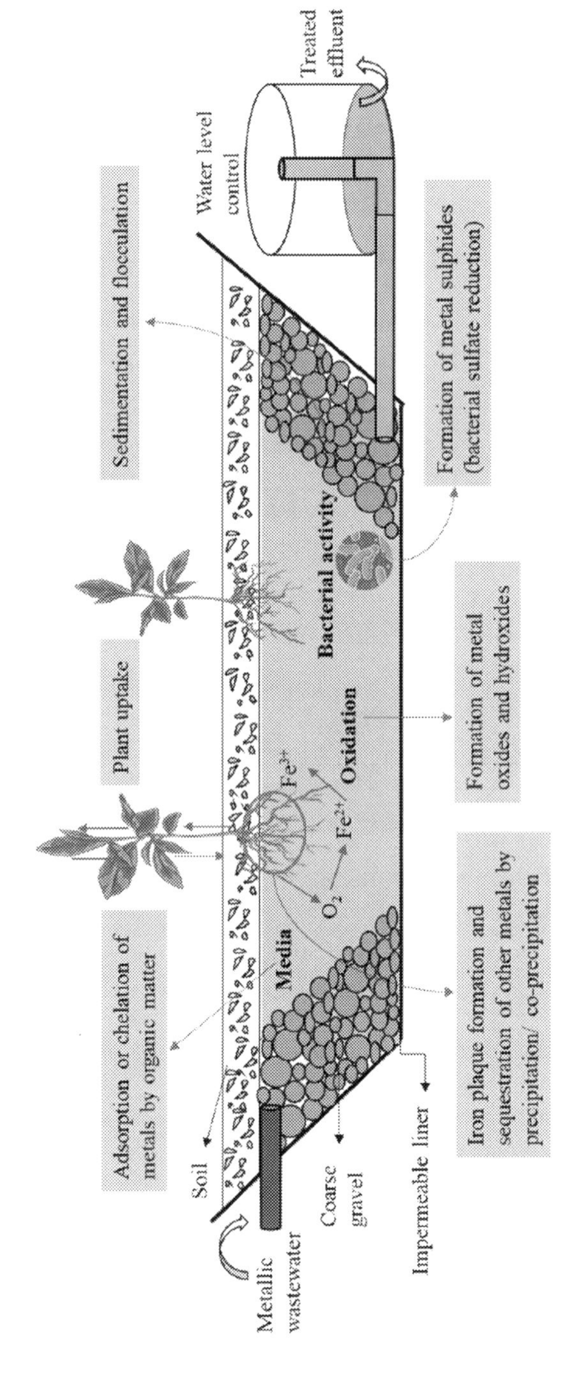

Fig. 6.3 *Metal removal mechanisms in typical horizontal sub-surface CWs.*

6.5.1.2 Sorption

The sorption mechanism of media (particularly SSF CWs) plays a dynamic role in removing metals and metalloids from wastewater. It may be identified either as short-term accumulation or long-term stabilization. Sorption defines a range of processes such as adsorption (surface phenomena where atoms/ions get attached to the surface of adsorbent), absorption (bulk phenomena) and precipitation reactions. The former is further classified as physisorption (physical processes with weak Van der Waals interactions) and chemisorption (chemical processes with strong covalent bonds). The adsorption process is predominant during the initial period of operation when the sorption sites of media are unsaturated. Cation exchange capacity (CEC) gives the measure of the capability of media to hold positively charged ions. The cations present on the media surface are exchanged with the metal ions or any other cations present in wastewater. The organic media containing many polar functional groups (such as acids and phenolics) impart high CEC and involved in chemical binding to form more stable chelate complexes as represented in Eq. (6.1) (Kadlec and Wallace, 2008):

$$2HP + M^{2+} \leftrightarrow MP_2 + 2H^+ \tag{6.1}$$

where '*P*' and '*M*' denote the organic material (peat) and a divalent metal ion, respectively.

The amount of metal-bound is given by Eq. (6.2) (Kadlec and Wallace, 2008):

$$K_{eq} = \frac{[MP_2][H^+]^2}{[HP]^2[M^{2+}]} \tag{6.2}$$

where the bracket signifies molar concentration of the species. It can be seen from Eqs. (6.1) and (6.2) that CEC increases with the organic matter content of the media, whereas it declines with decreasing pH.

The redox potential of media is also an essential factor deciding the removal of metals (like Fe and Mn) by providing reducing conditions. Several other factors such as properties of the concerned metals (valence, radius and coordination with oxygen), physicochemical environment (pH and redox), nature of the adsorbent, presence of other competing metals and soluble ligands influence the cation exchange adsorption (Matagi et al., 1998).

Metal adsorption can be explained as either Langmuir or the Freundlich adsorption isotherms models for various wetland media. The Langmuir isotherm is given by Eq. (6.3) (Kadlec and Wallace, 2008):

$$C_M = \frac{KC_W}{1 + aC_W} \tag{6.3}$$

where a is the Langmuir parameter (L mg^{-1}), C_M is the metal concentration in media (mg g^{-1}), C_W is the metal concentration in water (mg L^{-1}) and K is the Langmuir parameter (L g^{-1}).

Freundlich isotherms are often employed to explain heterogeneous adsorption as Eq. (6.4) (Kadlec and Wallace, 2008):

$$C_M = K_F C_W^n \tag{6.4}$$

where n is the Freundlich parameter (dimensionless) and K_F is the Freundlich parameter [(mg kg^{-1}) × (mg L^{-1})$^{-n}$].

Yadav et al. (2010) reported maximum removal of chromium (98.3%) and nickel (96.2%) using gravel media in wetland microcosms where the specific adsorption capacities of gravel were found to be 7–128 mg kg^{-1} for chromium and 10–89 mg kg^{-1} for nickel.

6.5.1.3 Precipitation and co-precipitation

Precipitation is a major driving mechanism for metal removal and co-precipitation reactions in wetland media. Precipitation of metals is facilitated when the solubility product (K_{sp}) of the concerned metal is exceeded, which is determined by on the solution pH and initial concentration of metal ions. The nature of the insoluble precipitate formed governs the stability and bioavailability of the precipitated metal ions into the aquatic environment. Secondary minerals, including (oxy-)hydroxides of metals such as Fe, Al and Mn assist the co-precipitation of other metals. Cu, Co, Ni, Pb and Zn are very often co-precipitated in the oxides of Fe and Mn (Matagi et al., 1998). The flocs of iron oxyhydroxides are formed under high pH and high dissolved oxygen conditions, whereas aluminium oxyhydroxides flocs may form under circumneutral pH conditions and do not require oxygen as represented by Eqs. (6.5)–(6.8), respectively (Kadlec and Wallace, 2008).

$$Fe^{2+} + H^+ + 0.25O_2 \rightarrow Fe^{3+} + 0.5H_2O \tag{6.5}$$

$$Fe^{3+} + 2H_2O \rightarrow FeOOH_{(sus)} + 3H^+ \tag{6.6}$$

$$FeOOH_{(sus)} \rightarrow FeOOH_{(sed)} \tag{6.7}$$

$$Al^{3+} + H_2O \rightarrow Al(OH)_3 \downarrow + 3H^+ \tag{6.8}$$

Additionally, many wetland plants are capable of maintaining oxidizing conditions by radial oxygen loss (ROL) from roots and formation of iron plaque at the root's surface, which are reported to retain Zn and As (Kiiskila et al., 2017). In reducing environment, pH controls the precipitation of metals as carbonates, hydroxides and sulfides.

6.5.1.4 Association of metals with oxides (oxidation) and hydroxides (hydrolysis)

Many metals may undergo oxidation or hydrolysis depending on pH, oxidation-reduction potential (ORP) and various anions present. The oxidation (abiotic or bacterially-mediated) or hydrolysis of Fe and Al are described in Eqs. (6.5)–(6.8). However, the abiotic oxidation of Mn is remarkably slow, as shown in Eq. (6.9) and oxidation starts only at pH > 8 making it the most difficult to remove (Sheoran and Sheoran, 2006).

$$Mn^{2+} + 0.25O_2 + 1.5H_2O \rightarrow MnOOH + 2H^+ \tag{6.9}$$

The redox potential of wetland media decides the stability of metal oxides and (oxy-)hydroxides as under reducing environment (below −100 mV), metals bound to Fe, Al and Mn flocs tend to remobilize and release back rapidly. Because oxidized iron acts as an electron acceptor, it can inhibit bacterial sulfate reduction by outcompeting SRBs for electron donors.

6.5.1.5 Precipitation as metal carbonates

The occurrence of metal carbonate precipitation is primarily governed by the water pH, partial pressure of CO_2 and dissolved carbonate concentration. It is an important mechanism in wetlands that drain limestone catchment areas (Matagi et al., 1998). The CO_2 utilization by plants and algal biomass may also increase pH via periphyton photosynthesis, leading to $CaCO_3$ supersaturation and facilitating the precipitation of calcitic minerals (Kadlec and Wallace, 2008). Carbonate precipitation of metals like Zn, Cu, Pb and Ni in wetlands is well documented in many studies (Kröpfelová et al., 2009) and it can be represented as shown in Eq. (6.10):

$$M^{2+}\begin{cases} SO_4 \\ Cl_2 \end{cases} + Na_2CO_3 \rightarrow MCO_3 \downarrow + Na_2\begin{cases} SO_4 \\ Cl_2 \end{cases} \tag{6.10}$$

6.5.1.6 Precipitation as metal sulfides

Metal sulfide precipitation is the most desired route for metal removal as most of the metal sulfide precipitates are relatively insoluble and stable than compared to oxide or hydroxide counterparts in reduced conditions over a wide pH range. The biological reduction of sulfate by SRBs under anaerobic conditions promotes the generation of hydrogen sulfide in wetlands, which further reacts with the metals present and metal precipitation as sulfides occurs (represented in Eqs. (6.11)–(6.12)) (Stumm and Morgan, 2012).

$$2CH_2O\,(organics) + SO_4^{2-} \rightarrow H_2S + 2HCO_3 \tag{6.11}$$

$$M^{2+} + H_2S + 2HCO_3 \rightarrow MS \downarrow + 2H_2O + 2CO_3 \tag{6.12}$$

Eq. (6.11) results in the production of mineral acidity from H_2S dissociation (shown in Eqs. (6.13)–(6.14)), but an equal release of HCO_3 simultaneously neutralizes this proton acidity.

$$H_2S \leftrightarrow HS^- + H^+, K_{p1} = \frac{[HS^-][H^+]}{[E_2S]}, pK_1 = 6.99 \tag{6.13}$$

$$HS^- \leftrightarrow S^{2-} + H^+, K_{p2} = \frac{[S^{2-}][H^+]}{[HS^-]}, pK_2 = 17.44 \tag{6.14}$$

The key requirements for sulfide precipitation are anaerobic conditions (ORP below −100 mV), presence of electron donors (simple carbon source) and microbial consortia (such as *Desulfovibrio* and *Desulfotomaculum*) capable of utilizing dissolved sulfate as electron acceptors. However, excessive generation of H_2S (or HS^-) can adversely cause toxicity to SRBs and may also increase metal mobility. Moreau et al. (2013) revealed the sequestration of metal (loid)s into biogenic FeS_2 associated with AMD contaminated sediments of natural estuarine wetland (Stege Marsh) in San Francisco Bay, USA, where various metal (loid)s such as As, Cu and Pb (~0.25–2 wt%) were sequestered in authigenic iron and zinc-sulfides.

6.5.1.7 Microbial metabolism

Microbial processes involved directly or indirectly in metal attenuation are classified into (a) metal assimilation into their cells, (b) biosorption, (c) microbial oxidation–reduction of metals, (d) methylation, and (e) biogeochemical cycle-assisted metal removal (Kosolapov et al., 2004). Biosorption is a passive adsorption process where metal ions are bound to the surplus extracellular electronegative charged materials or the cell walls. Biosorption comprises of several mechanisms like adsorption, ion exchange, chelation and entrapment interactions involving many functional groups, including carboxyl, sulfonate, phosphate, hydroxyl and amino moieties. The hydrophilic metals ions are assumed to be conveyed across the hydrophobic space of a bio-membrane where a receptor molecule binds a metal ion. The hydrophilicmetal-receptor complex then diffuses to the bulk of the membrane and releases the metal ion into the cytosol where it is trapped, and then the receptor diffuses back to the other surface of the membrane where it may collect another metal ion (Matagi et al., 1998). Anaerobic metal-reducing bacteria are accomplished to receive metals as terminal electron acceptors via respiration, and the resulting reduced metals may be less or more soluble, depending on the specific elemental chemistry. The immobilization of metals like chromium, selenium and uranium is reported through reduction processes bio-catalyzed by microorganisms (Green, 2011; Kosolapov et al., 2004). Iron and sulfate reduction driven by SRBs also facilitate the removal of metals as precipitates of hydroxides or sulfides, thereby increasing pH (discussed in Section 6.5.1.6).

6.5.1.8 Plant uptake

Metal accumulation in plants is an important biological process by phytostabalization, phytoaccumulation, and phytovolatization that may or may not contribute significantly to metal removal. Wetland plants can be generally divided into four groups: Emergent macrophytes (e.g.,*Carex rostrata, Phragmites australis, Scirpus lacustris,* and *Typha latifolia*), floating leaved macrophytes (e.g.,*Nymphaea odorata*, and *Nuphar lutea*), submerged macrophytes (e.g., *Myriophyllum spicatum, Ceratophyllum demersum*, and *Rhodophyceae*) and free-floating macrophytes (e.g., *Lemna minor, Spirodela polyrhiza*, and *Eichhornia crassipes*) (Williams and Kevern, 1964). Previous researchers presented quite conflicting results, expressing the insignificant to a considerably significant role of plants in metal elimination. Lee and Scholz (2007) reported the negative impact of *P. australis* in the removal of Ni and Cu, most likely by lowering pH due to the enhanced nitrification process. Many studies have reported negligible contribution of plants in metal uptake (1–3%) (Mays and Edwards, 2001; Singh and Chakraborty, 2020), while other studies have presented high uptake of metals (15–100%) and improved metal removal efficiency in the presence of plants (Batty, 2003; Younger et al., 2002).

The uptake rate and uptake routes of the metal vary largely depending on the type of plants, growth rate and initial metal concentration in plant tissue, e.g., metal uptake in emergent and surface-floating plants takes place through roots, where as roots as well as leaves are involved in the case of floating and submerged plants (Matagi et al., 1998). The passive movement of metals in the aqueous phase occurs by foliar absorption to the cell wall and then the plasmalemma. Plants may also contribute to metal trapping into the substrate via rhizodeposition and catalyze biochemical reactions (chelation) involving organic acids from root exudates such as citrate,oxalate, malate, malonate, fumarate, and acetate (Ryan et al., 2001). Secretion of protons and metal-reductases (for reduction of metal ions) facilitates the solubilization of metals by plant roots. Metal immobilization in the rhizosphere and storage in the below ground biomass (roots) is usually higher than its translocation to the above ground biomass (shoots), indicating the potential phytostabilization ability of the plants (Batool and Saleh, 2020). Apart from direct metal uptake in plant tissues, the indirect assistance is provided by means of ROL in the root zone, which results in iron plaque formation, where concentrations can reach 5–10 times the concentrations seen in the surrounding sediments (Sundby et al., 1998). However, there have been contradictory reports about whether plaque formation reduces or increases the metal-plant uptake. Leaf and stem litter also participate in metal accumulation and possibly over time it may become enriched with metals, due to cation adsorption or amalgamation of fine particles with adsorbed metals (Weis and Weis, 2004). The decomposing plant litter also acts as a carbon source for microbial functions. The decomposition of biomass may likely release or sink metals.

Plants adopt different mechanisms to tolerate metal toxicity by (a) metal sequestration in cellular sections, which are insensitive or tolerant to metals, (b) restricting metal translocation into shoots, (c) metal translocation to old leaves, (d) synthesis of biochemical biomarker chelators (phytochelatins, peptides) in plant tissues for metal chelation, (e) biomineralization of metals and (f) prompting the function of antioxidantenzymes (like glutathione peroxidase and reductase) (Weis and Weis, 2004).

Xu et al. (2020) investigated the transportation characteristics of metals in five different plant species (i.e., *Phragmites australis, Triarrhenalutarioriparia, Carex cinerascens Kukenth.,*

Artemisia selengensis, Phalarisarundinacea) and revealed Cd phytostabilization ability of *Triarrhena* and *Artemisia* with high bioaccumulation factor (8.26 and 4.31, respectively), whereas high translocation factor values of Cd (5.29) and Pb (3.25) in *Phragmites* indicated its ability for Cd and Pb phytoextraction. In another study, Batty and Younger (2004) indicated the reduction in plant growth (*Phragmites australis*) in the presence of excessive metal concentrations, possibly due to the inhibition of Ca uptake by competing for binding sites on roots.

6.5.2 Long-term performance feasibility

The long-term retention of metals in many wetlands has been reported. The bioavailability and migration of the metals are mainly determined by the pH, water chemistry, presence of organic matter and redox condition of the wetland. A 120-year-old natural wetland in Ireland receiving leachate from an abandoned lead-zinc mine has been estimated to retain 95% of Zn, 65% of As and so far, only 30% of the total capacity of the marsh has been utilized to retain metals, thus revealing the longevity of the marsh in the order of several centuries rather than decades (Beining and Otte, 1997). Jacob and Otte (2004) reported that in permanently flooded areas without live vegetation, the high organic matter content decreased Zn mobility and caused extremely high concentrations of acid-volatile sulfides (AVS), whereas, in areas with high organic matter content and living plants, Zn mobility increased and decreased concentrations of AVS, indicating root induced sediment oxidation or decreased sulfate-reduction. In many cases, wetlands used for wastewater treatment function only to store toxicants temporarily, but as there is increase in the pollutant load, the capacity of a wetland to incorporate more toxicants may be compromised and the wetland can potentially become a secondary source of toxicity. Szkokan-Emilson et al. (2014) investigated the potential metal release from the catchments of wetlands that once deposited an elevated amount of S and metals from smelter emissions in Sudbury, Canada. Results indicated that catchments containing wetlands have now became a major source of metal pollution to a lake as well as oxidation of sulfur during dry periods triggered the acidification followed by metal and sulfate release.

The precipitated solids need to be periodically removed and wetland bed may be refilled with new media for longer operation. However, a well-vegetated wetland can create adequate new removal sites (slow process) to enable the long-term treatment opportunity. Many studies suggest the accumulation rate of metal sludge in wetlands is about 1.25–1.75 in per year, indicating dikes that provide 3 ft of freeboard should offer sufficient volume for 25–50 years of treatment (Sheoran and Sheoran, 2006). Therefore, it is recommended to build systems either with larger freeboard. Disposal of the excavated sludges is not difficult or unduly expensive as it is not considered hazardous waste.

6.6 Conclusions

Metal enrichment of water bodies and sediments has increased many folds globally since the industrialization era. The deliberate discharge and flexible regulation policies in several parts of the world, especially developing countries, would suffer immediate wrath from metal contamination. The choice and suitability of a treatment process for metallic wastewater among a wide range of treatment technologies such as chemical precipitation, coagulation, flocculation, flotation, adsorption, ion exchange, membrane filtration and constructed wetlands will depend on the wastewater characteristics (such as pH and initial metal concentration), plant flexibility and reliability, economic aspects such as capital investment and operational costs (including energy consumption and maintenance) and finally its sustainability approach. CW technology has been widely used to remove metals from wastewater even at trace concentrations. Several researchers have reported the complete removal of metals in both natural and CWs. The critical factor for achieving success in wetland application is by understanding the fundamental mechanisms and processes controlling the metal removal within wetland and its interactions among various wetland components. Metal and metalloid from various wastewater are removed by combining the aspects of physical, chemical, and biological processes, which involves sedimentation, settling, filtration, adsorption, precipitation, co-precipitation, plant uptake and bacterial function to transform highly soluble metals into insoluble compounds. This book chapter entails details about the predominant mechanisms and other wetland functions in metal removal. However, more in-depth knowledge of the entire wetland process (living and non-living components) will help develop insight into its design, structure and feasibility. More studies are required to evaluate the metal removal performance and focus on the mitigation and restoration of wetlands.

References

Abdullah, N., Yusof, N., Lau, W., Jaafar, J., Ismail, A., 2019. Recent trends of heavy metal removal from water/wastewater by membrane technologies. J. Ind. Eng. Chem. 76, 17–38.

Aghel, B., Mohadesi, M., Gouran, A., Razmegir, M., 2020. Use of modified Iranian clinoptilolite zeolite for cadmium and lead removal from oil refinery wastewater. Int. J. Environ. Sci. Technol. 17 (3), 1239–1250.

Anaga, A., Abu, G.O., 1996. A laboratory-scale cultivation of Chlorella and Spirulina using waste effluent from a fertilizer company in Nigeria. Bioresour. Technol. 58 (1), 93–95.

Arivoli, A., Mohanraj, R., Seenivasan, R., 2015. Application of vertical flow constructed wetland in treatment of heavy metals from pulp and paper industry wastewater. Environ. Sci. Pollut. Res. 22 (17), 13336–13343.

Aslam, M.M., Malik, M., Baig, M., Qazi, I., Iqbal, J., 2007. Treatment performances of compost-based and gravel-based vertical flow wetlands operated identically for refinery wastewater treatment in Pakistan. Ecol. Eng. 30 (1), 34–42.

Aykol, A., Budakoglu, M., Kumral, M., Gultekin, A.H., Turhan, M., Esenli, V., et al., 2003. Heavy metal pollution and acid drainage from the abandoned Balya Pb-Zn sulfide Mine, NW Anatolia, Turkey. Environ. Geol. 45 (2), 198–208.

Babilas, D., Dydo, P., 2018. Selective zinc recovery from electroplating wastewaters by electrodialysis enhanced with complex formation. Sep. Purif. Technol. 192, 419–428.

Balladares, E., Jerez, O., Parada, F., Baltierra, L., Hernández, C., Araneda, E., et al., 2018. Neutralization and co-precipitation of heavy metals by lime addition to effluent from acid plant in a copper smelter. Miner. Eng. 122, 122–129.
Basha, C.A., Bhadrinarayana, N., Anantharaman, N., Begum, K.M.S., 2008. Heavy metal removal from copper smelting effluent using electrochemical cylindrical flow reactor. J. Hazard. Mater. 152 (1), 71–78.
Batool, A., Saleh, T.A., 2020. Removal of toxic metals from wastewater in constructed wetlands as a green technology; catalyst role of substrates and chelators. Ecotoxicol. Environ. Saf., 189.
Batty, L.C., 2003. Wetland plants-More than just a pretty face? Land Contamination and Reclamation 11 (2), 173–180.
Batty, L.C., Younger, P.L., 2004. Growth of *Phragmites australis* (Cav.) Trin ex. Steudel in mine water treatment wetlands: effects of metal and nutrient uptake. Environ. Pollut. 132 (1), 85–93.
Beining, B.A., Otte, M.L., 1997. Retention of metals and longevity of a wetland receiving mine leachate. In: Brandt, J.E., Galevotic, J.R., Kost, L. (Eds.), Proceedings of 14th Annual National Meeting—Vision 2000: An Environmental Commitment American Society for Surface Mining and Reclamation. Austin, Texas, 43–46.
Benhadji, A., Ahmed, M.T., Maachi, R., 2011. Electrocoagulation and effect of cathode materials on the removal of pollutants from tannery wastewater of Rouïba. Desalination 277 (1–3), 128–134.
Blesson, S., Pushparaj, A.N., Soda, S., 2021. Removal of Heavy Metals from Synthetic Mine Drainage in Laboratory Scale Constructed Wetlands. Trends in Civil Engineering and Challenges for Sustainability, 507–523.
Bruins, M.R., Kapil, S., Oehme, F.W., 2000. Microbial resistance to metals in the environment. Ecotoxicol. Environ. Saf. 45 (3), 198–207.
Burris, J., Gerber, D., McHerron, L., 1984. Removal of iron and manganese from water by Sphagnum moss Editor. In: Burris, J. (Ed.), Treatment of Mine Drainage by Wetlands. Pennsylvania State University, University Park, PA, pp. 1–13.
Campbell, K.M., Alpers, C.N., Nordstrom, D.K., 2020. Formation and prevention of pipe scale from acid mine drainage at iron Mountain and Leviathan Mines, California, USA. Appl. Geochem. 115, 104521.
Carvajal-Flórez, E., Cardona-Gallo, S.-.A., 2019. Technologies applicable to the removal of heavy metals from landfill leachate. Environ. Sci. Pollut. Res. 26 (16), 15725–15753.
Chandra, R., Yadav, S., Yadav, S., 2017. Phytoextraction potential of heavy metals by native wetland plants growing on chlorolignin containing sludge of pulp and paper industry. Ecol. Eng. 98, 134–145.
Chen, F., Yu, W., Qie, Y., Zhao, L., Zhang, H., Guo, L.-.H., 2019. Enhanced photocatalytic removal of hexavalent chromium through localized electrons in polydopamine-modified TiO_2 under visible irradiation. Chem. Eng. J. 373, 58–67.
Cimino, G., Ziino, M., 1983. Heavy metal pollution, Part VII. Emissions from Mount Etna Volcano. Geophys. Res. Lett. 10 (1), 31–34.
Deliyanni, E.A., Kyzas, G.Z., Matis, K.A., 2017. Various flotation techniques for metal ions removal. J. Mol. Liq. 225, 260–264.
Du, J., Zhang, B., Li, J., Lai, B., 2020. Decontamination of heavy metal complexes by advanced oxidation processes: a review. Chin. Chem. Lett. 31 (10), 2575–2582.
Engwa, G.A., Ferdinand, P.U., Nwalo, F.N., Unachukwu, M.N., 2019. Mechanism and health effects of heavy metal toxicity in humans. In: Karcioglu, O., Arslan, B. (Eds.), Poisoning in the modern world-new tricks for an old dog. BoD – Books on Demand, London, UK.
Fernández-González, R., Martín-Lara, M., Blázquez, G., Tenorio, G., Calero, M., 2020. Hydrolyzed olive cake as novel adsorbent for copper removal from fertilizer industry wastewater. J. Cleaner Prod. 268, 121935.
Gao, J., Zhang, J., Ma, N., Wang, W., Ma, C., Zhang, R., 2015. Cadmium removal capability and growth characteristics of *Iris sibirica* in subsurface vertical flow constructed wetlands. Ecol. Eng. 84, 443–450.
Gao, Y., Qiao, Y., Xu, Y., Zhu, L., Feng, J., 2021. Assessment of the transfer of heavy metals in seawater, sediment, biota samples and determination the baseline tissue concentrations of metals in marine organisms. Environ. Sci. Pollut. Res. 1–13.
Garrett, R.G., 2000. Natural sources of metals to the environment. Hum. Ecol. Risk. Assess. 6 (6), 945–963.
Ghosh, A., Ghosh Dastidar, M., Sreekrishnan, T., 2016. Recent advances in bioremediation of heavy metals and metal complex dyes. J. Environ. Eng. 142 (9), C4015003.

Green, S.J., 2011. Microorganisms and processes linked to uranium reduction and immobilization. In: Stolz, J.F., Oremland, R.S. (Eds.), Microbial metal and metalloid metabolism: advances and applications. ASM Press, pp. 117–138.

Hanchette, C., Zhang, C.H., Schwartz, G.G., 2018. Ovarian cancer incidence in the US and toxic emissions from pulp and paper plants: a geospatial analysis. Int. J. Environ. Res. Public Health 15 (8), 1619.

Hussain, Z., Arslan, M., Malik, M.H., Mohsin, M., Iqbal, S., Afzal, M., 2018. Treatment of the textile industry effluent in a pilot-scale vertical flow constructed wetland system augmented with bacterial endophytes. Sci. Total Environ. 645, 966–973.

Jacob, D.L., Otte, M.L., 2004. Long-term effects of submergence and wetland vegetation on metals in a 90-year old abandoned Pb–Zn mine tailings pond. Environ. Pollut. 130 (3), 337–345.

Kadlec, R.H., Wallace, S.D., 2008. Treatment wetlands. CRC press, Boca Raton, FL.

Kakoi, B., Kaluli, J.W., Ndiba, P., Thiong'o, G., 2017. Optimization of Maerua Decumbent bio-coagulant in paint industry wastewater treatment with response surface methodology. J. Cleaner Prod. 164, 1124–1134.

Kaushik, A., Singh, A., 2020. Metal removal and recovery using bioelectrochemical technology: the major determinants and opportunities for synchronic wastewater treatment and energy production. J. Environ. Manage. 270, 110826.

Kiiskila, J.D., Sarkar, D., Feuerstein, K.A., Datta, R., 2017. A preliminary study to design a floating treatment wetland for remediating acid mine drainage-impacted water using vetiver grass (*Chrysopogon zizanioides*). Environ. Sci. Pollut. Res. 24 (36), 27985–27993.

Kiran, M.G., Pakshirajan, K., Das, G., 2017. An overview of sulfidogenic biological reactors for the simultaneous treatment of sulfate and heavy metal rich wastewater. Chem. Eng. Sci. 158, 606–620.

Kjeldsen, P., Barlaz, M.A., Rooker, A.P., Baun, A., Ledin, A., Christensen, T.H., 2002. Present and long-term composition of MSW landfill leachate: a review. Crit. Rev. Environ. Sci. Technol. 32 (4), 297–336.

Kosolapov, D., Kuschk, P., Vainshtein, M., Vatsourina, A., Wiessner, A., Kästner, M., et al., 2004. Microbial processes of heavy metal removal from carbon-deficient effluents in constructed wetlands. Eng. Life Sci. 4 (5), 403–411.

Kröpfelová, L., Vymazal, J., Švehla, J., Štíchová, J., 2009. Removal of trace elements in three horizontal sub-surface flow constructed wetlands in the Czech Republic. Environ. Pollut. 157 (4), 1186–1194.

Kumar, N., Kumar, S., Bauddh, K., Dwivedi, N., Shukla, P., Singh, D., et al., 2015. Toxicity assessment and accumulation of metals in radish irrigated with battery manufacturing industry effluent. International J. Veg. Sci. 21 (4), 373–385.

Kumar, R., Bhatia, D., Singh, R., Rani, S., Bishnoi, N.R., 2011. Sorption of heavy metals from electroplating effluent using immobilized biomass *Trichoderma viride* in a continuous packed-bed column. Int. Biodeterior. Biodegrad. 65 (8), 1133–1139.

Kurniawan, T.A., Chan, G.Y., Lo, W.-H., Babel, S., 2006. Physico–chemical treatment techniques for wastewater laden with heavy metals. Chem. Eng. J. 118 (1–2), 83–98.

Lee, B.-H., Scholz, M., 2007. What is the role of *Phragmites australis* in experimental constructed wetland filters treating urban runoff? Ecol. Eng. 29 (1), 87–95.

Lim, P.E., Mak, K., Mohamed, N., Noor, A.M., 2003. Removal and speciation of heavy metals along the treatment path of wastewater in subsurface-flow constructed wetlands. Water Sci. Technol. 48 (5), 307–313.

Liu, J., Ye, S., Yuan, H., Ding, X., Zhao, G., Yang, S., et al., 2018. Metal pollution across the upper delta plain wetlands and its adjacent shallow sea wetland, northeast of China: implications for the filtration functions of wetlands. Environ. Sci. Pollut. Res. 25 (6), 5934–5949.

Liu, M., Li, X., He, Y., Li, H., 2020. Aquatic toxicity of heavy metal-containing wastewater effluent treated using vertical flow constructed wetlands. Sci. Total Environ. 727, 138616.

Lizama-Allende, K., Ayala, J., Jaque, I., Echeverría, P., 2021. The removal of arsenic and metals from highly acidic water in horizontal subsurface flow constructed wetlands with alternative supporting media. J. Hazard. Mater. 408, 124832.

Matagi, S., Swai, D., Mugabe, R., 1998. A review of heavy metal removal mechanisms in wetlands. Afr. J. Trop. Hydrobiol. Fish. 8, 23–35.

Mays, P.A., Edwards, G.S., 2001. Comparison of heavy metal accumulation in a natural wetland and constructed wetlands receiving acid mine drainage. Ecol. Eng. 16 (4), 487–500.

Moreau, J.W., Fournelle, J.H., Banfield, J.F., 2013. Quantifying heavy metals sequestration by sulfate-reducing bacteria in an acid mine drainage-contaminated natural wetland. Front Microbiol 4, 43.

Mungur, A., Shutes, R., Revitt, D., House, M., 1997. An assessment of metal removal by a laboratory scale wetland. Water Sci. Technol. 35 (5), 125–133.
Mustapha, S., Ndamitso, M., Abdulkareem, A., Tijani, J., Mohammed, A., Shuaib, D., 2019. Potential of using kaolin as a natural adsorbent for the removal of pollutants from tannery wastewater. Heliyon 5 (11), e02923.
Muthusaravanan, S., Sivarajasekar, N., Vivek, J., Paramasivan, T., Naushad, M., Prakashmaran, J., et al., 2018. Phytoremediation of heavy metals: mechanisms, methods and enhancements. Environ. Chem. Lett. 16 (4), 1339–1359.
Pandey, P.K., Sharma, R., Roy, M., Pandey, M., 2007. Toxic mine drainage from Asia's biggest copper mine at Malanjkhand, India. Environ. Geochem. Health 29 (3), 237–248.
Patel, T.L., Patel, B.C., Kadam, A.A., Tipre, D.R., Dave, S.R., 2015. Application of novel consortium TSR for treatment of industrial dye manufacturing effluent with concurrent removal of ADMI, COD, heavy metals and toxicity. Water Sci. Technol. 71 (9), 1293–1300.
Ramírez, S., Torrealba, G., Lameda-Cuicas, E., Molina-Quintero, L., Stefanakis, A.I., Pire-Sierra, M.C., 2019. Investigation of pilot-scale constructed wetlands treating simulated pretreated tannery wastewater under tropical climate. Chemosphere 234, 496–504.
Rao, J.R., Nair, B.U., Ramasami, T., 1997. Isolation and characterisation of a low affinity chromium (III) complex in chrome tanning solutions. J. Soc. Leather Technol. Chem. 81 (6), 234–238.
Ribeiro, C., Scheufele, F.B., Espinoza-Quiñones, F.R., Módenes, A.N., Vieira, M.G.A., Kroumov, A.D., et al., 2018. A comprehensive evaluation of heavy metals removal from battery industry wastewaters by applying bio-residue, mineral and commercial adsorbent materials. J. Mater. Sci. 53 (11), 7976–7995.
Runnells, D.D., Shepherd, T.A., Angino, E.E., 1992. Metals in water. Determining natural background concentrations in mineralized areas. Environ. Sci. Technol. 26 (12), 2316–2323.
Ryan, P., Delhaize, E., Jones, D., 2001. Function and mechanism of organic anion exudation from plant roots. Annu. Rev. Plant Biol. 52 (1), 527–560.
Sheoran, A.S., Sheoran, V., 2006. Heavy metal removal mechanism of acid mine drainage in wetlands: a critical review. Miner. Eng. 19 (2), 105–116.
Singh, S., Chakraborty, S., 2020. Performance of organic substrate amended constructed wetland treating acid mine drainage (AMD) of North-Eastern India. J. Hazard. Mater. 397, 122719.
Sochacki, A., Surmacz-Gorska, J., Faure, O., Guy, B., 2014. Polishing of synthetic electroplating wastewater in microcosm upflow constructed wetlands: effect of operating conditions. Chem. Eng. J. 237, 250–258.
Steinmüller, K., 2001. Modern hot springs in the southern volcanic Cordillera of Peru and their relationship to Neogene epithermal precious-metal deposits. J. South Amer. Earth Sci. 14 (4), 377–385.
Stumm, W., Morgan, J.J., 2012. Aquatic chemistry: chemical equilibria and rates in natural waters, 126. John Wiley & Sons, New York, pp. 1040.
Sundby, B., Vale, C., Caçador, Z., Catarino, F., Madureira, M.-J., Caetano, M., 1998. Metal-rich concretions on the roots of salt marsh plants: mechanism and rate of formation. Limnol. Oceanogr. 43 (2), 245–252.
Szkokan-Emilson, E., Watmough, S., Gunn, J., 2014. Wetlands as long-term sources of metals to receiving waters in mining-impacted landscapes. Environ. Pollut. 192, 91–103.
Tran, T.-K., Chiu, K.-F., Lin, C.-Y., Leu, H.-J., 2017. Electrochemical treatment of wastewater: selectivity of the heavy metals removal process. Int. J. Hydrogen Energy 42 (45), 27741–27748.
Wei, X., Kong, X., Wang, S., Xiang, H., Wang, J., Chen, J., 2013. Removal of heavy metals from electroplating wastewater by thin-film composite nanofiltration hollow-fiber membranes. Ind. Eng. Chem. Res. 52 (49), 17583–17590.
Weis, J.S., Weis, P., 2004. Metal uptake, transport and release by wetland plants: implications for phytoremediation and restoration. Environ. Int. 30 (5), 685–700.
Weissberg, B.G., Browne, P.R., Seward, T.M., 1979. Ore metals in active geothermal systems. Geochemistry of Hydrothermal Ore Deposits, 738–780.
Wildeman, T.R., Machemer, S.D., Klusman, R.W., Cohen, R.R., Lemke, P., 1990. Metal removal efficiencies from acid mine drainage in the Big Five constructed wetland, Proceedings of the 1990 Mining and Reclamation Conference and Exhibition. West Virginia University Morgantown, WV, pp. 417–424.
Williams, L.G., Kevern, N.R., 1964. Relative strontium and calcium uptake by green algae. Science 146 (3650), 1488–1488.
Wojciechowska, E., Waara, S., 2011. Distribution and removal efficiency of heavy metals in two constructed wetlands treating landfill leachate. Water Sci. Technol. 64 (8), 1597–1606.

Xu, J., Zheng, L., Xu, L., Wang, X., 2020. Uptake and allocation of selected metals by dominant vegetation in Poyang Lake wetland: from rhizosphere to plant tissues. Catena 189, 104477.

Yadav, A.K., Kumar, N., Sreekrishnan, T.R., Satya, S., Bishnoi, N.R., 2010. Removal of chromium and nickel from aqueous solution in constructed wetland: mass balance, adsorption-desorption and FTIR study. Chem. Eng. J. 160 (1), 122–128.

Younger, P.L., Banwart, S.A., Hedin, R.S., 2002. Mine water: Hydrology, Pollution, Remediation. Kluwer Academic Publications, London, UK.

Zeraatkar, A.K., Ahmadzadeh, H., Talebi, A.F., Moheimani, N.R., McHenry, M.P., 2016. Potential use of algae for heavy metal bioremediation, a critical review. J. Environ. Manage. 181, 817–831.

Zhou, Q., Yang, N., Li, Y., Ren, B., Ding, X., Bian, H., et al., 2020. Total concentrations and sources of heavy metal pollution in global river and lake water bodies from 1972 to 2017. Global Ecology and Conservation 22, e00925.

CHAPTER 7

A glance over current status of waste management and landfills across the globe: A review

Krishna Chaitanya Maturi[a], Aparna Gupta[b], Izharul Haq[a], Ajay S. Kalamdhad[a]
[a]Department of Civil Engineering, Indian Institute of Technology, Guwahati, Assam, India
[b]Department of Environmental Science, Central University of South Bihar, India

7.1 Introduction

India is one of the world's fastest-growing economies. For the first time in more than a century, India's economy has exceeded that of the United Kingdom and the sixth-largest economy by gross domestic product (GDP) and the third-largest economy by purchasing power parity (PPP). India's urban growth rate rose from 27.8 to 31.6 percent between 2001 and 2011, and it is projected that up to 50% of the Indian population will live in cities over the next decade (B. Gupta and Arora, 2016; Sharma and Jain, 2019). Expanding global human population, increased urbanization & technological advancement has increased generation of solid waste. Solid waste (SW) mismanagement is a worldwide concern in terms of environmental contamination, social inclusion, and economic sustainability, necessitating integrated assessments and holistic approaches to help address (Li et al., 2005). Solid wastes can contain any solid material found in nature, as well as numerous man-made items, making them the most diverse collection of substances possible.

Wastes are classified as solid, liquid, or gaseous depending on their physical state. Municipal wastes, hazardous wastes, medical wastes, and radioactive wastes are the four types of solid wastes. Planning, financing, construction, and operation of facilities for trash collection, transportation, recycling, and final disposal are all part of solid waste management. Municipal solid waste (MSW) is commonly characterized as waste materials generated in home, commercial, and public spaces, such as gardens, street sweeping, and drain silts, that may pose environmental and public health issues in the long term. MSW is classified into three categories, according on the source: Residential or household garbage generated by individual dwellings; commercial and/or institutional waste generated by larger sources of MSW such as hotels, office buildings, schools, and so on; municipal services waste generated by area sources such as roadways, parks, and so on. Food waste, paper, cardboard, plastics, textiles, glass, metals,

Biodegradation and Detoxification of Micropollutants in Industrial Wastewater
DOI: https://doi.org/10.1016/B978-0-323-88507-2.00001-4

wood, street sweepings, landscape and tree trimmings, and general waste from parks, beaches, and other recreational areas are commonly found in MSW. Other household garbage, such as batteries and consumer electronics, are occasionally mixed together with MSW.

7.2 Global scenario of landfilling

A landfill is one of the most common methods for disposing of municipal solid waste (MSW) (Lema et al., 1988; Renou et al., 2008). Approximately 95% of all MSW collected worldwide is disposed of in landfills (El-Fadel et al., 1997). In developed countries such as the United States, more than half of MSW is landfilled each year (Deng, 2009). In contrast, according to a 2006 United Nations waste management report, the Flemish area of Belgium landfilled 15% of industrial rubbish, or roughly 3 million tonnes. In 2006, China had 324 landfills in operation for the disposal of MSW, with around 43% of MSW being landfilled (Xu, 2008). According to a World Bank report, by 2050, with only 13.5 percent recycled and 5.5 percent composted, the globe would produce around 3.4 billion tonnes of MSW, up from two billion tonnes in 2018 (Kaza et al., 2018). According to the survey, between 30 and 40% of waste produced worldwide is destroyed or buried in the open rather than being properly managed. Brazil produced 79 million tonnes of MSW in 2018, an increase of 0.82 percent over 2017, with 92 percent of it gathered (an increase of 1.66 percent over 2017). This suggests that rubbish collection increased at a faster rate than waste generation.

However, 6.3 million tonnes of garbage went unchecked in the urban centers. Only 59.5 percent of the MSW collected was properly disposed of in landfills, with the remaining 40.5 percent dumped in inappropriate sites by 3001 municipalities. To put it another way, 29.5 million tonnes of MSW winds up in dumping sites or landfill because they lacked the necessary systems and safeguards to secure person's environment and human health (Abrelpe and Abrelpe, 2012; Penteado and de Castro, 2021).

According to a study, 60 percent of rubbish produced in developing countries is thrown carelessly, with 38 percent of it being landfilled, while incineration and recycling each contributed for 1%. Various countries have reported different landfill classifications (Idris et al., 2004; Kamaruddin et al., 2017). Various criteria such as the type of waste deposits, the type of liner used, landfill layout, and landfilling procurement are considered. On the other hand, the environmental consequences of improper MSW disposal endure and will continue to do so through inferior landfills. Landfill emissions can take many forms, including landfill gas, airborne particles, and leachate. Many African countries' landfill dumps have been discovered to have no or limited liner (Agamuthu, 2013; Bundhoo, 2018). According to Cossu and Piovesan (2007) in underdeveloped countries, unregulated urban waste dumping is the principal source of carbon sinks.

There are two sorts of landfills. The most common, and still in use in developing countries, are dumps in which MSW is dropped until it approaches a level deemed acceptable for aesthetic or technical reasons. After a landfill closes, some earth is placed on top of it. Over 6,500 landfills were used to store municipal solid waste, according to a report to Congress by the US Environmental Protection Agency (USEPA) in October 1988. Despite using a variety of environmental measures, these landfills posed substantial threats to ground and surface water resources, as well as health hazards from air and water pollution. To tackle these issues and standardize technical specifications for MSW landfills, the USEPA established new minimum federal criteria for MSWLFs on October 9, 1991 (CFR, 2009). Several of the minor landfills have closed as a result of the increasingly rigorous restrictions, leaving only 1767 landfills, according to the EPA (Agency, 2005). Huge landfill operations have taken advantage of economies of scale by serving large geographic areas and accepting various types of rubbish, such as commercial solid waste, non-hazardous sludge, and industrial non-hazardous solid wastes. In the year 2000, the United States' 75% municipal solid waste was anticipated to be disposed of in 500 large landfills.

The European standards for non-hazardous, Class II landfills are comparable to those in the US, and are based on French legislation passed in January 1996. The EU Landfill Directive of 1999, on the other hand (Directive, 1999), mandates that landfilling be limited to inert, non-biodegradable, and non-combustible materials in the near future. Nonetheless, it may take years for this law to be implemented in new EU member countries, thus landfill gas manufacturing will remain for the indefinite future. In contrast to the EU and Japan, landfilling remained the predominant method of MSW disposal in the United States. The overall generation of MSW each person in the advanced countries of the European Union and Japan is double that of the generation capacity of MSW each capita (i.e., before any recycling) (1.19 tonnes). This is to be expected, given that the US consumes 20–25 percent of global materials and fossil fuels, although accounting for only 5% of the world's population. The European Union (EU) and the bulk of the "golden billion" produce over 420 million tonnes of MSW, with at least 210 million tonnes (50%) being landfilled. According to waste management studies, the amount of MSW in poor countries, including Africa, is always greater than 0.2 tonnes per capita, with the bulk of the rubbish being food and yard waste, which is disposed of in landfills. For the 5.4 billion people in the developing world, this equates to 1080 million tonnes. When these values are added together, the total quantity of MSW that will be landfilled globally is expected to be about 1.5 billion tonnes.

The annual garbage production in SSA (Sub-Saharan Africa) is anticipated to be around 62 million tonnes. African cities are predicted to generate garbage at a rate of between 0.3 kg and 1.4 kg per capita per day, compared to an average of 1.22 kg per capita per day in developed countries (Dladla et al., 2016). In most regions of the world, landfills remain a frequent and inexpensive technique of managing municipal solid wastes. Several landfills in developing nations are outengineered, making it difficult to transform waste resources into commercially viable products. Furthermore, the ineffective systems provide a possible threat of pollution to groundwater resources. However, a study on framework for the development and operation of municipal solid waste management systems in hotter climates (Munawar and Fellner, 2013) clearly indicates that in developing countries with limited waste management budgets, such as Nigeria, Uganda, Ethiopia, and others, using Low-hydraulic-conductivity clay (less than 1×10^{-8} m/s) as a base liner may suffice.

According to previous studies in Nigeria, 68 percent of solid trash created by localities was thrown haphazardly, 21 percent was disposed of in suitable landfill sites, and 11 percent was burned (Adeniran et al., 2014; Regassa et al., 2011). Aurah (2013) According to reports, about 10% of all garbage generated in Nairobi thrown away in landfill. Additionally, two different studies done in Nigeria and Ghana showed that landfilling was used to dispose of only about 20% of solid waste in African countries (Aziale and Asafo-Adjei, 2013; Ogwueleka, 2009). The continually low numbers seen throughout African communities indicate that more public knowledge and awareness on waste management responsibilities is required. In many nations, urbanization has an influence on the entire rate of solid waste generation. In many large cities, critical concerns relating to municipal solid waste (MSW) collection, disposal techniques, and dumping sites remain unresolved. Problems linked with the decomposition of the organic fraction of trash offer the largest challenge in terms of river and groundwater pollution in many parts of Asia due to the relatively moist climate. The per capita Waste generation rate in Malaysia varies between 0.88 and 1.44 kg/day, based on the financial situation of the area. It's widely assumed to be 1.0 kg/ca/day. It is estimated that Kuala Lumpur produces almost 3000 tonnes of solid waste each day. The country produces over 6 million tonnes of waste each year (Idris, 2003; Nazeri Salleh, 2002). Solid waste is one of Thailand's most pressing environmental issues. Solid waste production in 2003 was roughly 40,165 tonnes per day, with 24 percent coming from the Bangkok Metropolitan Administration, 31% from municipalities, and 45% from rural areas. In 1993, Bangkok produced 9640 tonnes of rubbish each day. The collection service is thought to reach 60–80 percent of residents in Bangkok's municipal jurisdiction (Padungsirikul, 2003).

According to Laner et al. (2012), landfills disposed of up to 54% of the US's 250, 106 tonnes of municipal solid waste (MSW) in 2008. In addition, whereas 77 percent of MSW in Greece, 55 percent in the United Kingdom, and 51 percent in Finland was landfilled in 2008, almost 70 percent of MSW in Australia was diverted to landfills without being pretreated in 2002 (Laner et al., 2012). In Korea, Poland, and Taiwan, respectively, around 52 percent, 90 percent, and 95 percent of MSW is disposed of in landfills (Renou et al., 2008). The combined waste produced in India's four major cities of Mumbai, Delhi, Chennai, and Kolkata is about 20,000 tonnes per day, with the majority of it going to landfills (Chattopadhyay et al., 2009). The bulk of landfills across the globe are old and weren't built to keep toxic leachate from polluting the soil and groundwater underneath them. Although the United States has a population of 4.57 percent, it produces 13.98 percent of global MSW. Brazil's population is 2.96 percent of the global total, and its internal MSW production is 4.33 percent of the global total (Franco et al., 2021). According to a survey conducted by the US Environmental Protection Agency (USEPA), roughly 75 percent of United States' landfills pollute groundwater (Dresser and McKee, 2004). Water polluted by landfill leachate has been discovered all over the world, particularly in Europe, China and Australia (Ngo et al., 2009).

7.3 Indian scenario of landfilling

In the previous few decades, India's urban and semi-urban population has grown dramatically. The urban population is quickly growing as a result of fast industrial growth. According to the 2011 Census of India, the percentage of people living in cities climbed from 27.8% to 31.16 percent between 2001 and 2011. Its urban population increased by 31.8 percent in the last decade to 377 million, surpassing the total population of the United States, the world's third-largest country by population. Rapid population expansion combined with unplanned urbanization has resulted in many of the basic infrastructure issues in India's quickly rising cities, including water supply, storm water drainage, and sewerage, as well as solid waste management. The quantity of MSW has also expanded dramatically as people's lifestyles and social standing have improved in metropolitan areas (Sharholy et al., 2007). Yearly waste generation has been found to increase in tandem with population and urbanism, making disposal more problematic as more space is needed for ultimate disposal of such waste materials (Idris et al., 2004). During the past three decades, the National Environmental Engineering Research Institute (NEERI) carried out investigations in more than 50 Indian cities. According to the evaluation, MSW contains 30–45 percent organic waste, 6–10 percent recycle, and the rest is inactive materials. Organic matter in developing-country solid trash is substantially higher than in developed-country garbage (Bhide, 1983), and organic matter can be transformed into usable goods to relieve the pressure on existing landfills (Richard, 1992). The MSW (MoEF, 2000) Rules, 2000 recommends Source-specific waste collection and transportation, as well as suitable processing and disposal. Knowledge about the quantity and characteristics of MSW has been shown to aid in the development of a MSW management (MSWM) system lengthy strategy.

However, the vast majority of municipal governments lack accurate statistics and data. As a result, the CPCB of the Government of India in New Delhi determined that it was necessary to conduct a review of the MSWM system's current status in 59 cities throughout India, including 35 metro cities with populations greater than one million, as well as 24 state capitals and union territories. On the orders of the Honorable Supreme

Court of India, New Delhi, CPCB retained NEERI to assess the state of MSWM in these cities and towns. MSW is typically dumped in open areas in many Indian cities and towns, but this is not a practical waste disposal method due to the environmental concerns and ecosystem instabilities associated with dumpsites, including land, river, and poor air quality (Kansal, 2002). Over 90% of all MSW generated in India is disposed of in an unsatisfactory manner on land (Das et al., 1998). In India, 91.4 percent of collected garbage was landfilled, with the remainder being burned (6.4 percent) or composted (2.2 percent). Almost all cities have embraced open dumping for waste disposal, with the exception of Pune, which is developing a partially sanitary landfill, and Nashik, which disposes of waste in distinct cells that have chosen a sanitary landfilling process. Treatment of leachate following collection, as well as biogas extraction from landfills, are uncommon in the majority of cities. Partially covered cities include Mumbai, Kolkata, Chennai, Ahmedabad, Kanpur, Lucknow, Coimbatore, Nashik, Vadodara, Jamshedpur, Allahabad, Amritsar, Rajkot, Shimla, Thiruvananthapuram, and Dehradun. In 26 cities, compactors/bulldozers are used to compact garbage. Waste is disposed of along the valley ridges in mountainous regions cities. It was discovered that in a huge number of cities, rubbish is not weighed and the waste amount is approximated rely on the number of trips each day (Kumar et al., 2009).

In India, per capita garbage creation increased from 0.44 kg/day in 2001 to 0.5 kg/day in 2011, owing to changing lifestyles and rising spending power among urban Indians. In urban India, total MSW production is expected to be 68.8 million tonnes per year (TPY) or 188,500 tonnes per day (TPD). Within a decade, there has been such a dramatic increase in trash output that it has put a strain on all available natural, infrastructural, and fiscal resources. Smaller cities and towns collect less than half of the MSW generated, while larger cities and towns gather 70–90 percent. More than 91 percent of MSW collected formally is dumped on open land or dumped in landfills (Kumar et al., 2009). Around 2% of uncollected waste is estimated to be burned in the public on city streets. Approximately 10% of MSW collected is burned in the open or caught in landfill fires (Chakrabarti and Devotta, 2007). Every year, such open MSW burning and landfill fires emit 22,000 tonnes of pollutants into Mumbai's lower atmosphere. The contaminants include carbon monoxide (CO), carcinogenic hydro carbons (HC) (including dioxins and furans), particulate matter (PM), nitrogen oxides (NOx) and Sulphur dioxide (SO_2) (Chakrabarti and Devotta, 2007).

7.4 Land requirement for dumpsites

The quantity of area occupied by rubbish discarded post-independence, till 1997, was calculated in a 1998 research by TERI (The Energy Resources Institute, formerly Tata Energy Research Institute) titled "Solid Waste Management in India: Options and Opportunities." The study calculated the amount of area covered in football field multiples, resulting in 71,000 football fields of solid waste stacked 9 meters high. The analysis indicates that garbage created in 2001 might have taken up 240 sq km, or half the area of Mumbai, and waste generated in 2011 would have taken up 380 square kilometers, based on a 91 percent landfilling scenario, or 90 percent of Chennai, India's fourth largest city in terms of size; and trash created in 2021 would require 590 square kilometers, more than the size of Hyderabad (583 square kilometers), India's largest metropolis in terms of area (Gupta et al., 1998). According to the Ministry of Finance's Position Paper on the Solid Waste Management Sector in India, more than 1400 sq km of land will be required for disposal of solid waste by the end of 2047 if MSW is not adequately handled, which is comparable to the combined size of Hyderabad, Mumbai, and Chennai.

In India, the number of sanitary landfills (SLFs) is steadily increasing. SLFs now exist in eight cities, compared to zero in India's 74 cities. Pune, Ahmadabad, Surat, Jodhpur, Chandigarh, Navi Mumbai, Mangalore, and Nashik are the eight cities with SLFs. There are no preventative procedures in place at the Nashik waste complex to deal with landfill fires, which have been discovered to be a typical occurrence. In addition to the 8 cities having SLFs, another 13 (total 21) cities cover the wastes dumped with earth cover, and another 15 (total 24) cities compact or align the wastes. The frequency with which earth cover is applied to wastes is unknown. In Mumbai and Pune, LFG recovery from landfills has also been explored. Only seven landfills (in four locations) were judged to be economically viable in a study conducted by the USEPA's Methane to Markets programme. "Sanitation landfilling is ideally suited to developing countries (like India) as a means of managing waste disposal due to the flexibility and relative simplicity of the technology," according to the UNEP. This advice ignores the significant maintenance and running expenses of SLFs, as well as the requirement for SWM projects to be self-sustaining. Out of 59 cities assessed by the Central Pollution Control Board, 17 have suggested new landfill sites. 24 cities (23.4 million TPY) dump their garbage in 34 landfills that cover 1,900 hectares (Annepu, 2014).

Open, uncontrolled, and mismanaged dumping is common in many metropolitan cities. As a result, the environment has been severely harmed. In cities and towns, more than 90% of MSW is disposed of in an inadequate manner directly on land. Heavy metals have been rapidly leaking into coastal waters as a result of such dumping activity in numerous coastal municipalities. The daily cover procedures are inadequate, making leakage more likely. This is primarily due to a lack of information and skill on the side of local government officials. This pushes local governments to limit the use of even well-known protections and practises. Even though major cities such as Delhi, Mumbai, Kolkata, and Chennai are facing the problem of limited available land for waste disposal, there will undoubtedly be some improvement in the aspect of landfilling and final disposal of MSW and to ensure sanitary landfilling in India (Kaushal et al., 2012; Sharholy et al., 2006; Singh et al., 2007). The availability of land for waste disposal in major towns or cities, such as Delhi, is quite constrained (Das et al., 1998; Gupta et al., 1998; Kansal et al., 1998; Mor et al., 2006; Sharholy et al., 2006; Siddiqui et al., 2006). In most urban areas, MSW is disposed of by dumping it in close to the bottom areas outside the city without following sanitary landfilling rules. Most landfills lack a leachate collection process or surveillance and gathering equipment for landfill gas, and compressing and flattening of rubbish, as well as final soil coating, are rarely recorded practices (Bhide and Shekdar, 1998; Gupta et al., 1998). All trash, including hazardous wastes from hospitals, wind up at the disposal site since MSW is not segregated at the source. Industrial garbage is frequently dumped in landfills intended for household waste (Datta, 1997). For the final disposal of MSW, sanitary landfilling is a viable and recommended option. Because all other solutions produce some residue that must be disposed of through landfilling, it is a vital component of MSWM. However, landfilling appears to be the most extensively used practice in India in the next years, at which time various adjustments will be required to ensure sanitary landfilling (Das et al., 1998; Dayal, 1994; Kansal, 2002).

Landfilling is an important aspect of any MSWM system, and it is the last resort for MSW disposal once all other options have been exhausted (Aljaradin and Persson, 2012; Devi et al., 2016; Wilson et al., 2012). MSWM methods are primarily designed to address the health, environmental, aesthetic, land-use, resource, and economic challenges associated with inappropriate waste disposal (Ferronato et al., 2018; Marshall and Farahbakhsh, 2013). Nonetheless, MSW is disposed of in India's cities in an unscientific manner in low-lying areas or open dumps, without sensible precautions or management processes, causing harm to all aspects of the environment and human health (Alam and Kulkarni, 2016; Bundela et al., 2010; Devi et al., 2016; Gidde et al., 2008; Gupta and Arora, 2016; Kalyani and Pandey, 2014; Kaushal et al., 2012; Kumar and Samadder, 2017; Kumar, 2005; Mani and Singh, 2016; Narayana, 2009). The Solid Waste Management Rules, 2016, govern MSWM in India. Urban India is home to a third of the country's population and produces 54.75 million tonnes of municipal solid trash each year. The scientific repair of dumpsites has also been ordered by environmental litigation. The fact that many ancient dumpsites and landfills around the country constitute a hazard to human health and the environment mandates remediation of dumpsites and reclamation of degraded land. After ages of being used, these open dumps have grown in size and height, becoming major pollution sources. Leachate, a vile dark liquid concentration produced by waste decaying in these dumps, kills vegetation in its path and pollutes groundwater irrevocably. Methane, a greenhouse gas that produces 21 times more global warming than carbon dioxide, is also produced by the rubbish heaps. They also damage groundwater with leachate (liquid produced by airless waste). Air pollution is caused by fires that erupt frequently in dumpsites. The presence of these dumps encourages more garbage to be dumped there, despite the fact that they are already overburdened. Methane frequently auto-ignites in dumpsites, resulting in fires, smoke, and emissions, resulting in severe air pollution. In India, it is believed that over 10,000 hectares of urban land are enslaved in these dumpsites. These dumpsites produce methane (a greenhouse gas) and other landfill gases in the absence of oxygen, which contribute to global warming. According to the CPCB (2016) annual report of solid waste management, India has 2120 legacy waste dumpsites spread across 23 states: (according per MoHUA MIS data, 1,764 dumpsites are registered; the legacy waste dumpsites may include under construction and current SLF that do not require cleanup) (USEPA, 2020). Clearing these piles of years-old rubbish, also known as legacy waste, is one of the simplest and quickest ways to cut national emissions while also protecting nearby towns from polluted water sources, smoke, insects, and stink.

Despite the fact that cities with populations greater than 20 lakhs serve only 23% of the urban population, dumpsites in these cities account for nearly half (47%) of the total garbage found in dumpsites serving cities with populations greater than 1 lakh. This problem is exacerbated by the fact that these cities have grown the most in terms of area, and as a result, dumpsites that were once located on the outskirts of cities are now standing tall in the heart of them, exposing a large number of people to the health risks associated with open dumpsites. Furthermore, these dumpsites are expanding in size, becoming a nuisance for these cities and resulting in significant social, economic, and environmental costs. Cities will soon have to cope with the fact that these dumpsites will be unable to accept any more waste, and it is critical that they act quickly to address this problem. In response to the growth in trash generation, the amount of land required for garbage disposal is expected to rise.

7.5 MSWM in developed and emerging countries: a comparative analysis

According to gross national income, countries are classified as high-income countries (HICs), which include France, Germany, Spain, the United Kingdom, the United States, and Italy; upper-middle-income countries (UMICs), which include Brazil, Mexico, Poland, Lithuania, and Latvia; and lower-middle-income countries (LMICs), which include Algeria, China, India, Jordan, Turkey, Bulgaria, and T. While the majority of waste is collected and processed scientifically in HICs, LICs rely on open dumping. While HICs have successfully implemented the Five Rs, LICs and LMICs continue to struggle to collect 100% of MSW. HICs place a premium on waste reduction and recycling over landfilling and incineration of municipal solid waste, whereas LICs and LMICs prefer open dumping and sanitary landfilling. Germany has the highest overall recycling and composting rate (66.2%), followed by Nepal (55%), Lithuania (48%), Italy (45.1%), the United Kingdom (44.5%), Poland (44%), France (41.7%), the United States (34.6%), Bulgaria (31.8%), and Spain (31.8%). (31.8%). (29.7%). Despite billions of dollars invested in MSWM, India continues to lag behind LMIC, UMIC, and HIC countries in terms of economic performance. India's waste collection efficiency (82%) is the lowest among LMIC countries, trailing China (90%), Turkey (84%), Thailand (100%), Bulgaria (94%), Jordan (90%), and Algeria (90%). (92 percent). Only 36.66 percent of waste in India is treated, compared to 92% in China, 100% in Turkey and Bulgaria, 55%

in Jordan, and 46% in Thailand. Algeria, Pakistan, India, and Thailand have the highest rates of open and illegal waste disposal at 96%, 90%, 72%, and 54%, respectively. As a large waste-generating country, India needs to focus its MSWM heavily on high waste recycling, composting, and anaerobic digestion to reduce the load on waste disposal through incineration and sanitary landfilling (Sharma and Jain, 2019).

7.6 Conclusions

The quantity of garbage produced in India has risen substantially as a result of increasing urbanization, industrialization, population expansion, and economic development. Because of its reliance on limited waste infrastructure, the informal sector, and open waste disposal, India's MSWM system is ineffective. This paper includes an extensive assessment of the environmental impacts of waste management and waste disposal. The current situation of waste disposal in the world was attempted to summaries. Based on the discussion described above, it can be inferred that waste disposal remains a dominant form of urban waste management in many countries. Despite their relatively high potential to pollute the atmosphere, deposits continue to be one of the key waste disposal methods. Regular landfill surveillance is therefore important to detect and identify environmental hazards. The production of leachate is an unavoidable consequence of garbage disposal in landfills. Many environmental variables and operating procedures have an impact on landfill operations, causing temporal and geographical variations in landfill leachate amount and quality. To mimic these changes, as well as leachate production and transport processes in landfills, mathematical models have been created. Leachate quantity models have apparently been employed with some success, particularly in terms of bounding production rates, which can aid in the development of design requirements for leachate collecting and storage systems. As the importance of moisture (precipitation) is established in this study, it is suggestive through this study that some moisture additions can be designed in areas with low rainfall, such as arid zones. Although the volume of leachate was smaller, the strength of the resulting leachate was higher. Finally, it can be concluded that understanding the quality of leachate will aid in the planning and implementation of proper liner systems in sanitary landfill design and leachate treatment. Operators of contemporary landfills face a number of challenges, and the classification of existing dump sites is inconsistent in a number of countries. Appropriate garbage disposal is expected to remain one of the most pressing environmental and health concerns in Asia's emerging economies.

References

Abrelpe, E.A., Abrelpe, 2012. Panorama dos resíduos sólidos no Brasil. São Paulo Grappa.
Adeniran, A.A., Adewole, A.A., Olofa, S.A., 2014. Impact of Solid waste management on Ado Ekiti property values. Civ. Environ. Res. 6, 29–35.
Agamuthu, P., 2013. Landfilling in developing countries.

Agency, U.S.E.P., 2005. Municipal solid waste generation, recycling, and disposal in the United States. Off. Solid Waste, Facts Fig. 2003.
Alam, T., Kulkarni, K., 2016. Municipal solid waste management and its energy potential in Roorkee city, Uttarakhand, India. J. Inst. Eng. Ser. A 97, 9–17.
Aljaradin, M., Persson, K.M., 2012. Comparison of different waste management technologies and climate change effect—Jordan.
Annepu, R.K., 2014. Municipal Solid Waste Management in Asia and the Pacific Islands, MS Dissertation, Environmental Science and Engineering. Springer, Singapore. https://doi.org/10.1007/978-981-4451-73-4.
Aurah, C.M., 2013. Assessment of extent to which plastic bag waste management methods used in Nairobi City promote sustainability. Am. J. Environ. Prot. 1, 96–101.
Aziale, L.K., Asafo-Adjei, E., 2013. Logistic challenges in urban waste management in Ghana a case of Tema metropolitan assembly. Eur. J. Bus Manag. 5, 116–128.
Bhide, A.D., 1983. Solid waste management in developing countries. Indian National Scientific Documentation Centre.
Bhide, A.D., Shekdar, A.V., 1998. Solid waste management in Indian urban centers. Int. Solid Waste Assoc. Times 1, 26–28.
Bundela, P.S., Gautam, S.P., Pandey, A.K., Awasthi, M.K., Sarsaiya, S., 2010. Municipal solid waste management in Indian cities-A review. Int. J. Environ. Sci. 1, 591.
Bundhoo, Z.M.A., 2018. Solid waste management in least developed countries: current status and challenges faced. J. Mater. Cycles Waste Manag. 20, 1867–1877.
CFR, U.S., 2009. Code of Federal Regulations Title 40: protection of Environment, Part 136–Guidelines establishing test procedures for the analyses of pollutants, Appendix B to Part 136–definition and procedure for the determination of Method Detection Limit rev. 1.11.
Chakrabarti, T., Devotta, S., 2007. Scientist. National Environmental Engineering Research Institute.
Chattopadhyay, S., Dutta, A., Ray, S., 2009. Municipal solid waste management in Kolkata, India–A review. Waste Manag 29, 1449–1458.
Cossu, R., Piovesan, E., 2007. Modern role of landfill as geological sink for carbon and other elements, in: Sardinia 2007 Eleventh International Waste Management and Landfill Symposium.
CPCB, 2016. https://cpcb.nic.in/uploads/MSW/MSW_AnnualReport_2016-17.pdf.
Das, D., Srinivasu, M., Bandyopadhyay, M., 1998. Solid state acidification of vegetable waste. Indian J. Environ. Health 40, 333–342.
Datta, M., 1997. Waste disposal in engineered landfills. Narosa, New Delhi.
Dayal, G., 1994. Solid wastes: sources, implications and management. Indian J. Environ. Prot. 14, 669–677.
Deng, Y., 2009. Advanced oxidation processes (AOPs) for reduction of organic pollutants in landfill leachate: a review. Int. J. Environ. Waste Manag. 4, 366–384.
Devi, K.S., Swamy, A., Nilofer, S., 2016. Municipal solid waste management in India—An overview. Asia Pacific J. Res. 1, 118–126.
Directive, C., 1999. 99/31/EC of 26 April 1999. Off. J. L 182, 7.
Dladla, I., Machete, F., Shale, K., 2016. A review of factors associated with indiscriminate dumping of waste in eleven African countries. African J. Sci. Technol. Innov. Dev. 8, 475–481. https://doi.org/10.1080/20421338.2016.1224613.
Dresser, C., McKee, 2004. Guidelines for Water Reuse. U.S. Environmental Protection Agency, Washington, DC. In this issue.
El-Fadel, M., Findikakis, A.N., Leckie, J.O., 1997. Environmental impacts of solid waste landfilling. J. Environ. Manage. 50, 1–25. https://doi.org/10.1006/jema.1995.0131.
Ferronato, N., Gorritty Portillo, M.A., Guisbert Lizarazu, E.G., Torretta, V., Bezzi, M., Ragazzi, M., 2018. The municipal solid waste management of La Paz (Bolivia): challenges and opportunities for a sustainable development. Waste Manag. Res. 36, 288–299.
Franco, D.G. de B., Steiner, M.T.A., Assef, F.M., 2021. Optimization in waste landfilling partitioning in Paraná State. Brazil. J. Clean. Prod. 283, 125353. https://doi.org/10.1016/j.jclepro.2020.125353.
Gidde, M.R., Todkar, V.V., Kokate, K.K., 2008. Municipal solid waste management in emerging mega cities: a case study of Pune city, in: Proceedings of Indo Italian Conference on Green and Clean Environment, Pune, India (March 20-21).
Gupta, B., Arora, S.K., 2016. Municipal solid waste management in Delhi—The capital of India. Int. J. Innov. Res. Sci. Eng. Technol. 5, 5130–5138.

Gupta, Shuchi, Mohan, K., Prasad, R., Gupta, Sujata, Kansal, A., 1998. Solid waste management in India: options and opportunities. Resour. Conserv. Recycl. 24, 137–154.
Idris, A., 2003. Overview of municipal solid waste landfill sites in Malaysia, in: Proceedings of the 2nd Workshop on Material Cycles and Waste Management in Asia, December 2-3. Tsukuba, Japan, 2003.
Idris, A., Inanc, B., Hassan, M.N., 2004. Overview of waste disposal and landfills/dumps in Asian countries. J. Mater. cycles waste Manag. 6, 104–110.
Kalyani, K.A., Pandey, K.K., 2014. Waste to energy status in India: a short review. Renew. Sustain. energy Rev. 31, 113–120.
Kamaruddin, M.A., Yusoff, M.S., Rui, L.M., Isa, A.M., Zawawi, M.H., Alrozi, R., 2017. An overview of municipal solid waste management and landfill leachate treatment: malaysia and Asian perspectives. Environ. Sci. Pollut. Res. 24, 26988–27020.
Kansal, A., 2002. Solid waste management strategies for India. Indian J. Environ. Prot. 22, 444–448.
Kansal, A., Prasad, R.K., Gupta, S., 1998. Delhi municipal solid waste and environment-an appraisal. Indian J. Environ. Prot. 18, 123–128.
Kaushal, R.K., Varghese, G.K., Chabukdhara, M., 2012. Municipal solid waste management in India-current state and future challenges: a review. Int. J. Eng. Sci. Technol. 4, 1473–1489.
Kaza, S., Yao, L., Bhada-Tata, P., Van Woerden, F., 2018. What a waste 2.0: a global snapshot of solid waste management to 2050. World Bank Publications.
Kumar, A., Samadder, S.R., 2017. An empirical model for prediction of household solid waste generation rate–A case study of Dhanbad, India. Waste Manag. 68, 3–15.
Kumar, S., 2005. Municipal solid waste management in India: Present practices and future challenge. NEERI Rep.
Kumar, S., Bhattacharyya, J.K., Vaidya, A.N., Chakrabarti, T., Devotta, S., Akolkar, A.B., 2009. Assessment of the status of municipal solid waste management in metro cities, state capitals, class I cities, and class II towns in India: an insight. Waste Manag. 29, 883–895. https://doi.org/10.1016/j.wasman.2008.04.011.
Laner, D., Crest, M., Scharff, H., Morris, J.W.F., Barlaz, M.A., 2012. A review of approaches for the long-term management of municipal solid waste landfills. Waste Manag. 32, 498–512.
Lema, J.M., Mendez, R., Blazquez, R., 1988. Characteristics of landfill leachates and alternatives for their treatment: a review. Water. Air. Soil Pollut. 40, 223–250.
Li, B., Akintoye, A., Edwards, P.J., Hardcastle, C., et al., 2005. Critical success factors for PPP/PFI projects in the UK construction industry. Constr. Manag. Econ. 23 (5), 459–471. doi:10.1080/01446190500041537. In this issue.
Mani, S., Singh, S., 2016. Sustainable municipal solid waste management in India: a policy agenda. Procedia Environ. Sci. 35, 150–157.
Marshall, R.E., Farahbakhsh, K., 2013. Systems approaches to integrated solid waste management in developing countries. Waste Manag. 33, 988–1003.
MoEF, 2000. Municipal Solid Waste (Management and Handling) Rules 2000.
Mor, S., Ravindra, K., DeVisscher, A., Dahiya, R.P.P., Chandra, A., 2006. Municipal solid waste characterization and its assessment for potential methane generation: a case study. Sci. Total Environ. 371, 1–10. https://doi.org/10.1016/j.scitotenv.2006.04.014.
Munawar, E., Fellner, J., 2013. Guidelines for design and operation of municipal solid waste landfills in tropical climates. ISWA–the Int. Solid Waste Assoc.
Narayana, T., 2009. Municipal solid waste management in India: from waste disposal to recovery of resources? Waste Manag. 29, 1163–1166.
Nazeri Salleh, M., 2002. Physical and chemical characteristics of solid waste in Kuala Lumpur, Malaysia. Appropr. Environ. solid waste Manag. Technol. Dev. Ctries. 1, 461–468.
Ngo, H.-H., Guo, W., Xing, W., 2009. Applied technologies in municipal solid waste landfill leachate treatment. Encylopedia Life Support Syst 199.
Ogwueleka, T., 2009. Municipal solid waste characteristics and management in Nigeria. J. Environ. Heal. Sci. Eng. 6, 173–180.
Padungsirikul, P., 2003. Country report on solid waste management in Thailand, Proceedings of a Seminar on the Study of Safety Closure and Rehabilitation of Landfill Sites in Malaysia, September 18-19. Kuala Lumpur, 2003.

Penteado, C.S.G., de Castro, M.A.S., 2021. Covid-19 effects on municipal solid waste management: what can effectively be done in the Brazilian scenario? Resour. Conserv. Recycl. 164, 105152.

Regassa, N., Sundaraa, R.D., Seboka, B.B., 2011. Challenges and opportunities in municipal solid waste management: the case of Addis Ababa city, central Ethiopia. J. Hum. Ecol. 33, 179–190.

Renou, S., Givaudan, J.G., Poulain, S., Dirassouyan, F., Moulin, P., 2008. Landfill leachate treatment: review and opportunity. J. Hazard. Mater. 150, 468–493.

Richard, T., 1992. Municipal solid waste composting: physical and biological processing. Biomass Bioenergy 3, 163–180. https://doi.org/10.1016/0961-9534(92)90024-K.

Sharholy, M., Ahmad, K., Mahmood, G., Trivedi, R.C., 2006. Development of prediction models for municipal solid waste generation for Delhi city, Proceedings of National Conference of Advanced in Mechanical Engineering (AIME-2006), Jamia Millia Islamia. New Delhi, India, 1176–1186.

Sharholy, M., Ahmad, K., Vaishya, R.C., Gupta, R.D., 2007. Municipal solid waste characteristics and management in Allahabad, India. Waste Manag. 27, 490–496. https://doi.org/10.1016/j.wasman.2006.03.001.

Sharma, K.D., Jain, S., 2019. Overview of municipal solid waste generation, composition, and management in India. J. Environ. Eng. 145, 4018143.

Siddiqui, T.Z., Siddiqui, F.Z., Khan, E., 2006. Sustainable development through integrated municipal solid waste management (MSWM) approach–a case study of Aligarh District, in: Proceedings of National Conference of Advanced in Mechanical Engineering (AIME-2006), Jamia Millia Islamia. New Delhi, India, 1168–1175.

Singh, G., Siddiqui, T.Z., Jain, A., 2007. Sustainable Development through Integrated Municipal Solid Waste Management (MSWM) Approach–A Case Study of Indian School of Mines Campus, Proceedings of the International Conference on Sustainable Solid Waste Management. Chennai, India. Citeseer, pp. 5–7.

USEPA, 2020. Landfill Reclamation.

Wilson, D.C., Rodic, L., Scheinberg, A., Velis, C.A., Alabaster, G., 2012. Comparative analysis of solid waste management in 20 cities. Waste Manag. Res. 30, 237–254.

Xu, H., 2008. All the waste in China—The development of sanitary landfilling. Waste Manag World 4, 60–67.

CHAPTER 8

Micro and nanoplastic toxicity on aquatic life: fate, effect and remediation strategy

Md. Anwaruzzaman, Md. Irfanul Haque, Md. Nahidul Islam Sajol, Md. Lawshan Habib, M. Mehedi Hasan, Md. Kamruzzaman
Department of Applied Chemistry and Chemical Engineering, Bangabandhu Sheikh Mujibur Rahman Science and Technology University, Gopalganj, Bangladesh

8.1 Introduction

In the year of 1907, when first synthetic plastic was introduced, nobody thought about the upcoming danger from this type of inert materials. Plastic production wasn't that much high at the early years. The increasing growth of plastic was not recognized until 1950. After that for the next 70 years the manufacturing of plastic increased 200 folds and the amount reached 359 million tons in 2018. Now plastic products are everywhere. Plastics are used for food packaging, soft drinks are provided in plastic bottles, plastic utensils are used for cooking and drugs are also manufactured in plastic tubing. It becomes a part of our daily life and covers every demand in a simpler manner which makes them almost impossible to be replaced. Around 15 million tons of plastic are utilized for food packaging purposes only. Plastic covering protects from bacterial infections and other food borne diseases. On the other hand, migration of contaminant from packaging to food is considered as the major route of plastic contamination for human. Plastics comprises synthetic organic molecules like polyethylene (PE, high and low density), polystyrene (PS), polypropylene (PP), polyvinyl chloride (PVC), polyurethane (PUR) and polyethylene terephthalate (PET). To improve the properties like strength, coloration or flame retardation different types of additives are also used including stabilizers, plasticizers, flame retardants, pigments and fillers. These added extra chemicals are not firmly bound to the plastic. As a result, these toxic chemicals can easily leach out from main matrix into the environment. For better understanding of the microplastics (MPs) and nanoplastics (NPs) produced mainly from the above mentioned macromolecules it is obvious that extensive study of these polymers and additives are required (Revel, Châtel and Mouneyrac, 2018).

Huge production of plastic polymer leads to an enormous amount of plastic waste. These wastes are directly discarded in the lands or sewage system most often, which occupy a large area for landfills. These wastes have an adverse effect on natural terrestrial

Biodegradation and Detoxification of Micropollutants in Industrial Wastewater
DOI: https://doi.org/10.1016/B978-0-323-88507-2.00009-9

and aquatic environment and even on human health. When these plastics are indiscriminately discarded, they ended in the ocean where they can persist and accumulate. The inert nature and high durability of plastic products attract users a lot but these properties make the situation more complex to deal with the plastic wastes. Before 1980 incineration or recycling of plastic was neglected. Therefore 100% of plastic was discarded. Incineration of plastic after 1980 and recycling after 1990 increased by 0.7% every year. The recent trend of recycling plastic covers only around 25% of the whole plastic products which means the major portion (around 75%) of these materials find their destination in landfills or spread in the environment. Even in the highly developed countries recycling of plastic waste and new product fabrication from those waste is found to be problematic. Unavailable collection points, contaminated or mixed collection and limited market demand for the fabricated products are the major obstacles for recycling. As the recycling process isn't sufficient it ultimately boosts the production of micro- and nano-plastics (MNPs). The small fragment of plastic, called MNPs are produced by natural processes like ultraviolet radiation, heat or microbial activities and man-made actions like physical abrasion or chemical degradation. MNPs enters into aquatic system through waste water and runoff from urban areas. Even if the waste water is treated with the best treatment facilities, capable of reducing 98% of the plastics, it was found the 65 million MNPs escape the treatment facility on a daily basis worldwide (Murphy et al., 2016). Sludge from sewage is commonly used as fertilizer for lands and synthetic clothing fibers coming along with this sludge are found to be present even after many years.

When these tiny fragments enter into the freshwater network, they affect the aquatic flora and fauna such as phytoplankton, invertebrates, mollusks, and fishes. These particles get into the aquatic life either by direct consumption with food from water column or sediments or by eating the contaminated prey and thus entering into the food web. Once got in the food chain there is a high risk that these particles will also be bio-accumulated in human. Documentation for aquatic animals get ingested with MNPs includes turtles, seabirds, fish, crustaceans and worms (Wright, Thompson and Galloway, 2013). Researchers have found the capacity of microplastic ingestion in some other organisms and zooplankton (Cole et al., 2013). In recent years the presence of MPs and NPs has been reported in diverse aquatic ecosystem. Because of their nature of hydrophobicity these small fragments also act as an effective sorption medium for persistent organic pollutants (POPs) including polychlorinated biphenyls (PCBs), polycyclic aromatic hydrocarbons (PAHs) and organochlorine pesticides such as DDT which make them more toxic. Imposed cytotoxicity by these POPs are related to the amount of microplastic ingested by the respective organism. The bioavailability of those toxic chemicals increases after getting into the body because desorption rate gets increased in the body atmosphere. High number of POPs like PCBs can cause various problems such as endocrine disruption, teratogenicity, dysfunction of the liver and kidney, imbalanced hormone secretion and so on.

Macroplastics in marine environment possesses some aesthetic issues with economic repulsion for tourist industries and a hazard for other marine industries like fishing, shipping, power generation and so on. Plastics are also responsible for entanglement and marine equipment destruction. The impacts of large plastics on marine life involve the injury and even death of marine life, change of habitats of marine species, prevention of gas exchange by blocking the sea beds. Besides macroplastics, the smaller sized MNPs are drawing more attention recently. These small fragments are as consistent as their mother items. Moreover, because of their smaller size these particles can move easily from one place to another. These small elements can be easily taking up by wind, can escape the waste water treatment steps and finally ended into the ocean. As the surface area to volume ratio is higher for these particles they act as an active sorbent for organic wastes and microorganisms. Thus a new type of habitat for microbial assemblages emerges, called "plastisphere" (Zettler, Mincer and Amaral-Zettler, 2013). But sufficient studies aren't conducted regarding the microbial activity in these MP biofilms. Biofilms consist of microbial communities of very dynamic and diverse nature and extracellular polymeric substances (EPS) in which the microorganisms are set in. This EPS matrix along with the cell proximity protect the microbes from rough environment like UV radiation or physical abrasion. Genetic modification is also occurred within biofilms like metabolic diversity which is influenced by the increased horizontal gene transfer (HGT). Sharing of mobile genetic elements (MGE) also occurred which affects different biochemical functions. HGT is also responsible for the spreading of antibiotic resistance gene (ARG). So the biofilm formation on these increasing MPs not only possesses threat to the fate of the embedded microorganisms but also to the entire microbiomes and the result will be reflected on water quality, ecosystem, public health and every segments of the environment (Arias-Andres, Rojas-Jimenez and Grossart, 2019).

MPs can be further fragmented into NPs (diameter < 1 μm) by the same way as MPs are produced. The hazard of NPs should be considered separately as it is different from MPs in size. Detection and quantification of NPs aren't that easy which makes the situation more difficult. NPs can alter the metabolic functions of marine organisms. These particles possess potential health hazard and can induce toxicity in cells more adversely than MPs. The size of NPs allows them to permeate biological barriers and lipid membranes. NPs can affect the membrane structure and hinder the molecular diffusion and even gene expression (Sharma and Chatterjee, 2017).

The remediation of MNPs isn't an easy task. Different treatment procedures for MNP containing waste water are still at the intermediate level. Grit chamber/primary sedimentation, coagulation, sand filter, granular activated carbon filtration, membrane bioreactor, ozonation are the different physicochemical approaches utilized so far. Some ecofriendly green approaches and chemical methods are also chosen by some researches but the efficiency of these methods isn't that much satisfactory. While some methods can reduce only a specific shaped particle (spherical or fiber like), others required a lot of chemicals and energy. So, the most effective ways to reduce the MNP pollution are to control the sources, increase public awareness, proper collection and dumping of plastic waste and mostly increase the rate of recycling.

Owing very tiny size, MNPs can cause detrimental effect on aquatic ecosystem. Moreover detection and measurement of particles less than 10 μm is quite difficult (Kögel et al., 2020). So, the risk and possible harm upcoming from these particles are still not clear. As the uptake, retention and toxicity are directly related to the size it is high time to create awareness of the stakeholders about the impact of MNPs on our environment including human health.

8.2 Extensive use of plastic in our daily life

Marine waste includes a variety of processed or manufactured solid materials which are either directly discarded into the marine environment or transported there. This type of waste includes the debris of glass, paper, wood, textiles, rubber and plastics. Though paper, wood and natural fibers are easily biodegradable the others can persist for a long time in the aquatic environment. In a study of the floating debris on ocean surface it was found that plastics constitute the lion share (Law et al., 2010). There are also evidences that plastic exist in beach cleanups and even seafloor (Thiel et al., 2013). A survey, conducted in the South African beaches, showed that the densities of all type of plastic debris increased substantially within 5 years (Ryan and Moloney, 1990). In panama a cleaned beach for experiment was found to get the 50% of the previous wastes just in 3 months (Garrity and Levings, 1993). The highest abundance of plastic debris was found to be 258,408 items m^{-2} at Fan Lau Tung Wan beach, Hong Kong. More than 90% of the collected sample was constituted by MPs (<5 mm). Majority of the particles were of polystyrene type. The reason behind this huge plastic pollution would be the high population density and extensive use of plastic for food transportation (Fok and Cheung, 2015). Beside these anthropogenic actions some natural forces also contribute to plastic accumulation in the beaches. The abundance of plastic debris has been found to be increased after rainy season in comparison with the availability before that season. The direction of wind and current is also responsible for plastic accumulation in the beaches. Researchers has found that sea beaches in remote areas contain comparable number of wastes to the beaches near at highly industrialized places.

Plastics as well as the feedstock for plastic production, resins, are entering into the ocean. Spillage or mishandling causes the entrance of these small sized quasi-spherical resin pellets into the ocean. They were also included in the first report of marine plastic debris. Their durability isn't certain but they can persist around 5-10 years and the additives can exist 30-50 years (Gregory, 1978). Different plastic products converted from resin can enter the marine environment either unintentionally at the time of use or disposed as waste like plastic microbeads used in a variety of personal care products and synthetic fibers used in textiles.

The only reason for plastic pollution in marine environment is the uncontrolled production of plastic and the subsequent large number of plastic wastes in the land. The annual demand for plastic is increasing day by day. The largest sector of plastic production is packaging industry and then building and construction. Around 42% of total plastic produced in the year of 2015 constituted by only packaging sector. Light weight, transparency along with high strength make plastic a versatile material. As a packaging material none can beat plastic because of its low cost, excellent protecting properties from oxygen and moisture and bio-inertness. Plastic easily replaces other packaging materials like glass, metal and paper. In fact, plastic offers superior design than other materials. One third of the plastic resins produced are used for packaging material which include disposable one-time-use items. These single use items are largely used near the marine environments and finally disposed in the ocean (Andrady, 2003). In 2010 plastic waste generated within the 50 kms of coastal region was estimated to be 99.5 million tons.

For better understanding of the current situation sector wise plastic production and waste generated from the corresponding sector in the year of 2015 (Ritchie and Roser, 2018) is summarized in the Fig. 8.1. All the numbers provided in the chart are in million tons.

Ocean is frequently used for recreational purpose. Passengers and crew members of marine transport are responsible for around 10% marine waste. Dumping of plastic waste into the ocean has increased recently. Extraction of food from seas are also increasing day by day. These activities increased the availability of plastic waste in the ocean. Plastic gears are used in fishing net fleet globally. PE, PP and nylon are commonly used for the gear production. The broken or the lost gears find their destination into the ocean.

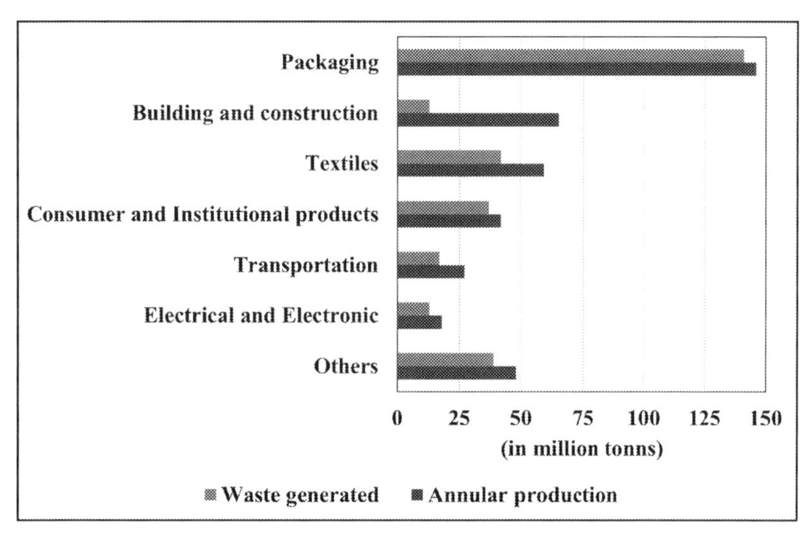

Fig. 8.1 *Plastic production and waste generation from the major contributing sectors in million tons. Data adopted from (Ritchie and Roser, 2018).*

Table 8.1 Types of plastic encountered most often in the ocean (Andrady, 2011).

Plastic class	Specific gravity	Percentage production	Products and typical origin
Low-density polyethylene	0.91–0.93	21%	Plastic bags, six-pack rings, bottles, netting, drinking straws
High-density polyethylene	0.94	17%	Milk and juice jugs
Polypropylene	0.85–0.83	24%	Rope, bottle caps, netting
Polystyrene and Foamed Polystyrene	1.05	6%	Plastic utensils, food containers Floats, bait boxes, foam cups
Nylon		<3%	Netting and traps
Thermoplastic Polyester	1.37	7%	Plastic beverage bottles
Poly(vinyl chloride)	1.38	19%	Plastic film, bottles, cups
Cellulose Acetate			Cigarette filters

Around 135,400 tons of plastic gear was discarded into the sea in the year of 1975. More than 18% marine plastic are coming from the fishing industry. Another contributor of plastic debris is aquaculture. Ocean transports also have significant contribution regarding plastic pollution. Virgin resin pellets from marine transport are entering into the ocean regularly. Common classes of plastic available in marine ecosystem are listed in Table 8.1 (Andrady, 2011). Beside these, the ever-increasing population is pushing immigration to coastal regions. Collectively these actions cause the increment of plastic influx into the ocean.

Plastics in marine environment are quantified by estimating the floating debris on the surface water which doesn't even close to the actual value. Plastic on surface water is estimated to be 10,000–100,000 tons. The mid water level MNPs or plastic debris mixed with sediments aren't determined by the common techniques. To float on the surface level the wastes have to have specific gravity lower than sea water which is 1.025. As seen from the Table 8.1 plastics having specific gravity higher than sea water will not rise up rather settled in the bottom of the ocean. Again plastic products when manufactured are mixed with different additives which may alter the specific gravity but ultimately denser plastics like nylon will submerge into coastal sediments (Andrady, 2011).

8.3 Characterizations of MPs and NPs

The problem regarding the characterization of MPs and NPs are mainly related to their size. Presence of both large and small particles in the wastewater arises problems associate with the masking of smaller particles by larger particles. Apart from size, because of the extensive use of composite plastic rather than single plastic, both the separation and characterization become difficult, costly and time consuming. Despite the limitations, several physical and chemical characterization techniques for the identification of microplastic and nanoplastic materials have been extensively studied. Among these techniques, some of them requires laborious sample preparation, while no sample preparation is needed for some. Both single and combination of two different characterization techniques have been used by research communities to overcome the problems regarding size and composition.

8.3.1 Physical characterizations

It is one type of preliminary analysis of the plastic particles. The particles can be separated based on their size, color, shape, density and opacity. Several techniques are available for the physical characterization of MPs and NPs.

Initially, because of relatively large size range of MPs visual inspection of plastics was done by naked eyes with forceps. It is easy, simple and MPs having a size range of 2-5 mm can be identified with naked eyes. However, though this technique doesn't require professional personal for completion, due to high probability of losing small particles during the sorting of large particles, this technique is not reliable for identification of particles less than 1 mm (McDermid and McMullen, 2004). However, MPs falls in the size range of micrometer is widely identified by optical microscopy (Nel and Froneman, 2015). Though size and appearance of MPs are identifiable with this technique, concrete conclusions cannot be drawn for the differentiation of natural and synthetic fibers. Besides, for transparent particles nearly 70% times results were not detectable (Song et al., 2015). To deal with the transparent and submicron MPs scanning electron microscopy (SEM) is extensively used nowadays. In this technique, instead of light, beam of high-speed electrons is projected on the surface of the sample. The backscattered electrons from the sample surface produce images with high resolution and fine texture. High magnification with high resolution images of plastic particles provided by SEM facilitates the discrimination of MPs from organic particles (Cooper and Corcoran, 2010). However, though these above mentioned techniques are suitable for MPs in different extents, none of them are suitable for the identification of NPs. On the other hand, atomic force microscopy (AFM) and hybrid AFM techniques are feasible and suitable for the NPs characterizations (Last et al., 2010). This technique relay upon the surface roughness of the sample and requires sample sizes and surface roughness to be as small as possible (<100 nm). This major condition of small surface and size limits the application of AFM for the identification of MPs. Hence, extensive research has been done on NPs with AFM rather than MPs (Fu et al., 2020).

Apart from appearance, size can also be evaluated with the help of above discussed microscopic techniques. However, as very small sample size is used for the microscopic analysis, these methods cannot provide the whole scenario of the particle sizes. Based on Brownian motion of particle, dynamic light scattering (DLS) and nanoparticle tracking analysis (NTA) techniques have been developed where the particle size of a sufficient amount of sample can be analyzed at a time to draw conclusions. In case of NTA, size distribution is done based on the relative speed of particles under the laser irradiation where larger particles move slower compare to smaller particles (James and Driskell, 2013). On the other hand, during DLS analysis, fluctuations of scattered light resulting from particle Brownian motion is measured (Yguerabide and Yguerabide, 1998). However, a comparative study of both DLS and NTA showed that DLS was more accurate in detecting larger particles, and NTA was better for small aggregates (Hou et al., 2018).

8.3.2 Chemical characterizations

Numerous spectroscopic and thermal analysis have been used for the chemical characterization of the MPs and NPs. Among them Fourier transform infrared spectroscopy (Shim, Hong and Eo, 2017), Raman spectroscopy (Collard et al., 2015), thermogravimetric (Dümichen et al., 2015), dynamic scanning calorimetry (Majewsky et al., 2016) analysis either solely or with junction of another technique are mostly used by research communities.

FTIR provides chemical information of the particles through the detection of chemical bond specific vibration modes of sample molecules at different infrared frequencies. Based on the results obtained from FTIR it is possible to distinguish carbon-based polymers from other organic and inorganic particles (Stuart, 2015). A well-established polymer spectrum library helps in the confirmation of plastics and identification of polymer type. However, if the sample contains hybrid plastics, obtained results are ambiguous as results can be varied due to instruments conditions and library doesn't contain most of the hybrid plastics (Shim, Hong and Eo, 2017). In this scenario an expert personal is necessary for the correct evaluation. Besides FTIR, Raman spectroscopy can also be used for the chemical identification of MPs and NPs. Non-contact of sample with the light source and smaller wavelength of laser source enables Raman spectroscopy for the identification of tiny particles (Cole et al., 2013). Combination of Raman and FTIR can be beneficial for the identification of complex MPs. However, Raman spectroscopy is costly and sensitive to pigments and additives which interfere with the identification of polymer types (Shim, Hong and Eo, 2017).

Apart from non-destructive spectroscopic methods, destructive thermal methods have also been used for the chemical and physical characterizations. The method requires for the reference materials of each polymer types as each polymer has different characteristics in DSC (Castañeda et al., 2014). Thermogravimetry combined with DSC had been used for the identification of different polyethylene and polypropylene based MPs (Majewsky et al., 2016).

In addition to these techniques several other hybrid techniques of both physical and chemical characterizations like AFM-IR (Cho et al., 2013), AFM-Raman (P Y Hung et al., 2015), gas chromatography - mass spectrometry (GC–MS) (Dümichen et al., 2017), thermal decomposition-GC–MS (P. Y. Hung et al., 2015), matrix assisted laser desorption-time of flight MS (Huppertsberg and Knepper, 2018), high performance liquid chromatography (HPLC) (Wang et al., 2017), SEM-energy dispersive X-ray spectroscopy (EDS) (Vianello et al., 2013), hyperspectral imaging (Serranti et al., 2018) have also been used for the MPs and NPs characterizations.

8.4 Environmental behavior and fate of MPs and NPs

8.4.1 Sources

The sources of MNPs are categorized into two major divisions: primary and secondary MNPs. Primary MNPs derive in the plastics industry from consumer goods, cosmetics and polymer raw materials such as polyethylene (PE), polystyrene (PS) and polypropylene (PP) (Tiwari, Santhiya and Sharma, 2020). For particular uses that require cosmetic abrasives, medicine vectors, and automotive and aerospace applications such as air blasting, primary MNPs are purposely made. Using sewage treatment systems, these MNPs are typically impossible to extract and will eventually accumulate in the atmosphere until they reach waste water. Secondary MNPs derive from larger plastics when various dynamic environmental factors such as wind, waves, temperature and UV light are increasingly broken into smaller parts (Guo et al., 2020).

In addition, repetitive use of plastic materials can also induce fragmentation and lead to secondary micro-NPs being created. Micro fibers from textile garments are a secondary source of MPs. Another key cause of MPs in the atmosphere is plastic pollution related to the movement of cars, including tire wear and tear, braking and road markings. The global average of microplastic emissions from road vehicle tire abrasion has been measured at 0.81 kg per capita per year. Wear and tear emitted from airplane tires accounts for nearly 2 percent of overall tire wear and tear emissions in the Netherlands, aside from road traffic. Furthermore, artificial turf also plays an important part in the secondary source of MPs, with artificial grass emissions ranging from 760 to 4500 tonnes a year being approximately estimated. Various types of MPs are thus being emitted into multiple natural environments and ecosystems (Guo et al., 2020).

Far different from ocean MPs sources, which primarily include land-based sources (contributing to 80%), coastal tourism, leisure, commercial fisheries (e.g. disposable fishing gear applications, etc., contributing to 18%), marine vessels and marine industries (e.g. aquaculture, oil rigs, etc.), MPs enter the soil through multiple sources, including landfills, soil amendments, land application of sewage sludge, wastewater-irrigation, compost and organic fertilizer, residues of agricultural mulching films, tire wear and tear, and atmospheric deposition etc as illustrated in Fig. 8.2 (Guo et al., 2020). In addition, biological processes of soil species, such as feeding operations, digestion, and excretion processes, can fragment plastic waste in the soil into MPs. The presence of MPs greatly decreases the fertility of the soil, and the movement and trophic transition of MPs to heavily polluted soils, especially those covered by waste water and plastic film, poses major ecological risks.

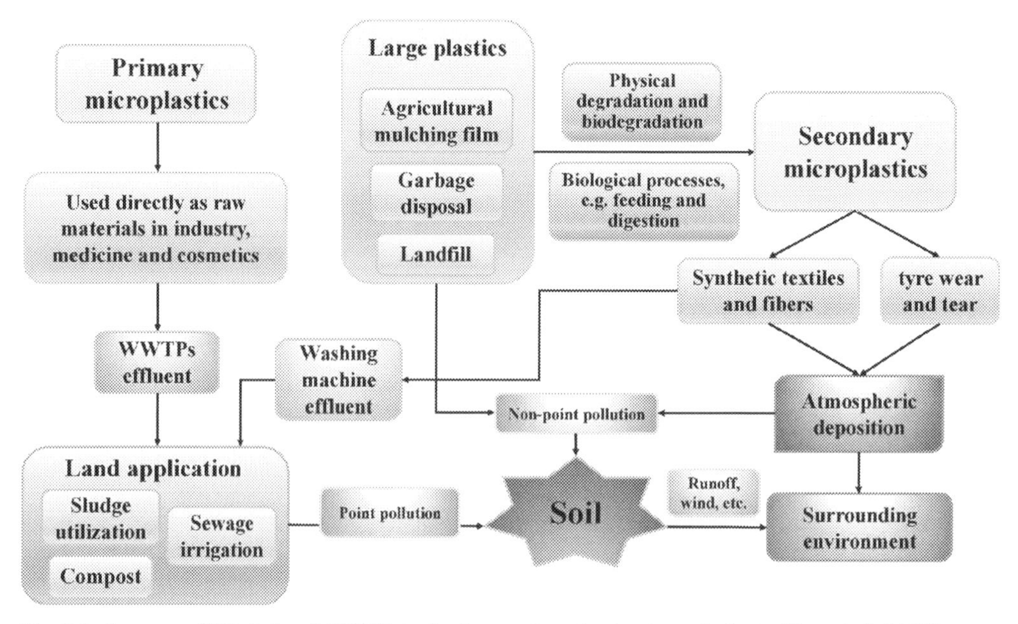

Fig. 8.2 ***Sources of MPs in land.*** WWTPs indicates wastewater treatment plants (Guo et al., 2020).

In terms of their bioavailability and deterioration, the scale, form, color and density of MPs all play a part. Usually, MPs are known as plastics <5 mm in size or between 1 μm and 5 mm; however, studies have identified them as <1 mm, <2 mm, 2-6 mm and <10 mm. The most popular types of MPs in the atmosphere include fragments, flakes, filaments and fibres, broken edges, granules and irregularly formed MPs. However, owing to their pervasiveness of everyday life, synthetic fibres and microbeads are more widely debated. Textiles, such as clothes, result in the release by ordinary use and laundering of synthetic fibers. NPs, too small to be seen with the human eye, are similar to MPs in that a specific definition of scale has not been provided. However, NPs are most commonly referred to as plastic particles of <0.1 μm in dimension (Lise Nerland et al., 2014).

8.4.2 Mechanism of microplastic conversion to NPs

A plastic particle's size is the prevailing attribute that decides its environmental destiny (e.g., migration). In addition, it may be size-dependent for bioaccumulation and toxicity. Considering that NPs derive largely from the breakup and conversion of larger plastic particles, it would be useful for a deeper understanding of NPs to explore the downsizing processes. By means of mechanical abrasion processes, NPs can be produced. Significant quantities of NPs are produced by the breakdown of regular polystyrene items by household blenders. Solid plastic waste fragmentation and MPs create NPs in the sewage system due to stream flow instability and mechanical systems in wastewater treatment plants (WWTPs). In the sea swash region, the natural breakup of larger plastic fragments can also be accomplished. Mechanical fragmentation of plastic pieces of macro and micro sizes is primarily caused by crack formation.

Another alternative mechanism for NP generation is hydrolysis (reacting with water), although it may not be the most efficient mechanism for minimizing plastic sizes. In contrast, UV irradiation-initiated degradation is a very successful downsizing process. The photodegradation of plastics is primarily due to reactive species of oxygen. The decrease in particle size may be due to the chain scission by attacks from free radicals, such as hydroxyl (•OH), alkyl (R•), alkoxyl (RO•) and peroxyl (ROO•) radicals produced from the UV light.

Biological destruction, and destruction by aquatic and terrestrial species of large plastic parts and MPs may also create environmental NPs. Zooplankton, squid, shrimps and other species have been known to eat plastic MPs and possibly NPs in aquatic ecosystems. Fragmentation or degradation of MPs into NPs has been recorded in polyethylene MPs (31.5 μm) exposed to Antarctic krill (*Euphausia superba*) along with algal foods. NPs of 150-500 nm size that were discovered in the digestive gland were generated after ingestion. In the common earthworm and snails, reduction of MPs into smaller sizes was observed, while fragmentation of MPs into NPs was not considered in these studies due to restriction of excess tools (Wang et al., 2021).

8.4.3 Migration of MPs and NPs

Owing to their unsystematic disposal, plastics enter the natural ecosystem and negatively impact the marine biota. This is a significant concern in recent decades, as the marine environment has the largest contribution to global primary productivity.

Mechanism of spreading of MNPs in aquatic environment is influenced not only by the properties of the MNPs, but also by the physical and chemical properties of the aquatic environment, as well as hydrodynamics, attachment, and uptake of aquatic species, both of which have a significant impact on the settling, re-suspension, and transportation distance of plastic particles, and thus on their environmental destiny. The majority of MNPs from land sources are discharged directly into rivers and lakes by wastewater treatment plant effluent or surface runoff. Tiny and medium plastic particles are mostly floating on the waters top, while big and hard plastic particles are concentrated at the bottom (Hoellein et al., 2019).

Furthermore, stream characteristics (water depth, flow, and obstructions in rivers such as large weirs) can influence microplastic migration in the freshwater environment, but there are few studies on the freshwater environment compared to the marine, and most of them have not paid attention to the inherent relationship between hydrological characteristics of rivers and microplastic migration. The deposited MPs/NPs tends to retain in the sediment for a short or long time, while the suspended plastics begin to flow to the ocean.

Another critical mode of transportation in the polar regions that is rarely discussed is sea ice. It uses a vertical pattern to mark polymer structure and size groups of microplastic particles in the drift trajectories of sea area during the growth of sea ice. However, in recent decades, global warming has accelerated the loss of sea ice, and these unquantified microplastic sinks are re-releasing significant amounts of MPs into the environment. Scientists would have to reconsider the dynamics, aggregation, and possible toxicological consequences of microplastic waste as a result of this (Huang et al., 2020).

8.4.4 Aggregation of MPs and NPs

The environmental fate of MNPs is primarily regulated by the processes of weathering and aggregation. The ageing of MNPs in the atmosphere is caused by different stressors (environmental factors), such as heat, water, UV irradiation, oxidants, micro-organisms, or by a mixture of these. As per the Arrhenius relationship, an increase in temperature will increase the weathering of NPs. Mechanical fragmentation (physical weathering) of NPs is caused by the shear forces of water (Section 8.4.2). Artificial aging using UV and O_3 co-exposure resulted in much rougher morphology and more oxygen-containing functional groups (e.g., hydroxyl, carbonyl, carboxyl) as compared with pristine NPs. During this abiotic oxidation process, reactive oxygen species such as hydroxyl radical (O•H), singlet oxygen ($^1O_2$) and superoxide radical ($•O_2^-$) induced the chain reactions, which degraded the structure of NPs (Wang et al., 2021).

In knowing the environmental fate of NPs, aggregation is a crucial challenge. Evidence has shown that NPs in environments can form milli-sized (mm-sized) aggregates. Furthermore, the development of heteroaggregates of inorganic colloids or organic matter contributes to NPs being either settled or migrated. The Derjaguin-Landau-Verwey-Overbeek (DLVO) theory has been generally adopted by numerous studies to explain the aggregation mechanism of NPs. The DLVO theory proposes that two independent forces (Eq. (8.1)), Van der Waals force (Eq. (8.2)) and the electrostatic double layer force (Eq. (8.3)) determine the stability of suspended particles:

$$V_T(d) = V_{vdw}(d) + V_{edl}(d) \tag{8.1}$$

$$V_{vdw}(d) = -\frac{A}{6}\left[\frac{2a^2}{d(4a+d)} + \frac{2a^2}{(2a+d)^2} + \ln\frac{d(4a+d)}{(2a+d)^2}\right] \tag{8.2}$$

$$V_{edl}(d) = \frac{32\pi\varepsilon\varepsilon_0 k^2T^2 a}{q_e^2 z^2} \times \tanh^2\left(\frac{zq_e\psi}{4kT}\right) \times e^{-xd} \tag{8.3}$$

where, $V_T(d)$ is the total interaction energy, $V_{vdw}(d)$ is the Van der Waals interaction energy, and $V_{edl}(d)$ represents the electric double layer interaction energy in Eq. (8.1). The parameters in Eq. (8.2) are defined as follows: A is the Hamaker constant for the NP dispersion system, whose value is dependent on the types of NPs and the aqueous media; a is the radius of NPs; d is the separation distance between NPs. As for the parameters in Eq. (8.3); ε is the dielectric constant of aqueous phase; ε_0 is the dielectric constant of vacuum; k is the Boltzman constant; T is the absolute temperature; q_e is the electron charge; z is the charge number; ψ is the surface potential of NPs (assumed to be equal to ζ-potential); κ is Debye length (Eq. (8.4)):

$$\kappa = \sqrt{\frac{2000 N_A q_e^2 I}{\varepsilon \varepsilon_0 k T}} \tag{8.4}$$

where N_A is the Avogadro constant, I is the ionic strength of the aqueous phase.

As shown in Eq. (8.3), an elevation in absolute value of ζ-potential ($|\psi|$) will lead to enhanced repulsive energy ($V_{edl}(d)$) and total interaction energy ($V_T(d)$), making it more difficult for NPs to form aggregates (Wang et al., 2021).

Conversely, a reduction in $|\psi|$ favors the aggregation process. For the interpretation of the environmental variables that control the aggregation process, the DLVO theory is critical. It has been verified by experimental findings that this principle is ideal for NP dispersion structures. In NP aggregation, various environmental variables, including pH, ion intensity, natural minerals and organic matter, play essential roles. It was observed that the negatively charged surfaces of polystyrene NPs possessed more negative ζ-potentials with the increase of solution pH. This resulted in an elevation in $|\psi|$, making NPs more stable according to DLVO theory. However, if the surface of NPs were positively charged at low pH conditions, an increase in solution pH may result in the aggregation due to a decrease in $|\psi|$ value. Inorganic ions also affect the aggregation process through changing the ionic strength of the solution. Higher concentrations of inorganic ions result in reduced $V_{edl}(d)$, favoring the aggregation of NPs. Natural minerals such as clay tends to form heteroaggregates with NPs due to electrostatic interactions. Natural organic matter protects NPs from aggregation by elevating the $|\psi|$ value (due to the formation of eco-corona). Additionally, the co-existence of natural organic matter and inorganic ions could contribute to the "bridging effect" if the concentration of organic matter is high enough (to allow the existence of un-adsorbed free organic matter in the system). Oxygen-containing groups in both NPs and organic matter could be bridged by inorganic metal cations (e.g. Ca^{2+}), resulting in heteroaggregation. In more complex structures, the traditional DLVO principle can be generalized to explain NP interactions (e.g., soil). Liu et al. adopted an expanded theory of DLVO to test the contact energy of porous media NPs (Liu et al., 2020). Three forces are taken into account in Total NPs-soil interaction energy: van der Waals force, electrostatic double layer force and hydrophobic effect (described by the Lewis acid-base interaction). Aged NPs are likely to have higher primary energy barriers, according to theoretical estimates, making them less likely to form aggregates. The theoretical estimate was consistent with the experimental results that aged NPs are more mobile in saturated porous media. Mao et al. assessed the aggregation behavior of NPs in the presence of extracellular polymeric substances (EPS) produced by microorganisms during biofilm formation on NPs (Mao et al., 2020). Steric repulsion has been introduced into the conventional DLVO theory in order to better explain the role of EPS in aqueous media. In the solutions with EPS, the energy barrier was larger, which was consistent with the observation that EPS prevented the aggregation of NPs by steric effects.

8.4.5 Deposition of MPs and NPs

In recent years, MPs and NPs have been detected in 690 species deposition of the marine environment, throughout the food chain, including fish, crustaceans, bivalves, mammals and plankton (Wang et al., 2019).

For microplastic absorption, phagocytosis and pinocytosis are two potential routes, but particle translocation is mostly size-dependent, with simpler internalization of smaller plastic particles. For example, 5-μm PS MPs were dispersed in the gills, liver, and gut in a study of the zebrafish *Danio rerio*, while 20-μm PS MPs collected only in the gills and gut. After exposure of this rotifer, researchers confirmed the dispersion of 50-nm MPs in different organs of *Brachionuskoreanus*, but 0.5- and 6-μm MPs localized only to its digestive tract. The distribution of MPs in marine animals is also influenced by surface charge. The deposition of MPs in embryos of the sea urchin *Paracentrotuslividus* with various surface coatings (carboxylated and amine polystyrene MPs) was studied and found that while carboxylated PS MPs were confined to its digestive tract, amine PS MPs were spread in the embryos (Xu et al., 2020).

A significant aspect which affects its toxicity is the accumulation of chemicals in an organism. Chorion zebrafish has pores for the transfer of oxygen and nutrients. In the embryo process, smaller NPs can penetrate these pores (0.5-0.7 μm diameter), whereas particles greater than pore size can bind to the chorion and be pulled inside by water flow or mistaken by larvae as food. Growth & development, immune system, oxidative stress, glucose level, energy metabolism and other biomarker response may adversely impact tissue accumulation of MPs/NPs. In various fluorescent-labelled colors and sizes, MPs/NPs are now available. Live simulation of the absorption and aggregation of fluorescent particles is due to the transparent nature of zebrafish embryos and larvae (Bhagat et al., 2020).

MPs/NPs are reported to accumulate in zebrafish embryo chorion, yolk sac, endotherm, muscle fibers, eye and spinal cord (Fig. 8.3). Sites recorded for its aggregation are larvae and adult fish, mouth, gut, head, blood, liver, heart, gills, muscle.

The absorption routes in fish, including oral, gill and skin absorption, are complex. Organ-specific accumulations of MNPs have been found in both larvae and adult zebrafish, with the intestinal tract being the primary and favored pathway. Studies have shown that MNPs can accumulate rapidly in zebrafish gills and intestines rather than in other regions. Increased accumulation of MPs in the presence of natural organic matter (NOM) was seen by the accumulation kinetics of MPs in zebrafish, with maximal accumulation within 72 hours (Bhagat et al., 2020).

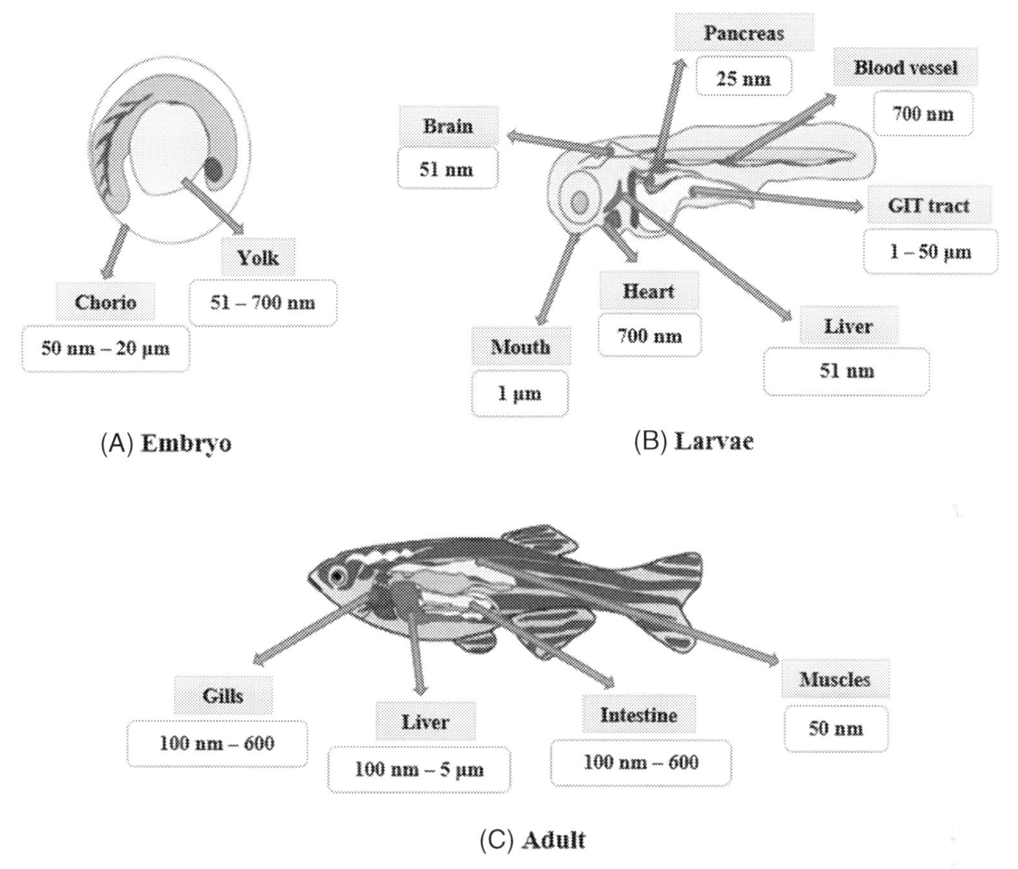

Fig. 8.3 Deposition of MNPs in different organs of (A) embryo, (B) larvae and (C) adult zebrafish. The numbers under each organ represents the size range of deposited MNPs found so far in the investigations in that respective organ (Bhagat et al., 2020).

8.5 Impact of MPs and NPs on aquatic life ecosystem

Contamination of plastics is one of the alarming issues in marine environments and the primary source of this contamination is small-scale plastic waste, such as MPs and NPs. An undeniable truth is the occurrence and deposition of MPs and NPs in the marine biota. Micro-sized plastics and plastic debris are distributed at various concentrations in aquatic ecosystems around the world. These MPs and NPs have various effects on the reproduction, creation, behavior, growth, and mortality of marine lives (Chae and An, 2017). Scientists are now realizing the detrimental effects of plastics on the aquatic environment, as shown by the growing number of studies focused mainly on marine biota in recent years. Accumulation of MP-associated contaminants may increase the potential risk of contaminant accumulation for higher trophic levels. As a result, it can create a range of effects on humans who are subjected to micro and nano plastics exposure. Till now the most studied taxonomic group was crustaceans (45%), followed by fish (21%), molluscs (18%), annelid worms (7%), echinoderms (7%) and rotifers (7%) (2%). Fishes are typically top predators and will eat MPs directly or through the consumption of microplastic-containing prey. The main consumers are planktonic rotifers and small crustaceans. Owing to their environment and feeding behavior, molluscs and other benthic species such as annelid worms are more susceptible to MPs. All of them occupy different roles in the aquatic food web and are exposed to different degrees of threat (de Sá et al., 2018).

8.5.1 Effects on aquatic microorganisms

The prevalence of MPs in marine ecosystems and their potential environmental impacts are capturing more focus recently. Less attention has been given to the relationship between microorganisms and plastic waste in aquatic ecosystems, despite long-standing evidence of the ability for floating plastic particles to function as microbial attachment sites and the subsequent production of plastic-associated biofilms. The potential interactions between MPs, chemicals associated with plastics, contaminants, microbial assemblages, and higher species are summarized in the following sections of this chapter. Evidence of the biological impact of plastic debris on microorganisms in aquatic environments is largely limited to information about bacterial and algae colonization and survival on polymer surfaces in seawater. The biodegradability of synthetic polymers depends on the shape and chemical properties of the material, the environment (e.g., seasonality and availability of oxygen), and metabolic interactions within the biofilms associated with the plastic. The majority of the research on the biotransformation of plastics focuses on microorganisms from terrestrial habitats but few studies have proved that microbial assemblages can use synthetic polymers as a nutrition source. A few experiments have been examined the potential of soil microorganisms in marine biota to biodegrade plastic waste. Due to the decreased supply of oxygen and light, the rates of oxidation of plastics in marine habitats are lower than their terrestrial counterparts. So it is unclear whether plastic is purposely destroyed by microbial organisms, treated by plastic-associated chemicals, or both (Harrison et al., no date).

Alarming discussions are going for a long time on the importance of human behaviors for microbial dispersal across the globe and intensified in the mid-20th century. Human involvement with Earth biomes induced major alterations in the composition and activity of microbial species, which could be seen as marked shifts in human microbiota or biogeochemical cycles of elements and delivery of particular genes. The last five years' data suggests that MP microbial species in the ecosystem have a distinct composition and function than most normal aquatic ecosystems. The exponential growth in the development of synthetic plastic polymers and their omnipresence in the atmosphere correlates, among other aspects, with the time of increasing industrialization and reported shifts in the microbiome of the earth. Moreover, in vitro studies by Ogonowski et al.

reported that the same initial aquatic microbial community results in different bacterial communities when grown on plastic polymers relative to other natural or inert substrates, suggesting a material-dependent sorting effect, especially during the early stages of particle colonization (Arias-Andres, Rojas-Jimenez and Grossart, 2019).

8.5.2 Effects on aquatic organisms

MPs are persistent in the aquatic ecosystem and are bioavailable to corals, zooplanktons, lobsters, worms, sea urchins, whales. For example, coral feeding studies have shown that corals confuse MPs as food and can eat up to ~50 μg cm^{-2} h^{-1} plastic; concentrations that are comparable to their consumption of plankton and organisms such as *Artemia nauplii*. Upon eating by aquatic species, they become bioaccumulated in the food chain and gradually reach higher tropical levels. MPs have been found in seabirds, turtles, crustaceans, and fish which lead to clogging of the intestinal tract, saturation suppression, inhibition of gastric enzyme secretion, imbalance of steroid hormone levels, delay in ovulation, and infertility (Sharma and Chatterjee, 2017).

Mechanisms referred to as might be necessary include digestive blockage, tissue abrasion, blockage of invertebrate feed appendices, tissue embedding, enzyme blockage, nutrient dilution, decreased feed stimulation, lower growth rate, and lower steroid hormone concentrations, and disrupted reproduction. The intensity of MPs-induced physical impacts depends on several variables. Plastics particles having high aggregation potential and translocation into tissues have a higher physical impact and are closely related to particle size. The sharp parts of ingested particles are more likely to inflict harm than round and fibers in the digestive tract are more likely to accumulate. The capacity of living beings to ingest microplastic is often considered important since this procedure can decide how easily a portion of the organism is ingested. Studies have shown that fish and other species can consume MPs due to their smaller size and less degradability which kills a variety of species, including plankton, vertebrates, and invertebrates (Sana et al., 2020).

8.5.3 Effect on large marine animals

Microplastic and NPs have many detrimental effects on species like water turtles, whales, harbor seals, and polar bears. 60.5% of sea turtles in Brazil have been ingested microplastic. Baleen whales were vulnerable to microplastic pollution as they were involved in filtering organisms that filter seawater and that facilitate the entry of MPs into their system. Also due to high fat and lipid content, whales are highly potential to ingest and accumulate MPs in the stomach and intestine. There have recently been several accounts of the death of stranded whales with a lot of microplastic litter in their intestines. Microplastic ingestion has been recorded in the stomach and intestines of harbor seals (*Phocavitulina*). The presence of microplastic particles was reported in the Hooker sea lions and fur seal scat. Polar bears are probably vulnerable to microplastic, although this issue has not been studied (Sharma and Chatterjee, 2017). While it has been shown that larger fish usually have more plastic parts in the intestine than smaller fish, no experiments have shown that microplastic translocation through the fish intestine presents a possible danger to human food welfare (Lise Nerland et al., 2014).

8.5.4 Effects on human health

The human can be exposed to MPs from primary MPs in cosmetics, toothpaste, scrubs, and hand wash. In addition to the toxicity effects of MPs, hazardous substances such as phthalates and PCBs inside MPs and pollutants adsorbed on the surface of MPs may contribute to human dietary exposure. Accumulation of MPs in seafood has consequences for human health. Marine species have been found to consume MPs and transfer them through the food network, such as mussels, oysters, crabs, marine cucumbers, and fish. There is no information available on the chemical composition, particle size, shape, or concentration of foodborne microplastic particles. The health risks resulting from the use of face washes, hand cleansers, and toothpaste containing PE microplastic particles have been measured by the German Federal Institute for Risk Control. In females, oestrogen-mimicking chemicals may cause breast cancer. The high proportion of microplastic pollutants in seafood is a big concern for food protection as well as humans are vulnerable to MPs because of their lifestyle. A detailed analysis of the potential health impact of MPs, coming from a range of foods in the general diet, should be carried out to determine the causative impact of contaminated aquatic foods on humans (Sharma and Chatterjee, 2017). NPs have typically raised questions about potential hazards and risks to the atmosphere and human health. Comprehension of potential human risk is essential for the detection of health hazards. Adverse effects of in vivo nanoparticles include cytotoxicity, inflammation, or the formation of reactive oxygen species (ROS). Many in vitro experiments using human cell lines have shown that polymer nanoparticles can stimulate the innate immune system, causing inflammatory reactions, or mediating oxidative stress. In vitro and in vivo results have demonstrated that nano plastic can cause cell response activation, in particular effects on the immune system. However, there are no long-term experiments of repeated exposures, as well as extensive substance research, including plastics such as PE, PET, and PP, since most of the studies are done with PS. Also, it is important to bridge the information gap between the ambient concentrations and the concentrations present in organisms to carry out appropriate studies on the effect on human health (Lehner et al., 2019).

8.6 Remediation of MPs and NPs from the aquatic system

8.6.1 Control of sources

In addition to other remediation methods such as physicochemical treatment, eco-friendly and green approaches, concentrate should be given on reducing the release of MPs and NPs in the environment. Recycling and management practices for plastic pollution control are not good enough. The release of plastic into the atmosphere is still unchecked and plastic disposal is still done improperly internationally or unintentionally. An estimate indicates that more than 80% of marine plastics come from land-based sources, such as stream and river-carried nanoparticles, coastal waste activities, bio-solid and compost applications, and excessive sewage disposal. Source control has been very important as plastics have been found in almost all media nowadays. As the large plastic particles are subjected to weathering and fragmentation until released into the atmosphere, they usually become dispersed in the media such as surface freshwater sediments, aquatic water, and the seabed, rivers, soil, and even in the atmosphere. The breakdown of the large plastic body into MPs and NPs results from natural forces such as mechanical forces of water, UV radiation, and biological metabolism. Marine-based sources include waste disposal from ships and fishing nets but many plastics come from our day to use. While research on marine plastics remains at the forefront, scientists have started to discuss the problem of NPs in other habitats (Wang et al., 2021).

8.6.2 Enhancement of removal efficiency

8.6.2.1 Physicochemical approach

Processes that are commonly used for the elimination of microplastic include coagulation, ozonation, rapid sand filter, skimming, sedimentation, dissolved air flotation, conventional activated sludge, and membrane bioreactor (MBR). The discussion includes the benefits and drawbacks of the processes as well as technical and economic aspects of each process are illustrated. MPs in the waste water could be eliminated in WWTPs by skimming, sedimentation, and tertiary filtration, but these processes are not designed for the removal of MPs. So, a large amount of tiny plastic particles in WWTPs could escape with the effluent and return to the receiving water system. In any full-scale WWTP, no particular treatment procedure has been implemented for the elimination of MPs and the MPs-targeted treatment technology is still at the preliminary testing level. If technology could be designed for household-scale MPs treatment to treat the waste water in the home, it is estimated that the total MPs in the waste water will be substantially decreased. For illustration, it is known that a high amount of MPs fiber is found in the effluent of washing machines. If techniques could be developed to reduce fibers releasing from washing machines the overall MPs in the wastewater are expected to be significantly reduced (Sun et al., 2019).

8.6.2.1.1 Grit chamber/primary sedimentation

The first phase of the wastewater treatment plant is the Grit chamber and primary sedimentation. MPs are mostly removed by surface skimming and sedimentation. Removal of the tiny plastic particle becomes easier due to the aeration mechanism at the rear of the grit chamber. Indeed, this process removes approximately 41% of MPs. In an analysis, scientists found that the concentration of MPs in influent was 79.9 MPs/L, and in effluent 47.4 MPs/L. Almost 57-64% efficiency for MP removal has been obtained at this stage. Average MPs decreased from 15.7 MPs/L to 3.4 MPs/L after passing this stage. During the primary process of the municipal wastewater treatment plant in Spain, about 74% of MPs were removed. However, in the primary stage of the largest wastewater treatment plant in Canada, 92% MP removal efficiency was achieved. Much of the MPs had fibrous shapes. Around 99% of MPs removal at this primary level was also reported where input concentration was 57.6 MPs/L. The high performance obtained was due to the fibrous shape of more than 96% particle. Overall, this treatment was effective in removing MPs and the remaining particles can be removed in the later stages. However, for the complete elimination of MPs, it is a must to use appropriate technology in secondary and tertiary treatment (Bui et al., 2020).

8.6.2.1.2 Coagulation

At the beginning of tertiary treatment, the coagulation was done with the help of chemical coagulants (ferrous and aluminum salts). These coagulants destabilize the surface charge and create flocks of MPs and other pollutants in wastewater and finally removed them by settling or skimming. Hidayaturrahman et al. studied the removal of MPs by coagulation by using polyaluminum chloride (PAC) as coagulant on influent with different MP concentrations (Hidayaturrahman and Lee, 2019). Their study revealed that the creation of MPs flocs was dependent on the concentration of MPs. Indeed, fewer flocks will be formed in the water with a lower MPs concentration with a certain dose of coagulant, resulting in a lower MPs removal quality. The ability to eliminate MPs was positively associated with the dosage of coagulants. The removal rate of MPs would appear to decline when the flocculant dose tended to increase. This is explained by the fact that when the coagulant dose increased excessively, the zeta potential of MPs reduced, resulting in trouble in creating flocks. Furthermore, the efficacy of the process of coagulation depends on the type of coagulant used. In an analysis both aluminum and ferric coagulants were concurrently investigated with the presence of polyethylene (PE), which was frequently found in higher percentage than other MPs. Aluminum coagulant was found more efficient in the removal of PE than the ferric coagulant (Bui et al., 2020).

8.6.2.1.3 Sand filter

In a sand filter tiny plastic particle may be removed by trapping in the sand grains or adhering to the surface of sand grains. Research on urban wastewater treatment in Italy showed that sand filtering and disinfection eliminated about 56% of MPs. The sand filtration device was used for water treatment in China for the elimination of MPs. The performance of sand filter removal by MPs was low (29%-44%). The findings showed that compared to the fragment type MPs, the number of sphere/pellet and fiber forms

MPs was much more retained, namely 31%-49% for pellets, 24%-51% for fibers, and 19%-28% for fragments. Thus, from the view of these scholars, this traditional filtration technology was not proposed as the primary treatment method for eliminating MPs (Bui et al., 2020).

8.6.2.1.4 Rapid sand filter

Many separate media layers make up the Rapid Sand Filter (RSF). Usually, three layers are consisting of grains of anthracite, silica powder, and gravel but the RSF may contain only sand sometimes. For example, a study by Hidayaturrahman and Lee used sand filters with a depth of 6.8 m, a sand grain size of 0.8-1.2 mm, and a hydraulic residence time of approximately 1.08 h (Hidayaturrahman and Lee, 2019). Around 74% of MPs were removed after passing through RSF where the concentration of MPs in the influent was 215 MPs/L. The efficiency of the removal of MPs by RSF containing 1 m of gravel and 0.5 m of quartz at the sewage treatment plant in Finland was assessed by Talvitie et al. (Bui et al., 2020). 97% elimination performance was obtained from 0.7 ± 0.1 MPs/L down to 0.02 ± 0.007 MPs/L influent. So. The sand filter was then deemed an appropriate technology for the removal of MPs for the low MPs concentration level.

8.6.2.1.5 Granular activated carbon filtration

In the aqueous environment, granular activated carbon (GAC) filtration has been used to handle many pollutants. Wang et al. tested MPs elimination capacity of the granular activated carbon filtration device in a drinking water treatment plant (Wang, Lin and Chen, 2020). 60.9% removal efficiency was found by this technology which was less than other traditional technologies such as sand filtration, RSF, coagulation, and ozonation. Moreover, in contrast to PP and PAM, PE is mostly accounted for the bulk of MPs removed. Contaminants are extracted in the GAC filtration by a combination of biodegradation and physical adsorption. However, the process for the removal of MPs from the GAC is still elusive so far. In the future, the GAC filtration mechanism will be an efficient technology for eliminating MPs, especially at low MP concentration ranges. It will be useful to analyze further surveys on cost-benefit analysis of distinct filtration rates and forms of filtration media (Bui et al., 2020).

8.6.2.1.6 Membrane disc-filter

For tertiary treatment disc-filter (DF) is used which is made up of large pore fiber membranes (10–20 μm). In Daegu, South Korea a group of scientists used DF with a pore size of 10 μm to eliminate MPs. The findings revealed that the DF system eliminated roughly 79% of MPs where influent concentration was in the range of 1444 MPs/L to 297 MPs/L. Talvitie et al. have observed that the DF decreased the concentration of MPs from 0.5 ± 0.2 to 0.3 ± 0.1 MPs/L (40%) with filters of 10 μm pore size and from

2.0 ± 1.3 to 0.03 ± 0.01 MPs/L (98.5%) with filters of 20 μm pore size. For greater removal performance, smaller-sized filters were usually expected. In this study opposite outcome was due to the disruption of previous stages of treatment which affected the sampling time. The DF demonstrated relatively poor effectiveness in the elimination of MPs compared to the literature analysis. This may be due to a large number of MPs adhering to the membrane surface, membrane fouling happened, the high-pressure backwashing procedure was applied to clean the disc filter and thereby allowed the MPs to flow through the membrane unintentionally. The secondary membrane layer was lost after backwashing, so MPs could be easily transferred during this early filtration stage (Bui et al., 2020).

8.6.2.1.7 Conventional activated sludge process

A common wastewater treatment technology based on biodegradation by activated sludge, which is then separated by a sedimentation tank, is the conventional activated sludge process (CASP). CASP has been used for many types of waste water to treat soluble/colloidal organic compounds and nutrients. Once the presence of MPs in surface water and waste water has been generally considered, the efficiency of CASP in the removal of MPs has also been evaluated. In this process, MPs could bind to suspended solids and be removed by settling and it could play a role as moving media for attachment growth. In the conventional activated sludge process, the removal efficiency of MPs was 98%. Hidayaturrahman and Lee found that the removal efficiency of MPs ranged from 42% to 77% (Hidayaturrahman and Lee, 2019). In the municipal waste water treatment plant (Spain), CASP has abolished 62% of MPs. Analysis of urban wastewater treatment schemes in Italy showed that the grid chamber and the CASP system extracted roughly 64% of MPs. In the anaerobic/anoxic/oxic (AAO) process, approximately 17% of MPs have been removed and converted into surplus sludge. Overall, the performance of CASP for MPs removal was not stable and differed widely. The leaching bisphenol A (BPA) from PVC MPs releases toxicity to inhibit the activity of CASP heterotrophic and nitrifying bacteria. Furthermore, the capacity to decompose MPs in CASP has not been presented in several tests. While the installation cost is low, major drawbacks of this technology are consuming an area and generating a lot of waste sludge (Bui et al., 2020).

8.6.2.1.8 Membrane bioreactor

Basically, the membrane bioreactor (MBR) is a combination of biological method and membrane separation technology. This technology is effective in the treatment and reuse of wastewater compared to the traditional activated sludge process (CASP). MBR technology also saves the area and cuts down the production of sludge and could be scaled up easily. Therefore, in the treatment of waste water containing antibiotics, pesticides, personal care products, pharmaceuticals, etc. as contaminants, MBR has been commonly known and effectively used. The cumulative information indicates that the MPs particle in surface water was greater than 300 μm. The MBR with the microfiltration membrane modules is very efficient to fully eliminate the particle of this size range. In recent years, MBR has been studied for the role of tiny plastic particle removal. Talvitie et al. used MBR with a pore size of 0.4 μm, including 20 submerged flat-sheet UF membranes. In this process MP concentration of influent was 6.9 ± 0.1 MPs/L (Talvitie et al., 2017). The findings revealed that when waste water goes through the MBR system, the majority of MPs were removed. MBR is highly efficient and stable for the reduction of MPs. Studies suggest that MBR is the most promising removal technology for the elimination of MPs. Besides, a potential analysis could explore the impact of MPs on membrane fouling. In future research, the deterioration and/or transformation of MPs in MBR should also be studied (Bui et al., 2020).

8.6.2.2 Chemical methods to treat MPs

To eliminate the polyethylene MPs in a stirred-tank batch reactor, researchers have used a robust and environmentally compatible electrocoagulation technique that enables sludge minimization, energy conservation, cost-effectiveness, and automation versatility. The reaction of metal ions such as Fe^{2+} and Al^{3+} emitted from sacrificial electrodes in a water stream with hydroxyl media anions initiates the in-situ production of metal hydroxide coagulants. The coagulants break up the colloids and stabilize the surface charges of the suspended microparticles, which enables the particles to get close enough to each other to interact through Van Der Waals force. Simultaneously, the coagulants form a sludge blanket to capture the suspended MPs in the sample of waste water. For both studies utilizing planned electrocoagulation, the findings suggest a removal efficiency greater than 90%. With a pH of 7.5 and a NaCl concentration of 0-2 g/L, the highest removal efficiency, 99.24%, was achieved. Moreover, to achieve the maximum removal rate, the lowest tested current density of 11 A/m^2, which is the best in terms of energy consumption, was the most efficient. Herbort et al. proposed agglomeration by sol-gel reactions based on alkoxy-silyl bond forming as a new sustainable MPs removal method derived from inert textile and cosmetic products (Herbort et al., 2018). Initially, functionalized molecular precursors were synthesized in an inert environment and then used for the formation of bio-inspired alkoxy-silyl. Micro-plastics, meanwhile, adhere together to create large three-dimensional agglomerates that can then be separated using cost-effective filtration techniques. The sol-gel produced in this manner is similar to hybrid organic-inorganic silica gels with a wide range of sensor and optical materials, medication, and corrosion resistance benefits and uses. The degradation processes for MPs are not completely understood. For breaking into components, microorganisms, catalysts, and photoactive materials, macroplastics have been studied, but there are not enough publications on microplastic degradation. The long-term aging behavior of polystyrene and polyethylene MPs processed in the marine environment by integrating heat-activated persulfate and Fenton methods was also studied. According to the results, the O/C ratio, the overall size of the MPs, and surface properties play a role in the ability to adsorb, which substantially impacts the rate of microplastic oxidation. A study of the structure and degradative abilities of the TiO_2-based micro & nanodevices for the photocatalytic treatment of MPs have been done. Bearing high capacity for water purification, remediation of environmental pollutions, and therapeutic usages, photo-active micromotors have drawn huge interest. The use of an Au-decorated TiO_2-based micromotor for the effective photocatalytic removal of polystyrene MPs from waste-water under UV exposure was reported. The micromotor propulsion is supplied by photochemical reactions in water and hydrogen peroxide initiated by electron-hole generation processes. Lack of selectivity was the most important disadvantage of this system besides this the need for low concentrations of H_2O_2 for promoting the phoretic activity makes this system useless in wastewater treatment (Padervand et al., 2020).

8.6.2.3 *Ecofriendly and green approach*

Microbial remediation is a practical, clean, affordable, and the most promising solution to degrade MNPs via a green route among all the existing methods. Several biochemical reaction pathways are involved in microbial plastic degradation. Mechanisms of degradation differ depending upon the chemical structure of MNPs. As microbial remediation processes depend on the factors like temperature, pH, oxidative stress, it is easy to bring a reduction in the plastic pollutants. Complete removal of MPs and NPs can be expected by using the respective carbon content as energy sources for the growth of microbes with the help of technological advancement. Microorganisms associated with the debasement/size decrease of plastic particles can likewise amass MNPs. The breakdown of MNPs can be clearly understood by examining changes in their physicochemical properties. For instance, by recognizing degradation results of plastics and comparing cell reactions. The normal job of microorganisms can be perceived through biodeterioration, bio fracture, osmosis, agglomeration, biosorption, and mineralization of MNPs (Tiwari, Santhiya and Sharma, 2020).

8.6.2.3.1 Adsorption on green algae

The existence of MPs and NPs is considered more harmful than other toxins in aquatic ecosystems owing to their adverse impacts on species like fish, mammals, marine birds, and reptiles. There persistency and non-degradability demand for elimination strategies. MPs are usually known as persistent materials, but based on their existence and chemical composition, they decay more and less. Plastic with a half-life less than the values defined in terms of REACH criteria for persistency is considered degradable. MPs and NPs adsorb a variety of contaminants on its surface, carry them and release them into other places. Their large surface area/volume ratio increases the chance of adsorption for most pollutants. The adherence properties of fluorescent MPs on the surface of marine microalgae called *Fucus vesiculosus* was observed. The size of the polystyrene MPs was

~20 μm in diameter, while the sorbent plant cells contained very small microchannels to limit the translocation of polystyrene MPs into the tissues. The findings showed a strong sorption rate of MPs (~94.5%), which is clarified by the position of alginate compounds released from cell walls in the cut regions. Indeed, alginate can enhance the binding of polystyrene particles to the surfaces of the seaweed due to the gelatinous properties of this anionic polysaccharide material. After studying the adsorption of 20-500 nm polystyrene particles on *Pseudokirchneriella subcapitata*, it was found that positively charged polystyrene MPs are more easily adsorbed on the surface of the algae than those with negative charges. Overall, the sorting of MPs on the surface of the algae greatly relies on the surface charge of the particles (Padervand et al., 2020).

8.6.2.3.2 Enzyme technology

For the processing, separation, purification, and degradation of plastics enzyme technology can be explored. Many non-toxic and biodegradable enzymes can be used for the removal of micro and NPs. A few plastic polymer chains (PE, PP, PS, and PVC) have been exposed to degradation by a different group of enzymes in the last decade. Several enzymes like esterases, proteases, cutinase, and laccase have demonstrated positive effects on the degradation of plastics. Interestingly, a bacterium *Ideonella sakaiensisis* can use polyethylene terephthalate as its primary energy and carbon source. Using two enzymes (PETase and MHETase), this bacterium converts PET into terephthalic acid and ethylene glycol monomers. Although many enzymes are known and their plastic degradation behavior is understood, they hardly have a discussion. This is due to the lack of technology for improving the performance of these plastic degrading enzymes and the development of them. Recent research on the enzymatic degradation of plastics has created significant interest in improving enzyme function by protein/enzyme modification. Compared to wild-type strains, an engineered PETase mutant from *Ideonella sakaiensis* showed an improvement in the three mutants (R61A, L88F, and I179F) by 1.4-fold, 2.1-fold, and 2.5-fold. This research indicates that through logical protein engineering and by changing core hydrophobic grooves of substrate binding sites, enzyme activity can be increased. Interestingly, the role of protein-engineered enzymes in the degradation of MPs has been shown in a recent paper. The study notes that polyurethane microplastic degradation starts with polymer binding peptides, such as anchor peptides, which act as a synthetic polymer binding tool. The observed decrease in polyurethane shelf life from 41.8 to 6.2 h, roughly 6.7 fold, is attributed to the increased degradation efficiency of the polyurethane protein-engineered enzyme Tachystatin A2 (anchor peptide) relative to Tcur1278-3 WTT (wild type) (Tiwari, Santhiya and Sharma, 2020).

8.6.2.3.3 Metagenomics

In the investigation of "hidden" genetic characteristics, metagenomics has played a promising role. The development of biotechnological applications for polymer degradation is assisted by the discovery of new DNA sequences, enzymes, metabolic mechanisms, and metabolically active molecules with better metabolic functions. It is possible to split the identification of novel genes and DNA sequences into sequence-based and functional-based screening approaches. The functional-based screening methods can reveal these novel functions of identified enzymes. To produce a metabolic process for the elimination of polyethylene waste, the discovery of new microbial polyester hydrolases and construction of an assembly of metabolically active enzymes remains a major concern. A study on shotgun metagenomic sequencing of biofilms fouling plastic and bioplastic microcosms has disclosed that plastic biofilms are influenced by sulfate-reducing microorganisms. Study of bio-plastic disclosed improvement in activity of adenylyl sulfate reductase (aprBA), esterase, depolymerase, and dissimilatory sulfite reductase (dsrAB). Novel species of *Desulfovibrio, Desulfobacteraceae, and Desulfobulbaceae* have been recognized by metagenomic genome reconstruction among the huge SRMs, and these genomes bear genes that are important for both bioplastic degradation and sulfate reduction. These findings indicate that bioplastics will facilitate a remarkable change in the benthic microbial diversity and genomes, preferring microbes that are highly active in bioplastic degradation and sulfate reduction (Tiwari, Santhiya and Sharma, 2020).

8.6.2.3.4 Nano technology

The principal difference between plastics and MPs is nothing but a specific size range. Micro & NPs either generated from different sources or are produced from large plastic pieces because of biotic and abiotic factors. Step by step increment of MNP poisons in the environment and the comparing toxicological effect on different life forms alerts to restrict plastic production and utilization. A superior method to lessen MNPs in the biological system is to produce and use eco-friendly plastics to quicken their degradation. Nanotechnology could be a promising field to investigate appropriate techniques to control the release of MPs & NPs in the environment. There are several methodologies by which nanotechnology can contribute to improving the efficiency of the plastic breakdown process. The biodegradation of plastics is improved by the addition of nanoparticles in microbial cultures. The presence of SiO2 nanoparticles at various concentrations has been reported to affect the growth of plastic-degrading bacteria. Usually, Fullerene 60 nanoparticles are known to be poisonous at higher doses but it indicates an improvement in the plastic deterioration mechanism when used at a reduced concentration. The presence of nanoparticles in the bacterial culture media influences their growth thus biodegradation productivity of a potential polymer-degrading microbial consortium increases significantly. These investigations propose the significance of nanoparticles and bacterial associations. Investigating other nanoparticles for their capacity to advance the microbial remediation cycle can help in the commercialization of plastic degradation (Tiwari, Santhiya and Sharma, 2020).

8.7 Conclusion

Plastic pollution along with climate change and other manmade actions are hampering the marine biodiversity though most of the outcomes considered from plastic pollution are hypothetical. To understand the actual threat long term research on the route of ingestion, bioaccumulation, metabolism and biomagnification should be conducted. Evidences say that MPs are entering into the marine food chain. Thus, it can be easily assumed that human being is also susceptible to micro and NPs. Beside this MNPs are used in different daily care products like cosmetics, toothpastes etc. which are considered as the primary sources of MPs. Though many first world nations already banned the use of MNPs in any kind of products, it isn't stopped worldwide. If the primary sources can be eliminated completely the MNP pollution will be cut down significantly. Like primary MNPs, some restrictions are also given on the use of plastic products especially on the one-time-use products which is considered to be the major secondary source. One of the most important legislations is the 1978 protocol for the prevention of pollution from marine vessels. It was considered that those water vehicles possess remarkable and controllable number of hazards for marine environment.

Ingestion and entanglement are the two major routes of plastic pollution for aquatic lives. Ingested macroplastics and MPs are very detrimental to the health of marine organisms. Though it doesn't cause immediate mortality, sub-lethal and chronic effects have long term consequences. Ingested MPs sometimes causes the decrease of food consumption. When MPs are uptaken by the sea birds these plastic fragments are accumulated in the gastrointestinal tract and result in the blockage there. This ultimately affects the feeding stimuli and the level of activity. Several fish species of different aquatic environments are found to be ingested with macro and MPs. Into 504 fish species of English Channel high presence of synthetic plastic was observed and 92.4% of them were categorized as microplastic. Beside fish and seabirds, sea turtles are also highly susceptible to microplastic pollution. From 1985 to 2012 ingestion of plastic had been observed to be increased by 20%. Though the harm possessed by entanglement draws less attention it is actually more hazardous than ingestion. Mammals, sea turtles, sea birds and crustaceans are very vulnerable to the entanglement and this can lead to death. Drowning, suffocation, reduced activity including predation and the probability of getting caught are the outcomes of entanglement process. The lost or the intentional disposal of fishing materials, mostly composed of plastic, are the main reason of entanglement (LI, TSE and FOK, 2016).

Decomposition in the natural environment or biodegradation of plastic is dependent on the nature of the material. Most of the high molecular weight plastics are resistant to natural decomposition. As a result, the smaller fragments of these macromolecules can also persist for a long time. Recently scientists are trying their best to produce biodegradable plastics which will be easily degraded in the natural environment. Photodegradation is an effective way of plastic fragmentation by natural process. Ultraviolet light from sun causes the incorporation of oxygen into plastic which ultimate helps the light induced degradation mechanism. This process doesn't work in the ocean that much because the temperature cannot get so high but high water pressure facilitate plastic degradation inside the ocean. High temperature and humidity are found to accelerate the process of differentiation of plastic debris.

Ubiquity of microplastic makes the situation worst. It is almost impossible to remove plastic from marine ecosystem. Beside rules and regulations, investment for the development of plastic collection technologies should be increased. Manta trawler, a specific net usually placed at the back part of ocean vehicles, is used for skimming out the floating debris from water bodies. Then comes the plastic eating drones which can also capable of removing plastic from ocean. Solid floating boom can also be used to collect plastic from the ocean. To protect the aquatic animals from being entangled in the net sonic transmitter can be utilized. It was estimated that 7.25 million ton of plastic can be removed from the ocean (LI, TSE and FOK, 2016).

Increasing public awareness is a key to reduce the MNP pollution of ocean. If the consumption behavior can be changed by making people aware of the hazards of plastic pollution, a huge change can be expected. It should be added in the agendas of different international conferences related to this field. Campaigns on microplastic pollution should be carried out globally as most of the people are not concerned about that. Plastic industries should be pressurized to modify their products and manufacture biodegradable plastic. Starch or pullulan like materials can be utilized for that purpose. As tertiary recycling is well known now, conversion of larger molecules to the smaller molecules can help with the reduction of plastic wastes as these processed products can be utilized as feedstock for new material production.

References

Andrady, A.L., 2003. Plastics and the Environment. John Wiley & Sons.

Andrady, A.L., 2011. Microplastics in the marine environment. Mar. Pollut. Bull. 62 (8), 1596–1605.

Arias-Andres, M., Rojas-Jimenez, K., Grossart, H.P., 2019. Collateral effects of microplastic pollution on aquatic microorganisms: an ecological perspective. TrAC - Trends in Analytical Chemistry, Elsevier B.V., 234–240. doi:10.1016/j.trac.2018.11.041.

Bhagat, J., et al., 2020. Zebrafish: an emerging model to study microplastic and nanoplastic toxicity. Sci. Total Environ. 728, 138707. doi:10.1016/j.scitotenv.2020.138707.

Bui, X.T., et al., 2020. Microplastics pollution in wastewater: characteristics, occurrence and removal technologies. Environmental Technology and Innovation. Elsevier B.V. doi:10.1016/j.eti.2020.101013.

Castañeda, R.A., et al., 2014. Microplastic pollution in St. Lawrence River sediments. Can. J. Fish. Aquat. Sci. 71 (12), 1767–1771. doi:10.1139/cjfas-2014-0281.

Chae, Y., An, Y.J., 2017. Effects of micro- and nanoplastics on aquatic ecosystems: current research trends and perspectives. Mar. Pollut. Bull. 124 (2), 624–632. doi:10.1016/j.marpolbul.2017.01.070.

Cho, H., et al., 2013. Improved atomic force microscope infrared spectroscopy for rapid nanometer-scale chemical identification. Nanotechnology 24 (44), 444007. doi:10.1088/0957-4484/24/44/444007.

Cole, M., et al., 2013. Microplastic Ingestion by Zooplankton. Environ. Sci. Technol. 47 (12), 6646–6655. doi:10.1021/es400663f.

Collard, F., et al., 2015. Detection of Anthropogenic Particles in Fish Stomachs: an Isolation Method Adapted to Identification by Raman Spectroscopy. Arch. Environ. Contam. Toxicol. 69 (3), 331–339. doi:10.1007/s00244-015-0221-0.

Cooper, D.A., Corcoran, P.L., 2010. Effects of mechanical and chemical processes on the degradation of plastic beach debris on the island of Kauai, Hawaii. Mar. Pollut. Bull. 60 (5), 650–654. https://doi.org/10.1016/j.marpolbul.2009.12.026.

Dümichen, E., et al., 2015. Analysis of polyethylene microplastics in environmental samples, using a thermal decomposition method. Water Res. 85, 451–457. doi:10.1016/j.watres.2015.09.002.

Dümichen, E., et al., 2017. Fast identification of microplastics in complex environmental samples by a thermal degradation method. Chemosphere 174, 572–584. https://doi.org/10.1016/j.chemosphere.2017.02.010.

Fok, L., Cheung, P.K., 2015. Hong Kong at the Pearl River Estuary: a hotspot of microplastic pollution. Mar. Pollut. Bull. 99 (1–2), 112–118.

Fu, W., et al., 2020. Separation, characterization and identification of microplastics and nanoplastics in the environment. Sci. Total Environ. 721, 137561. doi:10.1016/j.scitotenv.2020.137561.

Garrity, S.D., Levings, S.C., 1993. Marine debris along the Caribbean coast of Panama. Mar. Pollut. Bull. 26 (6), 317–324.

Gregory, M.R., 1978. Accumulation and distribution of virgin plastic granules on New Zealand beaches. N.Z. J. Mar. Freshwater Res. 12 (4), 399–414.

Guo, J.J., et al., 2020. Source, migration and toxicology of microplastics in soil. Environ. Int. 137 October 2019, 105263. doi:10.1016/j.envint.2019.105263.

Harrison, J.P. et al. (no date) P A P E R Interactions Between Microorganisms and Marine Microplastics: a Call for Research.

Herbort, A.F., et al., 2018. Alkoxy-silyl induced agglomeration: a new approach for the sustainable removal of microplastic from aquatic systems. J. Polym. Environ. 26 (11), 4258–4270.

Hidayaturrahman, H., Lee, T.-.G., 2019. A study on characteristics of microplastic in wastewater of South Korea: identification, quantification, and fate of microplastics during treatment process. Mar. Pollut. Bull. 146, 696–702.

Hoellein, T.J., et al., 2019. Microplastic deposition velocity in streams follows patterns for naturally occurring allochthonous particles. Sci. Rep. 9 (1), 1–11.

Hou, J., et al., 2018. Nanoparticle tracking analysis versus dynamic light scattering: case study on the effect of Ca2+ and alginate on the aggregation of cerium oxide nanoparticles. J. Hazard. Mater. 360, 319–328. https://doi.org/10.1016/j.jhazmat.2018.08.010.

Huang, D., et al., 2020. Microplastics and nanoplastics in the environment: macroscopic transport and effects on creatures. J. Hazard. Mater. Elsevier B.V. doi:10.1016/j.jhazmat.2020.124399.

Hung, P.Y., et al., 2015. Potential application of tip-enhanced Raman spectroscopy (TERS) in semiconductor manufacturing, in Proc.SPIE. doi:10.1117/12.2175623.

Hung, P.Y., et al., 2015. Potential application of tip-enhanced Raman spectroscopy (TERS) in semiconductor manufacturing, Metrology, Inspection, and Process Control for Microlithography XXIX, 9424 (March 2015), 94241S. doi:10.1117/12.2175623.

Huppertsberg, S., Knepper, T.P., 2018. Instrumental analysis of microplastics—Benefits and challenges. Anal. Bioanal. Chem. 410 (25), 6343–6352. doi:10.1007/s00216-018-1210-8.

James, A.E., Driskell, J.D., 2013. Monitoring gold nanoparticle conjugation and analysis of biomolecular binding with nanoparticle tracking analysis (NTA) and dynamic light scattering (DLS). Analyst 138 (4), 1212–1218. doi:10.1039/C2AN36467K.

Kögel, T., et al., 2020. Micro-and nanoplastic toxicity on aquatic life: determining factors. Sci. Total Environ. 709, 136050.

Last, J.A., et al., 2010. The Applications of Atomic Force Microscopy to Vision Science. Invest. Ophthalmol. Vis. Sci. 51 (12), 6083–6094. doi:10.1167/iovs.10-5470.

Law, K.L., et al., 2010. Plastic accumulation in the North Atlantic subtropical gyre. Science 329 (5996), 1185–1188.

Lehner, R., et al., 2019. Emergence of Nanoplastic in the Environment and Possible Impact on Human Health. Environ. Sci. Technol. American Chemical Society. doi:10.1021/acs.est.8b05512.

Li, W.C., Tse, H.F., Fok, L., 2016. Plastic waste in the marine environment: a review of sources, occurrence and effects. Sci. Total Environ. 566–567, 333–349. https://doi.org/10.1016/j.scitotenv.2016.05.084.

Lise Nerland, I. et al., 2014.'Norwegian Institute for Water Research Negative environmental impact', in, p. 55.
Liu, K., et al., 2020. Terrestrial plants as a potential temporary sink of atmospheric microplastics during transport. Sci. Total Environ. 742, 140523.
Majewsky, M., et al., 2016. Determination of microplastic polyethylene (PE) and polypropylene (PP) in environmental samples using thermal analysis (TGA-DSC). Sci. Total Environ. 568, 507–511. doi:10.1016/j.scitotenv.2016.06.017.
Mao, Y., et al., 2020. Nanoplastics display strong stability in aqueous environments: insights from aggregation behaviour and theoretical calculations. Environ. Pollut. 258, 113760.
McDermid, K.J., McMullen, T.L., 2004. Quantitative analysis of small-plastic debris on beaches in the Hawaiian Archipelago. Mar. Pollut. Bull. 48 (7–8), 790–794. doi:10.1016/j.marpolbul.2003.10.017.
Murphy, F., et al., 2016. Wastewater treatment works (WwTW) as a source of microplastics in the aquatic environment. Environ. Sci. Technol. 50 (11), 5800–5808.
Nel, H.A., Froneman, P.W., 2015. A quantitative analysis of microplastic pollution along the south-eastern coastline of South Africa. Mar. Pollut. Bull. 101 (1), 274–279. doi:10.1016/j.marpolbul.2015.09.043.
Padervand, M., et al., 2020. Removal of microplastics from the environment. A review. Environ. Chem. Lett., 807–828. doi:10.1007/s10311-020-00983-1.
Revel, M., Châtel, A., Mouneyrac, C., 2018. Micro(nano)plastics: a threat to human health?. Current Opinion in Environmental Science and Health, Elsevier B.V., 17–23. doi:10.1016/j.coesh.2017.10.003.
Ritchie, H., Roser, M., 2018. Plastic pollution. Our World in Data.
Ryan, P.G., Moloney, C.L., 1990. Plastic and other artefacts on South African beaches: temporal trends in abundance and composition. S. AFR. J. SCI./S.-AFR. TYDSKR. WET. 86 (7), 450–452.
de Sá, L.C., et al., 2018. Studies of the effects of microplastics on aquatic organisms: what do we know and where should we focus our efforts in the future? Sci. Total Environ. 1029–1039. doi:10.1016/j.scitotenv.2018.07.207.
Sana, S.S., et al., 2020. Effects of microplastics and nanoplastics on marine environment and human health. Environmental Science and Pollution Research 27 (36), 44743–44756. doi:10.1007/s11356-020-10573-x.
Serranti, S., et al., 2018. Characterization of microplastic litter from oceans by an innovative approach based on hyperspectral imaging. Waste Manage. (Oxford) 76, 117–125. https://doi.org/10.1016/j.wasman.2018.03.003.
Sharma, S., Chatterjee, S., 2017. Microplastic pollution, a threat to marine ecosystem and human health: a short review. Environmental Science and Pollution Research 24 (27), 21530–21547. doi:10.1007/s11356-017-9910-8.
Shim, W.J., Hong, S.H., Eo, S.E., 2017. Identification methods in microplastic analysis: a review. Anal. Methods 9 (9), 1384–1391. doi:10.1039/c6ay02558g.
Song, Y.K., et al., 2015. A comparison of microscopic and spectroscopic identification methods for analysis of microplastics in environmental samples. Mar. Pollut. Bull. 93 (1–2), 202–209. doi:10.1016/j.marpolbul.2015.01.015.
Stuart, B., 2015. Infrared SpectroscopyKirk-Othmer Encyclopedia of Chemical Technology. Major Reference Works, pp. 1–18. https://doi.org/10.1002/0471238961.0914061810151405.a01.pub3.
Sun, J., et al., 2019. Microplastics in wastewater treatment plants: detection, occurrence and removal. Water Res., 21–37. doi:10.1016/j.watres.2018.12.050.
Talvitie, J., et al., 2017. Solutions to microplastic pollution–Removal of microplastics from wastewater effluent with advanced wastewater treatment technologies. Water Res. 123, 401–407.
Thiel, M., et al., 2013. Anthropogenic marine debris in the coastal environment: a multi-year comparison between coastal waters and local shores. Mar. Pollut. Bull. 71 (1–2), 307–316.
Tiwari, N., Santhiya, D., Sharma, J.G., 2020. Microbial remediation of micro-nano plastics: current knowledge and future trends. Environ. Pollut. 265, 115044. doi:10.1016/j.envpol.2020.115044.
Vianello, A., et al., 2013. Microplastic particles in sediments of Lagoon of Venice, Italy: first observations on occurrence, spatial patterns and identification. Estuarine Coastal Shelf Sci. 130, 54–61. https://doi.org/10.1016/j.ecss.2013.03.022.
Wang, L., et al., 2017. A Simple Method for Quantifying Polycarbonate and Polyethylene Terephthalate Microplastics in Environmental Samples by Liquid Chromatography–Tandem Mass Spectrometry. Environ Sci Technol Lett 4 (12), 530–534. doi:10.1021/acs.estlett.7b00454.

Wang, L., et al., 2021. Environmental fate, toxicity and risk management strategies of nanoplastics in the environment: current status and future perspectives. J. Hazard. Mater. 401. doi:10.1016/j.jhazmat.2020.123415.
Wang, W., et al., 2019. The ecotoxicological effects of microplastics on aquatic food web, from primary producer to human: a review. Ecotoxicol. Environ. Saf. 173 (November 2018), 110–117. doi:10.1016/j.ecoenv.2019.01.113.
Wang, Z., Lin, T., Chen, W., 2020. Occurrence and removal of microplastics in an advanced drinking water treatment plant (ADWTP). Sci. Total Environ. 700, 134520.
Wright, S.L., Thompson, R.C., Galloway, T.S., 2013. The physical impacts of microplastics on marine organisms: a review. Environ. Pollut. 178, 483–492.
Xu, S., et al., 2020. Microplastics in aquatic environments: occurrence, accumulation, and biological effects. Sci. Total Environ. 703, 134699. doi:10.1016/j.scitotenv.2019.134699.
Yguerabide, J., Yguerabide, E.E., 1998. Light-Scattering Submicroscopic Particles as Highly Fluorescent Analogs and Their Use as Tracer Labels in Clinical and Biological Applications: II. Experimental Characterization. Anal. Biochem. 262 (2), 157–176. https://doi.org/10.1006/abio.1998.2760.
Zettler, E.R., Mincer, T.J., Amaral-Zettler, L.A., 2013. Life in the "Plastisphere": microbial Communities on Plastic Marine Debris. Environ. Sci. Technol. 47 (13), 7137–7146. doi:10.1021/es401288x.

CHAPTER 9

Endocrine-disrupting pollutants in domestic and industrial wastewater: occurrence and removal by advanced treatment system for wastewater reuse

P Snega Priya[a], M Kamaraj[b], J Aravind[c], S Sudhakar[d]
[a]Department of Medical Microbiology, SRM Medical College Hospital and Research Center, Kattankulathur, Tamil Nadu, India
[b]Department of Biotechnology, College of Biological and Chemical Engineering, Addis Ababa Science and Technology University, Addis Ababa, Ethiopia
[c]Dhirajlal Gandhi College of Technology, Omalur, Tamil Nadu, India
[d]Department of Biotechnology, PGP College of Arts and Science, Namakkal, Tamil Nadu, India

9.1 Introduction

The importance of good-quality drinking water is inevitable for human existence, and life will be difficult without the availability of safe drinking water (Swaminathan et al., 2013). There is growing fear of water resource shortage and it is becoming an important topic as a severe scarcity of water has been seen all over the globe. Added to this, there is an emerging concern about the potentially harmful substances present in water bodies. These emerging micropollutants (MPs) have shown to be present in both industrial and domestic wastewater in unnoticeable quantities (ngL^{-1} - mgL^{-1} scales) (Abbas et al., 2019). A major group in such compounds is Endocrine Disrupting compounds (EDCs) that are capable to elicit ill effects on the endocrine systems of humans and wildlife. The endocrine system plays an important role in regulating various functions of the human body and maintains the relative stability of the internal environment (Marty et al., 2018). Any malfunction of the endocrine system shall cause serious health concerns in the human body. Since EDCs are capable to disrupt the endocrine system, they are considered as serious pollutants in water pollution. The US Environmental Protection Agency (USEPA) defines an EDC as: "An exogenous agent that interferes with the synthesis, secretion, transport, binding, action, or elimination of natural hormones in the body that are responsible for the maintenance of homeostasis, reproduction, development, and/or behavior" (USEPA, 1997). According to the World Health Organization (WHO) definition, an EDC is "an exogenous substance or mixture that can alter normal hormonal functions in humans and animals, and consequently affects the endocrine system of living organisms." Estrogen and endocrine-disrupting phenolic compounds

Biodegradation and Detoxification of Micropollutants in Industrial Wastewater
DOI: https://doi.org/10.1016/B978-0-323-88507-2.00006-3

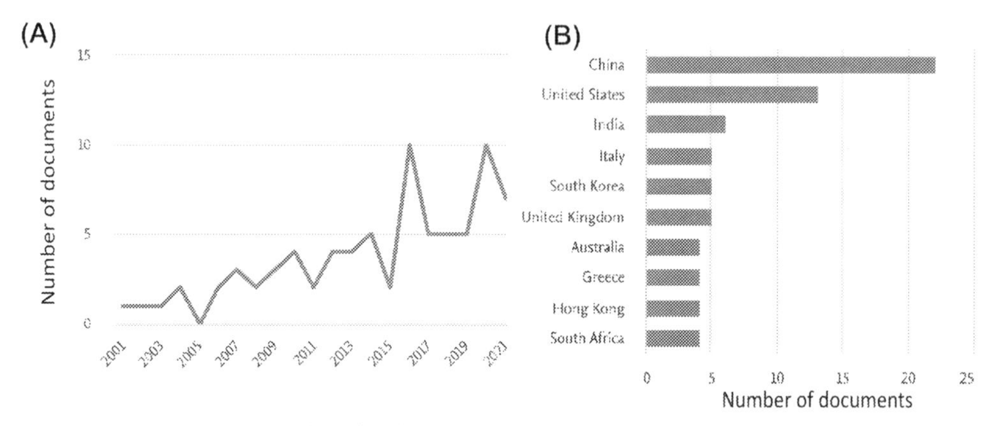

Fig. 9.1 ***Scopus database analysis for the research output in the keyword of "Endocrine-disrupting pollutants in wastewater and removal treatment" for the period of 2011-2021 (Scopus database 06/02/2021).*** (A) documents for year-wise, (B) documents by Top 10 country.

(Li et al., 2013) are the two major EDCs in the aquatic environment. Estrogen is found in sewage at low concentrations (ngL^{-1}) but has been reported to have a high activity of estrogen, including steroidal estrogens which are natural such as 17β estradiol (E2) and 17α-ethynylestradiol (EE2) which is a synthetic contraceptive (Ternes et al., 1999). The endocrine-disrupting phenolic compounds like nonylphenol (NP) and bisphenol A (BPA) have the characteristics of low estrogenic activity. Their concentration is relatively high in wastewater, reaching micrograms per liter (Zegura et al., 2017). The distribution of EDCs is very important to understand the source and to formulate the methods of removal of EDCs. Many research efforts to eradicate this problem, hence there are various studies are carried out in this aspects in the last decade (Fig. 9.1). The solution for the eradication or minimization of the problems associated with the EDCs is overcome by proper identification of EDCs in the environment, the accurate measurement of EDCs in water systems, and formulation of methods for removal of EDC with a high success rate.

9.2 Sources, fate, and interaction of EDCs with biota

Several natural and anthropogenic routes act as an entry point of EDCs to the environment which creates the interaction of EDCs with humans and wildlife. The chemical characteristics, exposure level, and hazards of EDCs toward the environment differ widely based on the nature of sources. The literature reports that ubiquitous occurrence of the wide variety of EDCs and their inadvertent transformation of by-products are found in most type of environmental media including soil and sediments, water (surface, ground, and marine water) with the analytical limits (usually ngL^{-1}) (Aris et al., 2020).

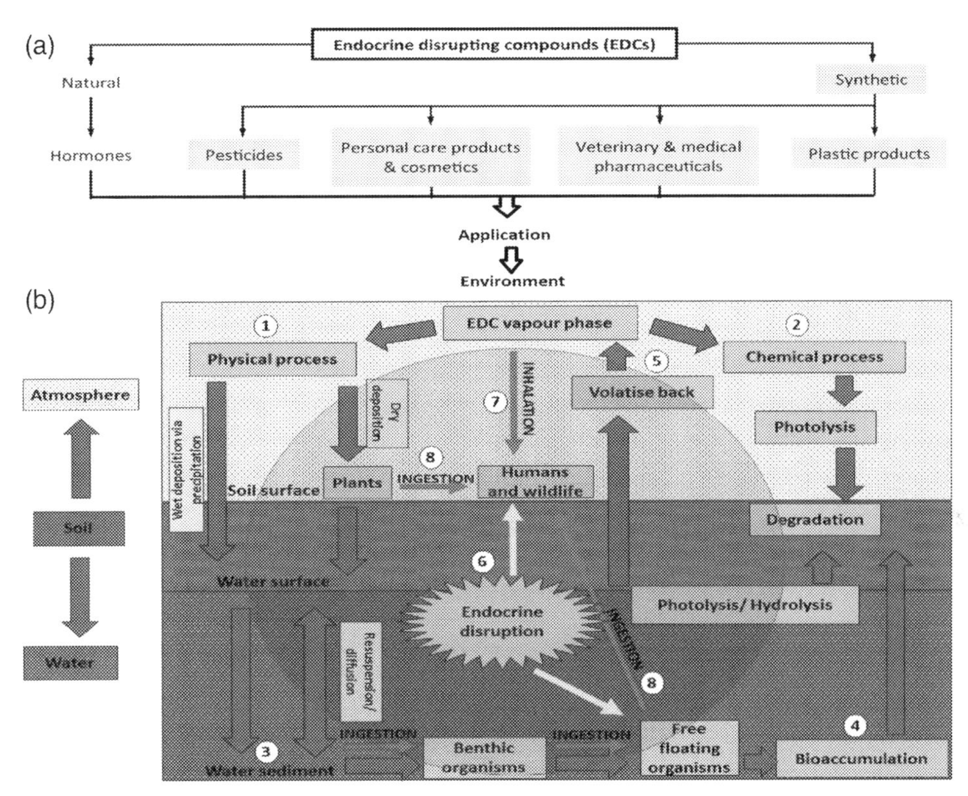

Fig. 9.2 Sources of EDCs (a) and Occurrence of EDCs in the environment and interaction with biota (b): From the atmosphere, EDCs in the vapor phase are transferred to soil surface either by physical process (1) or by chemical process (2). The physical process involves wet deposition and dry deposition while the chemical process involves photolysis. Both of the processes may ultimately lead to degradation or further transfer of EDC to water bodies (3) where resuspension or diffusion occurs; certain EDCs are adsorbed to the sediments. Bioaccumulation of persistent EDC among aquatic organisms occurs (4) and certain EDCs are volatilized back to the atmosphere (5). In this cycle, human and wildlife exposure is threatened to endocrine disruption (6) via inhalation of EDC from the atmosphere (7) and consumption of EDC deposited primary producers and bioaccumulated tissues of secondary consumers (8). *(Reproduced after modification with permission from Annamalai and Namasivayam V, 2015 Copyright © 2015 Published by Elsevier Ltd)*

Natural EDCs contain estrogens, androgens, and phytoestrogens that come from plants. The synthetic EDCs can be classified into diverse chemical groups such as phenolics, pesticides, polyhalogenated, bisphenol, phthalates, personal care, and pharmaceutical products, etc. Synthetic EDCs are frequently utilized for the manufacture of various day to day use products such as cleaners, detergents, adhesives, surfactants, lubricant, fragrances, fire retardants, additives, dietary intake, textiles, furniture, paints, plastics, disinfectants, electrical appliances, agricultural products, etc. (Omar et al., 2018). An overview of the source of EDCs to the environment is depicted in Fig. 9.2A.

In general, the EDCs enter the environments through the disposal of EDCs in an original state, metabolites, or byproducts either directly or indirectly. In the context of this, there are two major sources of EDCs occurrence in the environment are referred to as "nonpoint" and "point" sources. The points refer to the direct discharge of EDCs from a single source (i.e. Shaft, pipe, well, floating craft, discrete fracture, and intensive activities) of polluted water. The nonpoint refers to the entry of EDCs through multiple sources such as usage of chemicals in various waste management streams (i.e. Industrial effluents, Urban effluents, solid waste, animal waste, agricultural waste, and accidental leakage, etc.). Solid waste from urban and industries is a huge pile of waste that contributes to the occurrence of EDCs in the soil and water environment. The rainfall further leads to the leaching of these EDCs to the surrounding soil, surface water stream, and groundwater. The waste generated from the industrial waste which contains more volume of EDCs is considered as a major source of EDCs entrance to the soil and water system when it's released without proper waste remediation procedures. Besides, industrial waste is widely considered a major source of water and air pollution. The waste, which contains toxic chemicals, is released into the environment without appropriate procedures Furthermore, the domestic solid waste, sewage, and wastewater that are produced by household activities from the day to day usage of personal and pharmaceutical products containing the EDCs are also being a source for the EDCs occurrence in the environment (Aris et al., 2020).

The physicochemical properties and the nature of the environment where the EDCs occur will impact the fate and half-life of EDCs. The EDCs distributions in the environmental compartments are governed by three equilibrium portioning coefficients: air-water, water-octanol, and octanol-air. The EDCs mainly occur in the gaseous phase in the atmosphere, while a few sorbs to particles and a few sorbs onto suspended particles due to their semi-volatile nature. It can be transported by wet and dry gaseous vapor deposition, sorption, dissolution, volatilization, resuspension, sedimentation, and erosion in the environment. Both biotic and abiotic process induces the natural removal of the EDCs from the atmosphere (Fig. 9.2B). The biotic route involves microbial degradation in water, soil, and sediments; where the abiotic process involves photolysis (direct and indirect), hydrolysis, and reduction/oxidation reactions. EDCs that occur in the soil, sediment, and water mostly enter the atmosphere by volatilization when levels in the air are reduced. EDCs in the water separated into particles and dissolved phases are taken up by aquatic biota or dump to the bottom sediments and further transported back to the water stream via resuspension of diffusion. Certain EDCs enter the food chain during the environmental cycling and bioaccumulate in tissues through ingestion and inhalation causing endocrine disruption of the human and animal system (Annamalai and Namasivayam, 2015). A list of some EDCs along with their sources, type, and examples is given in Table 9.1.

Table 9.1 Selected sources, category (type), and examples of substances that have been reported as potential endocrine disruptors (data obtained from https://www.ccohs.ca/oshanswers/chemicals/endocrine.HTML).

Sources	Category	Substances
Agricultural runoff / Atmospheric transport	Organochlorine Pesticides (found in insecticides, many now phased out)	DDT, dieldrin, lindane
Agricultural runoff	Pesticides currently in use	Atrazine, trifluralin, permethrin
Municipal effluent Agricultural runoff	Natural Hormones (Produced naturally by animals); synthetic steroids (found in contraceptives)	Estradiol, estrone, and testosterone; ethynyl estradiol
Incineration, landfill	Polychlorinated Compounds (from industrial production or by-products of mostly banned substances)	Polychlorinated dioxins, polychlorinated biphenyls
Industrial and municipal effluents	Alkylphenols (Surfactants - certain kinds of detergents used for removing oil - and their metabolites)	Nonylphenol
Industrial effluent	Phthalates (found in plasticizers)	Dibutyl phthalate, butyl benzyl phthalate
Pulp mill effluents	Phytoestrogens (found in plant material)	Isoflavones, lignans, coumestans
Consumer products	Cosmetics, personal care products, cleaners, plastics	Parabens, phthalates, glycol ethers, fragrances, cyclosiloxanes, bisphenol A (BPA)
Harbors	Organotins (found in antifoulants used to paint the hulls of ships)	Tributyltin

9.3 Removal of EDCs via physical and chemical treatment

9.3.1 Physical treatment

9.3.1.1 Membrane technology

The basic types of membrane separation process include Pressure-driven and electrically driven. The pressure-driven filtration process involves Microfiltration (MF), Ultrafiltration (UF), Nanofiltration (NF), and Reverse osmosis (RO). These techniques use hydraulic pressure for forcing water into the membrane (Walha et al., 2007). The electrically driven membrane process involves the movement of ions across the membrane is achieved by passing electric current. It leaves the purified water behind the membrane. The Membrane treatment is very helpful in the achievement of a higher quality. They are widely used for the treatment of solids and hard substances in the aquatic system. Many research works suggest that the membrane process can be widely used for the removal of EDCs. To add on the processes like NF and RO can remove Phytoestrogens (70–93%) and surfactants (92–99%) from the wastewater released from industries. The agricultural waste products like herbicides, pesticides, etc. in water bodies can be efficiently removed by NF or MF (Bodzek, 2015). The high removal efficacy of the RO and NF process was also noted in the removal of phthalates which is 99.9%. Phenolic estrogenic compounds are removed by NF technology (Bodzek et al., 2004). There is an alarming concentration rise in the synthetic hormone in the aquatic environment which is released by pharmaceuticals, hospital waste, and from households. The synthetic hormones include mestranol, α-ethinylestradiol, and diethylstilbestrol. The membrane process has been shown high efficacy in the treatment of synthetic hormones even the low molecular weight pollutants can also be removed by RO and NF (Bodzek and Dudziak, 2006).

9.3.1.2 Absorption by activated carbon

The activated carbon (AC) is the most widely used process for removing various contaminants in water. It is available in a powdered form (powder activated carbon, PAC) and granular form (granular activated carbon, GAC). The adsorption and filtration provided by GAC can be used instead of media made up of anthracite in a conventional technique. The GAC can also be used as an adsorbent bed in post-conventional filtration (Snydera et al., 2007). Many types of research have proved the efficacy of AC (PAC and GAC) for the treatment of organic water pollutants (Zhou et al., 2007). EDC can be removed by AC and it was proven by many articles (Fukuhara et al., 2006). The Endocrine substances like estrone (E1) and 17β-estradiol (E2) have also been removed by AC treatment. However, both PAC and APC are found to be effective in the removal of emerging contaminants from wastewater (Budimirovic et al., 2017). The activated carbon adsorption works on the principle that the pollutants (adsorbates) move from the aqueous phase to the solid phase (adsorbent) (Rodriguez et al., 2017). It is a widely used method since it has a broad spectrum of adsorption, high porosity, large surface area, and higher surface interaction (Rizzo et al., 2019).

9.3.2 Chemical treatment

9.3.2.1 Advanced oxidation

Advanced oxidation processes (AOPs), can be used as an efficient alternative for removing estrogens that are in trace amounts. The chemical advanced oxidation (CAO) is defined as the process which uses strong oxidants to transform contaminants or pollutants in the aquatic system to a less toxic or non-toxic substance by the reduction-oxidation system (Redox). The general mechanism of CAO involves ozonation (O_3), Chlorine (Cl_2), Hydrogen peroxide (H_2O_2), or their combination with UV, etc. Ozone can be effectively used for cleaning drinking waters and this is how it has entered into the degradation of several organic micropollutants. Ozone can react selectively, as an oxidant in its molecular form (O_3). Hence it leads to ozonation reactions (Von Gunten,

2003) depending on the structure of the organic substrate, steroid derivatives can be degraded to lower molecular weight compounds.

The Fenton process is based on the use of ferrous ions in association with hydrogen peroxide in acidic media (pH = 3). This process coupled with the oxidation process to coagulation/flocculation, with the latter process occurring later when the pH is raised to neutralize the final effluent (Petruzzelli et al., 2007). This treatment appears to be promising in the degradation of estrogens because of the self-regenerating cycle operated by MnO_2, thus proving to be cost-effective in the long run. Ferrate ion was investigated as a viable oxidant and coagulant. Although the oxidation potential of ferrate ions is greater than that of ozone, the acidic conditions of operation strongly limit its use in full-scale installations. Both chlorine and chloramines have the potential to react with various EDCs. (Lee et al., 2005). Besides, the EDC removal can be achieved at a higher rate by the use of O_3 (Tijani et al., 2013). Researchers suggested that the Bisphenol A (BPA) and E2 can be removed effectively by the O_3 with a concentration of 0.1 mmol/L. Moreover, when O_3 is combined with UV, the efficacy is found to be high. This process is pH-dependent because the removal efficacy was higher in pH of 6.6 than 8 (Wu et al., 2012). The incomplete chlorination reaction by using chlorine is also helpful in the removal of EDCs. It was found that chlorine can remove BPA which is a byproduct produced by many industries. Overall the CAO in the removal of EDC has high efficiency and requires technical stringency e.g., pH. also, the cost of setup is high (Gao et al., 2020).

9.3.2.2 Heterogeneous photocatalysis

This technique has emerged recently technique for degrading hazardous and non-biodegradable substances in an aquatic environment and drinking water such as CO_2, H_2O, and inorganic ions. The major challenge, despite the useful functions, the post-separation of the used TiO_2 photocatalysts from wastewater treatment remains difficult. This disadvantage can be overcome by immobilization of TiO_2 upon many supports such as silica, activated carbon, clay, zeolites, and many others (Yao et al., 2010). These support materials are chosen because of their higher surface area. Several types of research have proven the use of Tio_2 photocatalysts for the degradation of a wide variety of EDCs (Gmurek et al., 2017).

9.3.2.3 Homogeneous advanced oxidation processes

This is a reaction process in which both the reactant and photocatalysis are in the same phase. The very commonly used homogeneous photocatalysis is Ozone, Metal oxide, and Hydrogen peroxide. The mineralization or decomposition of emerging micropollutants like EDCs in water pollution by homogeneous advanced treatment, therefore, uses ozone, hydrogen peroxide, ultraviolet light. The combination process with UV light /ferric ions can also be used to disrupt toxic pollutants. Some of the previous research conducted using either a single or combined system to destroy the contaminants of emerging concern and proven to be effective (Xu et al., 2007). Li et al. 2011 suggested the technique of irradiation with xenon lamp in presence of the Bismuth silicate oxide metal (BSO) will aid in the degradation of Pentachlorophenol. Giri et al. 2010 used seven different advanced oxidation methods (UV, UV/TiO_2, UV/H_2O_2, O_3, O_3/UV, O_3/ TiO_2, O_3/ TiO_2/UV and found that the combined ozonation is found to be very effective. Another research combined the membrane reactor with TiO_2 to decontaminate the pharmaceutical wastewater and found it has 95% removal efficacy of Carbamazepine pollutant (Chong and Jin. 2012).

9.4 Nanomaterial assisted removal of EDCs

Nanofiltration uses membranes that are denser that have good features that give noteworthy removal efficacy than that of another process. The pore size of nanofiltration membranes is very small. The Ionisable groups present in the nanofillers typically make them negatively charged and have higher charge density. The latter can be removed very efficiently (>90%) and can stop the mono and multivalent ions because of their electrostatic interaction between the charged group (Youcai, 2018). Thus this type of filter provides the separation based on the charge and size of the substance. This process is relatively cost-efficient and operates under low pressure than osmosis. It has low molecular weight cut off than another filtration process, hence it can be used as a barrier for a wide spectrum of compounds.

9.4.1 Carbon nanotubes

Carbon nanotubes have extraordinary adsorption properties and can be used widely in the removal of contaminants from wastewater (Savage and Diallo, 2005) They have received wide attention due to their capabilities of water treatment. The removal of contaminants by carbon nanotubes has higher efficacy than the carbon-based adsorbents (Saleh et al., 2008) It is due to their large surface area. CNTs can be effectively used against Endocrine disrupting compounds. The carbon nanotubes (CNT) are hollow cylindrical tubes, high pore volume, hydrophobic walls, and good conductivity of electricity, chemical properties, and large surface area. They effectively absorb diverse pollutants like heavy metals, phenols, EDCs, and other chemicals (Ma et al., 2016) CNT has elucidated the removal of EDCs like BPA, E1, and E2. In combination with membrane filters, the CNT demonstrates high potential (60.4–95.2%) to remove EDCs from wastewater (Wang et al., 2015). Another interesting combination of CNT/ TiO_2 that works in the principle of photocatalytic degradation has shown increased adsorptive efficacy of removal of EDCs from wastewater (Kurwadkar et al., 2019).

9.4.2 Nanocomposites

At present, the majority of the photocatalysis reported are heterogenous oxide semiconductor materials of which have pure and composite forms. Nano Tio_2 and ZnO are the most studied photocatalysts in the last decades in the individual metals and as composites for the EDCs degradation in aqueous solutions due to the high ultraviolet sensitivity and chemical stability. Kamaraj et al. (2014a) studied the Bisphenol-degradation by the $Ce_xZn_{1-x}O$ nanocomposites which were prepared at the doping and composite of cerium and Zno under sunlight irradiation and achieved near-complete mineralization. Some researchers have studied the TiO_2 composites with precious metals such as gold and silver. Mostly the gold nanoparticles are given preference due to their stability, nontoxicity, and biocompatibility. Sornalingam et al. (2017) reported that the gold loaded TiO_2 composites are more competent than p25 TiO_2 in the degradation of estrone (E1) under UVA and visible LEDs irradiation. Besides, silver orthophosphate (Ag_3PO_4) and Lanthanum cobaltite ($LaCoO_3$) has also tested for the degradation of organic pollutant due to their high catalytic activity. A study reported that the $Ag_3PO_4/LaCoO_3$ composites have mineralized the BPA up to 77.27% in 40 min (Guo et al., 2016).

9.4.3 Zero-Valent iron

Nanocatalysts like zero-valence metal, semiconductor materials, and bimetallic nanoparticles can be used for the degradation of pollutants/contaminants in wastewater. Their properties were shaped dependent and have a higher surface area (Zhao et al., 2011). Many articles suggest that magnet-based nanosorbents have higher efficacy in the removal of organic contaminants (Campos et al., 2012). Iron oxide nanomaterials also have good capabilities of removing contaminants (Li et al., 2003).

9.5 Polymer-based removal of EDCs

The innovative alternative to conventional methods is molecularly imprinted polymers (MIP) and non-imprinted polymers (NIP) because they do not lead to the formation of transformed products. The main difference between MIP and NIP is their specificity. MIP can selectively target and remove targeted compounds whereas NIP can remove different organic compounds. NIP has non-specific binding is accredited to hydrophobic interactions between polymer and organic contaminants. The development of new, effective, and affordable technologies for the removal of EDCs from water without producing any toxic degradation products has become a priority. MIP and NIP have been studied extensively for removal of EDCs during water and wastewater treatment (Meng et al., 2005; Lin et al., 2008). Polymers are suitable for the removal of EDCs from the water medium, hence widely used due to their advantages such as increased pore size, large surface area, volume distribution, malleable surface chemistry and mechanical strength, the possibility of regeneration under suitable conditions. The polymeric adsorbent has the achievability of adapting and changing the physicochemical and mechanical characteristics of material via different polymerization conditions

(Vieira et al., 2020). The adsorption behavior of polymeric adsorbent [poly(EGDMA-MATrp) beads] is tested against the widespread diethyl phthalate by Özer et al. (2015). Lee and Kwak (2020) tested the new developed cross-linked polymer for the removal of BPA from wastewater.

9.6 EDC bioremediation technologies

The wastewater treatment technology which is conventional methods depends on the microbes to degrade EDCs. In these treatment plants, wastewater is sent to a reactor that contains a microbial mixture. This reactor is filled with to provide aeration, which helps in the degradation of EDCs. This technique helps maintain the biomass of microbes as flocs. This helps in exposure to the maximum between the contaminated water and flocs. This technique also helps the adsorption of contaminants of organic matter. It is rapid and helps in the efficient separation of the wastewater (Scholz, 2016). The maintenance of these flocs relies on the characteristics and synthesis of Extracellular polymeric substances (EPS). In addition, the distribution of polysaccharides, proteins, nucleic acid, and lipids in the EPS can help in the facilitation of nutrient accumulation from the environment and the organic contaminant adsorption (Flemming and wimmenger, 2010). The sludge that is activated is unsteady and does not achieve EDC degradation completely (Eio et al., 2014). To overcome the disadvantage there should be a treatment plant with high adsorption capability, cost efficiency, stability, and solubility (Khan et al., 2019). In context to this, the conventional sludge system can be enhanced by specialized microbes, algae, and fungi (Haq and Raj, 2019; Gao et al., 2020). The list of selected microorganisms employed for the degradation of EDCs is given in Table 9.2.

Table 9.2 Microorganisms utilized for the degradation of EDCs.

EDCs	Microorganism	Removal efficiency (%)	References
Bisphenol-A (BPA)	*Pseudomonas* sp. strain KU1, *Pseudomonas* sp. strain KU2, and *Bacillus* sp. strain KU3	81 ± 3, 78 ± 4 and 74 ± 2 respectively	Kamaraj et al., 2014
4-*nonylphenol* (4-t-NP), 4-tert-octylphenol (4-t-OP), 4- cumylphenol (4-CP)	*Umbelopsis isabellina*	90	Janicki et al., 2016
EDCs	White-rot fungus *Pleurotus ostreatus* HK35	>90	Křesinová et al., 2018
17-β estradiol (E2) and 17-α ethinylestradiol (EE2)	Microalgae, *Selenastrum capricornutum* and *Chlamydomonas reinhardtii*	88 to 100	Hom-Diaz et al., 2015

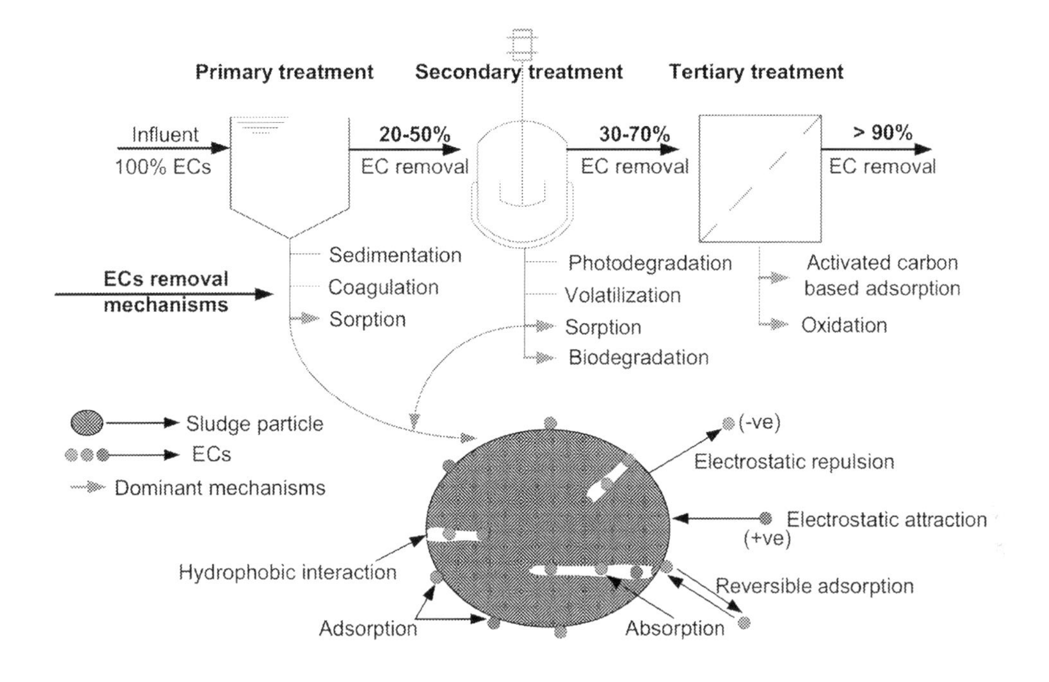

Fig. 9.3 ***Outline of emerging contaminants (EC) removal mechanism in a typical wastewater treatment plant.*** *(Reproduced with permission from Rout et al., 2021, Copyright © 2021 Published by Elsevier Ltd.)*

The important microbes that can be used techniques help mitigate the contaminants. It is done by the production of enzymes and secondary metabolites of microbes (Renneberg et al., 2017). The EDC degrading organisms include Bacteria, Cyanobacteria, Microalgae, and fungi. Their mechanism of degradation of EDCs includes adsorption, intracellular and extracellular mechanism, and accumulation (Xiong et al., 2018). They have many advantages being recyclable and provide efficient treatment and synthesize biomass that can be used in the production of biofuels and animal feeds (Hashemian et al., 2019). The combination treatments such as microalgae- microalgae and microalgae-bacteria are more efficient (Xiong et al., 2018). The microalgae combined with the conventional sludge system are also found to be efficient in the treatment of EDCs. It also helps in the reduction of energy consumptions for providing oxygen since microalgae can synthesis the oxygen naturally (Ariza, 2018). The schematic overview of mechanisms employed in a typical wastewater treatment plant for the removal of emerging contaminants is depicted in Fig. 9.3.

9.7 Conclusion and recommendation

EDCs in the environment are now a worldwide concern due to their prevalent occurrence, persistence, and bioaccumulation. However, the EDCs occurred in the water resources may not have an extensive impact on human health, but it might lead to significant health risks due to their long-term exposure (accretion of EDCs from low to high concentration and the combined effects of EDCs metabolites and by-products). The removal of EDCs using the existing and traditional treatment technology has challenges in elimination efficiency, cost-effectiveness, and sustainable development for the entire treatment and remediation system. Despite the progressive developments in the EDCs removal technologies, considerable limitations still exist which demand future research for the utilization of holistic EDCs removal technique. The major challenges and associated research directions are as follows:

1. The legislation needs to be emphasized the usage and release of EDCs in a industries
2. Public awareness needs to be created on the occurrence of EDCs in the common environment and their adverse effects
3. The developments of new procedures for the extraction of EDCs and detection of their accurate occurrence in the environment
4. The proper techniques need to identify the transformation products of EDCs which are not usually monitored for their persistency and toxicity due to the lack of knowledge.
5. In the advanced oxidation process, hydroxyl radicals may mineralize the EDCs or create by-products by the transformation. Hence, the knowledge of intermediates is necessary to take necessary actions to minimize or prevent the possible hazards.
6. Although the EDCs removal efficiency is high in AC-mediated adsorption, still it's a partial solution as it eliminates the EDCs from one phase and accumulates in another. Hence, further research is recommended to find subsequent techniques for the reusability or remediation of EDCs adsorbed AC.
7. Generally, the biodegradation of EDCs are been tested under an individual microorganism with limited parameters. Therefore the dynamic studies using microbial consortia need to be carried out for the degradation of EDCs.
8. In MBRs, membrane fouling is a major issue that restricting their usage to some extent. However, the EDCs can be retained and removed subsequently by interacting with the fouling layer. Hence, the negative and positive impacts of fouling are needed to be studied in detail for overcoming the limitation.
9. Although the nanomaterials with their enhanced photocatalytic and adsorptive properties are resulting in extensive EDCs removal application, still their costs of production are high. Hence, future research is recommended to find a cost-effective way for the production of nanomaterials.

10. The membranes and polymer utilization for the removal of EDCs are found suitable due to their thermal and mechanical properties. However, there is a need to explore various materials suits to prepare the membrane and polymer with enhanced hydrophilicity and EDCs removal efficiency.
11. The integration of one or multiple technologies into a single system for EDCs removal is highly recommended. This integration can harmonize with each other to overcome the challenges of each technique and led to a more efficient EDCs removal process.

Conflict of interest disclosures

The authors declare they have no conflicts of interest.

Funding

No fund was received from any funding agency or organization towards making this manuscript

Acknowledgments

The authors wish to acknowledge all who had been instrumental in the creation of this review article.

References

Abbas, A., Schneider, I., Bollmann, A., et al., 2019. What you extract is what you see: optimising the preparation of water and wastewater samples for in vitro bioassays. Water Res. 152, 47–60. https://doi.org/10.1016/j.watres.2018.12.049.

Annamalai, J., Namasivayam, V., 2015. Endocrine disrupting chemicals in the atmosphere: their effects on humans and wildlife. Environ. Int. 76, 78–97. http://dx.doi.org/10.1016/j.envint.2014.12.006.

Aris, A.Z., Mohd Hir, Z.A., Razak, M.R., 2020. Metal-organic frameworks (MOFs) for the adsorptive removal of selected endocrine disrupting compounds (EDCs) from aqueous solution: a review. Applied Materials Today 21, 100796. https://doi.org/10.1016/j.apmt.2020.100796.

Ariza, A., 2018. Photo-Activated Sludge: a novel algal-bacterial biotreatment for nitrogen removal from wastewater 2018. Wageningen University, IHE Delft Institute for Water Education, The Netherlands. Dissertation. https://library.wur.nl/WebQuery/wda/abstract/2247355.

Bodzek, M., Dudziak, M., Luks–Betlej, K., 2004. Application of membrane techniques to water purification. Removal of phthalates. Desalination 162, 121–128. https://doi.org/10.1016/S0011-9164(04)00035-9.

Bodzek, M., Dudziak, M., 2006. Elimination of steroidal sex hormones by conventional water treatment and membrane processes. Desalination 198, 24–32. https://doi.org/10.1016/j.desal.2006.09.005.

Bodzek, M., 2015. Application of membrane techniques for the removal of micropollutants from water and wastewater. Copernican Letters 6, 24–33. http://dx.doi.org/10.12775/CL.2015.004.

Budimirovic, D., Velickovic, Z.S., Djokic, V.R., Milosavljevic, M., Markovski, J., Levic, S., et al., 2017. Efficient As (V) removal by α-FeOOH and α-FeOOH/α- MnO2 embedded PEG-6-arm functionalized multiwall carbon nanotubes. Chem. Eng. Res. Des. 119, 75–86. https://doi.org/10.1016/j.cherd.2017.01.010.

Campos, A.F.C., Aquino, R., Cotta, T., Tourinho, F.A., Depeyrot., J., 2012. Using speciation diagrams to improve synthesis of magnetic nanosorbents for environmental applications. Bull. Mater. Sci. 34 (7), 1357–1361. https://doi.org/10.1007/s12034-011-0328-5.

Chong, M.N., Jin, B., 2012. Photocatalytic treatment of high concentration carbamazepine in synthetic hospital wastewater. J. Hazard. Mater. 199–200, 135–142. https://doi.org/10.1016/j.jhazmat.2011.10.067.

Eio, E.J., Kawai, M., Tsuchiya, K., Yamamoto, S., Toda, T., 2014. Biodegradation of bisphenol A by bacterial consortia. Int. Biodeterior. Biodegradation 2014 (96), 166–173. https://doi.org/10.1016/j.ibiod.2014.09.011.

Flemming, H., Wingender, J., 2010. The biofilm matrix. Nat. Rev. Microbiol. 8, 623–633. https://doi.org/10.1038/nrmicro2415.

Fukuhara, T., Iwasaki, S., Kawashima, M., Shinohara, O., Abe, I., 2006. Adsorbability of estrone and 17β -estradiol in water onto activated carbon. Water Res. 40, 241–248. https://doi.org/10.1016/j.watres.2005.10.042.

Gao, X., Kang, S., Xiong, R., Chen, M., 2020. Environment-Friendly Removal Methods for Endocrine Disrupting Chemicals. Sustainability 12 (18), 7615. https://doi.org/10.3390/su12187615.

Giri, R.R., Ozaki, H., Ota, S., Takanami, R., Taniguchi, S., 2010. Degradation of common pharmaceuticals and personal care products in mixed solutions by advanced oxidation techniques. Int. J. Environ. Sci. Technol. 7 (2), 251–260. https://doi.org/10.1007/BF03326135.

Gmurek, M., Olak-Kucharczyk, M., Ledakowicz, S., 2017. Photochemical decomposition of endocrine disrupting compounds – A review. Chem. Eng. J. 310, 437–456. http://dx.doi.org/10.1016/j.cej.2016.05.014.

Guo, J., Dai, Y.Z., Chen, X.J., Zhou, L.L., Liu, T.H., 2016. Synthesis and characterization of Ag3PO4/LaCoO3 nanocomposite with superior mineralization potential for bisphenol A degradation under visible light. J. Alloys Compd. 696, 226–233. https://doi.org/10.1016/j.jallcom.2016.11.251.

Haq, I., Raj, A., 2019. Endocrine-Disrupting Pollutants in Industrial Wastewater and Their Degradation and Detoxification Approaches. In: Bharagava, R., Chowdhary, P. (Eds.), Emerging and Eco-Friendly Approaches for Waste Management. Springer, Singapore. https://doi.org/10.1007/978-981-10-8669-4_7.

Hashemian, M., Ahmadzadeh, H., Hosseini, M., Lyon, S., Pourianfar, H., 2019. Production of microalgae-derived high-protein biomass to enhance food for animal feedstock and human consumptionAdvanced bioprocessing for alternative fuels, biobased chemicals, and bioproducts. Woodhead Publishing, Sawston, pp. 393–405. https://doi.org/10.1016/B978-0-12-817941-3.00020-6.

Hom-Diaz, A., Llorca, M., Rodríguez-Mozaz, S., Vicent, T., Barceló, D., Blánquez, P., 2015. Microalgae cultivation on wastewater digestate: β -estradiol and 17α-ethynylestradiol degradation and transformation products identification. J. Environ. Manage. 155, 106–113. https://doi.org/10.1016/j.jenvman.2015.03.003.

Janicki, T., Krupi´nski, M., Długo´nski, J., 2016. Degradation and toxicity reduction of the endocrine disruptors nonylphenol, 4-tert-octylphenol and 4-cumylphenol by the non-ligninolytic fungus *Umbelopsis isabellina*. Bioresour. Technol. 200, 223–229. https://doi.org/10.1016/j.biortech.2015.10.034.

Křesinová, Z., Linhartová, L., Filipová, A., Ezechiáš, M., Mašín, P., Cajthaml, T., 2018. Biodegradation of endocrine disruptors in urban wastewater using Pleurotus ostreatus bioreactor. New Biotechnol. 43, 53–61. https://doi.org/10.1016/j.nbt.2017.05.004.

Kamaraj, M., Ranjith, K.S., Sivaraj, R., Rajendrakumar, R.T., Salam, H.A., 2014a. Photocatalytic degradation of endocrine disruptor Bisphenol-A in the presence of prepared CexZn1-XO nano composites under sunlight irradiation. J. Environ. Sci. 26, 2362–2368. https://doi.org/10.1016/j.jes.2014.09.022.

Kamaraj, M., Sivaraj, R., Venckatesh, R., 2014b. Biodegradation of bisphenol A by the tolerant bacterial species isolated from coastal regions of Chennai, Tamil Nadu, India. Int. Biodeterior. Biodegradation 93, 216–222. https://doi.org/10.1016/j.ibiod.2014.02.014.

Khan, N.A., Khan, S.U., Ahmed, S., Farooqi, I.H., Yousefi, M., Mohammadi, A.A., et al., 2019. Recent trends in disposal and treatment technologies of emerging-pollutants—A critical review. TrAC, Trends Anal. Chem. 122, 115744. https://doi.org/10.1016/j.trac.2019.115744.

Kurwadkar, S., Hoang, T.V., Malwade, K., Kanel, S.R., Harper, W.F., Struckhoff, G., 2019. Application of carbon nanotubes for removal of emerging contaminants of concern in engineered water and wastewater treatment systems. Nanotechnol. Environ. Eng. 4 (1), 12. https://doi.org/10.1007/s41204-019-0059-1.

Lee, J.H., Kwak, S.Y., 2020. Branched polyethylenimine-polyethylene glycol-β-cyclodextrin polymers for efficient removal of bisphenol A and copper from wastewater. J. Appl. Polym. Sci. 137, 1–9. https://doi.org/10.1002/app.48475.

Lee, Y., Yoon, J., Von Gunten, U., 2005. Kinetics of the oxidation of phenols and phenolic endocrine disruptors during water treatment with ferrate (Fe (VI)). Environ. Sci. Technol. 39(22), 8978-8984. https://doi.org/10.1021/es051198w.

Li., P., Miser, D.E., Rabiei, S., Yadav, R.T., Hajaligol, M.R., 2003. The removal of carbon monoxide by iron oxide nanoparticles. Appl. Catal. B 43 (2), 151–162. https://doi.org/10.1016/S0926-3373(02)00297-7.

Li, J., Jiang, L., Liu, X., Lv, J., 2013. Adsorption and aerobic biodegradation of four selected endocrine disrupting chemicals in soil–water system. Int. Biodeterior. Biodegradation 76, 3–7. https://doi.org/10.1016/j.ibiod.2012.06.004.

Li, Y., Niu, J., Yin, L., Wang, W., Bao, Y., Chen, J., et al., 2011. Photocatalytic degradation kinetics and mechanism of pentachlorophenol based on superoxide radicals. J. Environ. Sci. 23 (11), 1911–1918. https://doi.org/10.1016/S1001-0742(10)60563-3.

Lin, Y., Shi, Y., Jiang, M., Jin, Y., Peng, Y., Lu, B., et al., 2008. Removal of phenolic estrogen pollutants from different sources of water using molecularly imprinted polymeric microspheres. Environ. Pollut. 153 (2), 483–491. https://doi.org/10.1016/j.envpol.2007.08.001.

Ma, Z., Yin, X., Ji, X., 2016. Evaluation and removal of emerging nanoparticle contaminants in water treatment: a review. Desalin. Water Treat. 57, 11221–11232. https://doi.org/10.1080/19443994.2015.1038734.

Marty, M.S., Borgert, C., Coady, K., Green, R., Levine, S.L., Mihaich, E., et al., 2018. Distinguishing between endocrine disruption and non-specific e_ects on endocrine systems. Regul. Toxicol. Pharmacol. 99, 142–158. https://doi.org/10.1016/j.yrtph.2018.09.002.

Meng, Z., Chen, W., Mulchandani, A., 2005. Removal of estrogenic pollutants from contaminated water using molecularly imprinted polymers. Environ. Sci. Technol. 39 (22), 8958–8962. https://doi.org/10.1021/es0505292.

Omar, T.F.T., Aris, A.Z., Yusoff, F.M., Mustafa, S., 2018. Occurrence, distribution, and sources of emerging organic contaminants in tropical coastal sediments of anthropogenically impacted Klang River estuary, Malaysia. Mar. Pollut. Bull. 131, 284–293. https://doi.org/10.1016/j.marpolbul.2018.04.019.

Özer, E.T., Osman., B., Kara, A., et al., 2015. Diethyl phthalate removal from aqueous phase using poly(EGDMA-MATrp) beads: kinetic, isothermal and thermodynamic studies. Environ. Technol. 36, 1698–1706. https://doi.org/10.1080/09593330.2015.1006687.

Petruzzelli, D., Boghetich, G., Petrella, M., Dell'Erba, A.L., Abbate, P., Sanarica, S., 2007. Pre-treatment of industrial landfill leachate by Fenton's oxidation. Glob. Nest J. 9, 51–56. https://doi.org/10.30955/gnj.000353.

Renneberg, R., Berkling, V., Loroch, V., 2017. Green biotechnology. Biotechnology for beginners, 2nd Ed. Academic Press, New York, pp. 233–279.

Rizzo, L., Malato, S., Antakyali, D., et al., 2019. Consolidated vs new advanced treatment methods for the removal of contaminants of emerging concern from urban wastewater. Sci. Total Environ. 655, 986–1008. https://doi.org/10.1016/j.scitotenv.2018.11.265.

Rodriguez-Narvaez, O.M., Peralta-Hernandez, J.M., Goonetilleke, A., Bandala, E.R., 2017. Treatment technologies for emerging contaminants in water: a review. Chem. Eng. J. 323, 361–380. https://doi.org/10.1016/j.cej.2017.04.106.

Rout, P.R., Zhang, T.C., Bhunia, P., Surampalli, R.Y., 2021. Treatment technologies for emerging contaminants in wastewater treatment plants: a review. Sci. Total Environ. 753 (20), 141990. https://doi.org/10.1016/j.scitotenv.2020.141990.

Saleh, N.B., Pfefferle, L.D., Elimelech, M., 2008. Aggregation kinetics of multiwalled carbon nanotubes in aquatic systems: measurements and environmental implications. Environ. Sci. Technol. 42 (21), 7963–7969. https://doi.org/10.1021/es801251c.

Savage, N., Diallo, M.S., 2005. Nanomaterials and water purification: opportunities and challenges. J. Nanopart. Res. 7 (4–5), 331–342. https://doi.org/10.1007/s11051-005-7523-5.

Scholz, M., 2016. Activated sludge processesWetlands for water pollution control, 2nd ed. Elsevier, pp. 91–105.

Snydera, S.A., Adhamb, S., Reddingc, A.M., et al., 2007. Role of membranes and activated carbon in the removal of endocrine disruptors and pharmaceuticals. Desalination 202, 156–181. https://doi.org/10.1016/j.desal.2005.12.052.

Sornalingam, K., Mcdonagh, A., Zhou, J.L., Mah, J., Ahmed, M.B., 2017. Photocatalysis of estrone in water and wastewater: comparison between Au-TiO2 nanocomposite and TiO2, and degradation by-products. Sci. Total Environ. 610–611, 521–530. https://doi.org/10.1016/j.scitotenv.2017.08.097.

Swaminathan, M., Muruganandham, M., Sillanpaa, M., 2013. Advanced oxidation processes for wastewater treatment. Int. J. Photoenergy 683682. https://doi.org/10.1155/2013/683682.

Ternes, T.A., Stumpf, M., Mueller, J., Haberer, K., Wilken, R.D., Servos, M., 1999. Behavior and occurrence of estrogens in municipal sewage treatment plants–I. Investigations in Germany, Canada and Brazil. Sci. Total Environ. 225, 81–90. https://doi.org/10.1016/S0048-9697(98)00334-9.

Tijani, J.O., Fatoba, O.O., Petrik, L.F., 2013. A review of pharmaceuticals and endocrine-disrupting compounds: sources, effects, removal, and detections. Water Air Soil Pollut. *224* (11), 1–29. https://doi.org/10.1007/s11270-013-1770-3.

United States Environmental Protection Agency (USEPA), 1997. Special report on environmental endocrine disruption: an effects assessment and analysis. office of Research and Development, Washington, DC.

Vieira, W.T., de Farias, M.B., Spaolonzi, M.P., et al., 2020. Removal of endocrine disruptors in waters by adsorption, membrane filtration and biodegradation. A review. Environ. Chem. Lett. 18, 1113–1114. https://doi.org/10.1007/s10311-020-01000-1.

Von Gunten, U., 2003. Ozonation of drinking water: part I. Oxidation kinetics and product formation. Water Res. 37, 1443–1467. https://doi.org/10.1016/S0043-1354(02)00457-8.

Walha, K., Amar, R.B., Firdaous, L., Quemeneur, F., Jaouen, P., 2007. Brackish groundwater treatment by nanofiltration, reverse osmosis and electrodialysis in Tunisia: performance and cost comparison. Desalination 207, 95–106. https://doi.org/10.1016/j.desal.2006.03.583.

Wang, W.L., Wu, Q.Y., Wang, Z.M., 2015. Adsorption removal of antiviral drug oseltamivir and its metabolite oseltamivir carboxylate by carbon nanotubes: efects of carbon nanotube properties and media. J. Environ. Manage. 162, 326–333. https://doi.org/10.1016/j.jenvman.2015.07.043.

Wu, Q., Shi, H., Adams, C.D., Timmons, D., Ma, Y., 2012. Oxidative removal of selected endocrine disruptors and pharmaceuticals in drinking water treatment systems, and identification of degradation products of triclosan. Sci. Total Environ. 439, 18–25. https://doi.org/10.1016/j.scitotenv.2012.08.090.

Xiong, J., Kurade, M., Jeon, B., 2018. Can microalgae remove pharmaceutical contaminants from water? Trends Biotechnol. 36 (1), 30–44. https://doi.org/10.1016/j.tibtech.2017.09.003.

Xu, B., Gao, N.Y., Rui, M., Wang, H., Wu, H.H., 2007. Degradation of endocrine disruptor bisphenol A in drinking water by ozone oxidation. J. Hazard. Mater. 138, 526–533. https://doi.org/10.1007/s11783-007-0060-y.

Yao, S., Li, J., Shi, Z., 2010. Immobilization of TiO2 nanoparticles on activated carbon fiber and its photodegradation performance for organic pollutants. Particuology 8, 272–278. https://doi.org/10.1016/j.partic.2010.03.013.

Youcai, Z., 2018. Pollution control Technology for Leachate from municipal solid waste: landfills, incineration plants, and transfer stations. Butterworth-Heinemann, 65–522. https://doi.org/10.1016/C2017-0-03224-X.

Zegura, B., Klemencic, A.K., Balabanic, D., Filipic, M., 2017. Raw and biologically treated paper mill wastewater effluents and the recipient surface waters: cytotoxic and genotoxic activity and the presence of endocrine disrupting compounds. Sci. Total Environ. 74, 78–89. https://10.1016/j.scitotenv.2016.09.030.

Zhao, X., Lv, L., Pan, B., Zhang, W., Zhang, S., Zhang, Q., 2011. Polymer-supported nanocomposites for environmental application: a review. Chem. Eng. J. 170 (2–3), 381–394. https://doi.org/10.1016/j.cej.2011.02.071.

Zhou, J.H., Sui, Z.J., Zhu, J., Li, P., Chen, D., Dai, Y.C., 2007. Characterization of surface oxygen complexes on carbon nanofibers by TPD, XPS and FT-IR. Carbon N Y 45, 785–796. https://doi.org/10.1016/j.carbon.2006.11.019.

CHAPTER 10

Use of microalgae for the removal of emerging contaminants from wastewater

Sunipa Deb, Soma Nag
Department of Chemical Engineering, National Institute of Technology Agartala, Tripura, India

10.1 Introduction

A large number of pollutants are generated in the waste water by various industrial and every day human activities. Use of the pesticides, pharmaceuticals, medicines, personal care products, shampoos, detergents, food additives, preservatives in daily life and many more organic and inorganic compounds released from industrial activities are considered as Emerging Contaminants (ECs). The municipal wastewater is generated from household and also from hospitals and commercial places. In the modern era, emphasis is given not only for the health of human beings but also for the environment and this directs the countries to dispose the wastewater wisely. Nowadays, clean water availability is menacing due to the increased use of these life style products. The emerging contaminants are micro pollutants and present in large numbers and have varied compositions, resulting most of them don't have any standard regulatory measures. They are poisonous to the environment and living beings and can affect deadly even in small concentrations. Their bio-accumulation creates high risk for human health. The wastewater is highly bio degradable especially for the organic pollutants.

Algae-based treatment is more effective than chemical-based treatment in extracting nutrients and heavy metals from waste water. There are so many microalgae-based pre-treatment systems for nutrient removal from waste water, some of which include primary decantation of effluent, active sludge, oxidation ditches, etc. In these systems, microalgae supply aerobic bacteria with oxygen to promote the breakdown of organic matter and carbon dioxide is absorbed by bacterial respiration. Biological fixation by microalgae is considered a promising technology for the post-combustion of carbon dioxide. The above method is an inexpensive alternative to traditional waste water aerobic treatment technologies. Micro algae provide a low cost option for environmentally friendly option compared to the traditional waste water treatment methods (Amenorfenyo et al., 2019).

Microalgae are highly flexible to poisonous organic pollutants and are very good in adsorbing heavy metal and emerging contaminants. Several research studies have been

Biodegradation and Detoxification of Micropollutants in Industrial Wastewater
DOI: https://doi.org/10.1016/B978-0-323-88507-2.00002-6

published to justify the use of microalgae in treatment of emerging contaminants at laboratory conditions and found to be effective in the removal of popular emerging contaminants such as caffeine, ibuprofen, galaxolide and other personal care products.

Photosynthetic micro-organisms have the capacity to convert solar energy to chemical energy which is essential for carbon-di-oxide fixation. Microalgae are prokaryotic or eukaryotic photosynthetic organisms. They can quickly grow and can live in adverse conditions. Cyanobacteria are called prokaryotic micro algae, green algae and diatoms are found in eukaryotic microalgae, which are also classified as Chlorophyta and Bacillariophyta. Under microalgae, a diverse category of autotrophic species are represented, which are resilient organisms able to thrive under challenging conditions in aquatic environments. There are also photosynthetic complexes consisting of molecules such as Chlorophyll a, b, c, d, e and bacteriochlorophylls, etc. Green micro algae also have a special property, they have a simple structure, a broad surface/volume ratio of cells, which help them to grow faster and their biomass would be doubled in less than a day (Aketo et al., 2020).

10.1.1 Selection of micro algae

A good number of micro-algae strains are available for handling of waste water and proper selection of the micro algae strain is a very critical step for the effective elimination of different compounds and nutrients from the waste water. In the photosynthesis process, a significant amount of CO_2 is consumed by algae which later on convert to biomass and oxygen released in this process help to extract urea from water, which enhances their activity for bio-conservation. Light, sugar, nitrogen, CO_2, phosphorus and potassium are needed for the growth of microalgae. Some specific micro algae species can generate the related bio-compounds. Microalgae often can synthesize a significant amount of lipids, proteins, and carbohydrates which are used in production of bio-fuels and other valuable products.

10.1.2 Cultivation

Algal cultivation is very trendy with the aim of economic growing large amounts of lipids and biomass (Lewis, 2013). There are a variety of factors that have a significant effect on the cost of algal harvesting, such as pH, temperature, source of carbon and light. Large variations in temperature and pH can impede algal growth. The most widely used reactors for algal cultivation are open air systems, closed system, and system for biofilm.

The purpose and method of cultivation depends on the species of micro-algae, where different chemical compounds, including pigments, polysaccharides, fatty acids and vitamins, are used to extract biomass. Open and closed systems are two conventional methods for micro algae farming. Each system has its own benefits and drawbacks. The open micro algae system has the higher pollution risk and it also has a higher need for process control. The initial development cost is high in the case of a closed system, but it is the easiest system to control cultivation parameters. In the bioremediation process, the micro algae are used in a sustainable way by biomass exploration method. Among the constraints faced in biomass harvesting, separation processes and biomass processing are the main two obstacles (Abdel-Raouf et al., 2012).

A variety of micro-bolic pathways can be used by microalgae. In phototrophic metabolism process, light is used as energy source and carbon dioxide as a source for carbon for the upstream cultivation of microalgae. On the other hand, heterotrophic metabolism is the process in which only organic compounds are used as energy. Some micro-algae have the ability to switch between phototrophic metabolism and heterotrophic metabolism in some environmental conditions (Escapa et al., 2019).

10.2 Microalgae-based mechanism for degradation of pollutants

A schematic diagram of micro algae based degradation technique is shown in Fig. 10.1. In this technique, biodegradation, photo-degradation, volatilization and biomass sorption are the important pathways for the remediation of the contaminants. Among these, biodegradation and photo-degradation are the most important steps, while sorption and volatilization are only significant for hydrophobic compounds (Norvill et al., 2017; Matamoros et al., 2015; Bilal et al., 2018). These collective pathways guide in effective designing of bioreactors for efficient removal of pollutants. Enzymes and biofilms also have consistent role in bioremediation.

10.2.1 Biodegradation

Biodegradation by micro algae is a sustainable, eco-friendly technique for emerging contaminants removal from wastewater. In this process the enzymes produced by the microorganisms help in breaking down the organic chemicals and are considered as the pathways of metabolic degradation or co-metabolism. On the other way, the degradation of micro pollutants is dependent on non-specific enzymes present within the environment in order to catalyse the metabolism of other substrates for co-metabolism. Carbon and energy for the metabolic degradation are supplied by the organic chemicals present in wastewater.

Co-metabolism is primarily responsible for the degradation of toxins as the response of the micropollutants is poor to sustain the growth of microalgae. Biodegradation is efficiently performed even with the increase of toxicity level of the pollutants; because of acetaminophen metabolites waste water is not completely free of poisonous substances. Due to this, it is required to examine their influence on the living organisms in order to evaluate the performance of microalgae-based treatment methods. For the low strength of contaminants, biodegradation can be motivated by using the adsorption mechanism (Delrue et al., 2016; Escapa et al., 2016; Matamoros et al., 2016).

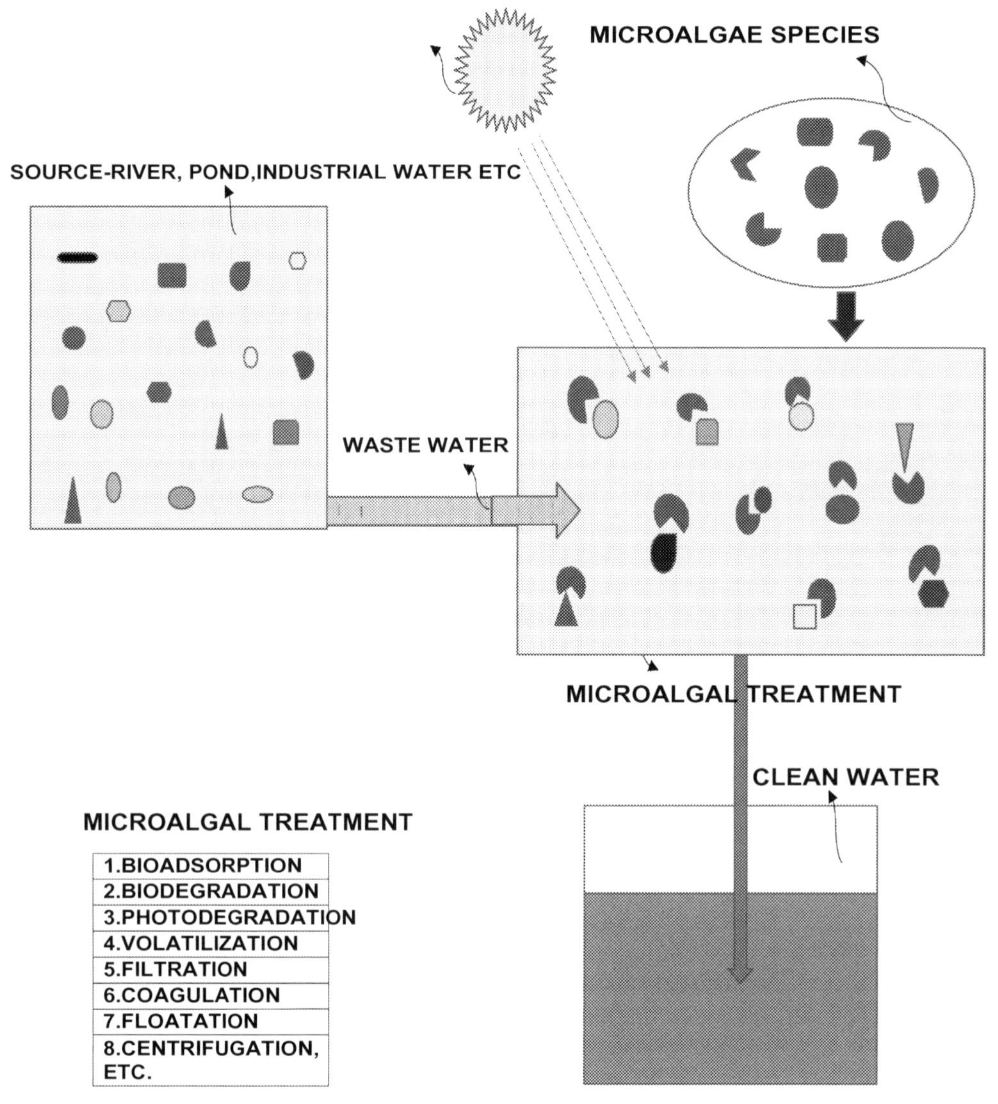

Fig. 10.1 ***Microalgae-based mechanism for degradation of pollutants.***

10.2.2 Biosorption

Biosorption and biodegradation are very effective tools for eradication of emerging contaminants. Biosorption is a combination of physical and chemical methods in which biomaterials are applied to take out the impurities from the solution. The sorption depends upon the structure of the contaminants, nature of the algae and also the surrounding conditions. The presence of leading functional groups, e.g., carboxyl, phosphoryl, amide etc. make the cellular wall of microalgae negatively charged and draws the pollutants and all other elements with cationic groups by the electrostatic interaction (Fomina and Gadd, 2014). Specific micro algae can remove up to 100 percent emerging contaminants and have the ability to biodegrade the pollutants using extracellular enzymes. Biosoption of estrogen and antibiotics is also taken place by specific algal biomass.

10.2.3 Photo-degradation

In a micro algae-based treatment method, illumination is required for photosynthesis. So, photo-degradation plays a crucial role in the reduction of micro contaminants and the process mostly depends upon the solar energy. Photosynthesis is also a mechanism by direct or indirect UV photodegradation for micropollutant removal. In a direct photo-degradation process, the pollutants adsorb the light directly resulting chemical reactions. The shape of the micro-pollutant determines by whether radiation absorption occurs and in that case, the molecular electricity is also increased, leading to the breakage and degradation of the bond. However, in the case of indirect photolysis, the dissolved natural materials present in waste water consume a small amount of energy and creates reactive oxygen species which ultimately degrades the aimed pollutants. The mechanism of photodegradation is based on the molecular structure of micro-pollutants and environmental conditions. The drawbacks of this removal mechanism are that, it incorporates the unfinished degradation of micro-pollutants, in addition to the constructive dependence on the chemical bonds of impurities (Gruchlik et al., 2018; Norvill et al., 2016; Norvill et al., 2017).

10.2.4 Volatilization

The elimination of unstable micro pollutants in a primarily open micro algae-based treatment system can be contributed to a particular process which is known as volatilization. The removal of trace pollution occurs by using volatilization in the microalgae-primarily based manner. The removal is influenced by the air stripping sensitivity and temperature. However, volatilization is not a preferred solution for the treatment of waste water as it does not degrade the pollutants into smaller molecules.

10.3 Removal strategies in micro algae based treatment systems

Algae-based treatments are considered to be more efficient than chemical methods for elimination of nutrients and heavy metals from waste water. The micro algae can remove the nutrients, like, nitrogen, phosphorus, and carbon as well as the heavy metals, personal care products, antibiotics etc. and produce high grade treated water. The microalgae-based systems have unique configurations, ranging from open ponds to closed photo-bio-reactors (PBR), also apply suspended cells, immobilized cells, consortia of micro algae and bacteria for better performance (Agüera et al., 2020). These methods can recover and reuse the algal bio mass for many other purposes, such as, production of protein-rich food, bio fuel, fertilizer etc.

10.3.1 Open ponds

Open ponds are the mostly chosen reactors for large scale cultivation of microalga, due to the lower construction cost, less power requirement, and simple cleaning process as compared to the closed photo-bio-reactors. The common types of open ponds are circular ponds, tanks and algal ponds. Microalgae and microorganisms develop through symbiosis and do not require aeration as a consequence. In case of high rate algal ponds, there is possibility of bacterial contamination due to shallow heights, however, this facilitates better penetration of light. The open pond system offers an extra benefit as compared to the traditional activated sludge system. The depth of high rate algal ponds vary within 100 cm whereas that of the activated sludge process is in the range of 2.4–6 m (Norvill et al., 2016). In conventional waste water treatment plants, some of the analgesics and anti-inflammatory drugs are removed a bit better compared to the algal pond, however, they are more efficient in biodegrading other pharmaceutical products, such as, diclofenac and antibiotics. Biodegradation is governed by some physic-chemical properties, such as functional groups, pH, temperature, hydrophobicity etc. and also the types of pollutants. The benefit of algal ponds is that, without the need for aeration, they can produce microalgae biomass with comparable removal efficiencies. One of the main disadvantages of enforcing algal ponds is that they require a large surface area, for enhancing the efficiency of pollutant removal and productivity of biomass (Matamoros et al., 2015; Norvill et al., 2017).

10.3.2 Closed photo bioreactors

Close photo bioreactors (PBRs) require excessive expenses for setting-up and have high maintenance cost. The systems are more efficient in terms of light and gas distribution, and also have highmass-transfer efficiency and low contamination compared to the high rate algal pond. Closed photo bioreactors are better than open systems at removing the emerging pollutants. In closed photo bioreactors, micro-pollutant elimination is performed by suspended microalgae. Various types of closed PBRs are used for wastewater treatment and microalgae farming. Closed PBRs provide controlled environment for studying how isolated microalgae can eliminate pollutants and that helps to identify the right species of micro algae for better elimination of the targeted pollutant. Microalgae-based treatment systems can eradicate multiple forms of emerging contaminants, such as endocrine disrupting compounds (EDCs), pharmaceuticals and personal care products (PPCPs). Closed PBR structures have been more effective in removing micro-pollutants, even at excessive concentrations, compared to microalgae-primarily based ponds. Closed PBRs in association with the activated sludge process, can eliminate the insecticides efficiently (Ahmed et al., 2017; Chang et al., 2017; Escapa et al., 2016).

10.3.3 Immobilized cells

One of the most significant disadvantages of fully microalgae-based wastewater remedial process, is the harvesting and separation of algal biomass from the treated water. Due to very small size of micro algae (2–50 micro meter), they remain suspended within the medium and their negative charge prevents their accumulation. Their concentration is also low within the cultures. Due to smaller size and low concentration, it is very difficult to collect the micro algae from the culture and the mechanical, electrical and chemical procedures applied for separation are very costly. Sludge of microalgae, afterwards, requires extra processing. The simplest low-value techniques are required for handling huge volumes of wastewater and biomass. Immobilization technology seems be a better option for collecting microalgae biomass from the reactors. Prominently retained cell density and cell catalytic action; survival in extreme environments; clean separation and reusability, are some of the advantages of the immobilization technology; which makes them a better choice over the other techniques. For more versatility within the reactor architecture than the conventional suspension systems, immobilization has been taken into account. Six types of immobilization techniques have been identified for the immobilization of microalgae, such as, adsorption, semi-permeable capture, membrane, immobilization of affinity, liquid-liquid emulsion containment, covalent coupling, and polymer capture. Synthetic and natural polymer are used widely for immobilization of microalgae, where they are static and alive within the gel matrix. The primary issues with immobilized microalgae systems are: (1) mild use (2) consistency of the gel (3) penetration of the substrate in the gel matrix; which have impact on the efficiency and economy of the process. The immobilized algae growth is not affected by the contaminants present in the wastewater; moreover they are very tough to the pollutants and can be recovered easily. Municipal, sewage or domestic waste water can be treated nicely by this process (Ferrando and Matamoros, 2020; Torres et al., 2008; Chojnacka, 2010).

10.3.4 Consortia

The growth of micro algae is challenged in real wastewater as they are to face continuous threat from the other micro-organisms present in the system and also to adjust with the environment. However, a grouping between algae and bacteria is seen to be much less vulnerable to the changes in ecological situations and more challenging to the contamination. A natural bio-flocculation occurs by the easily settled flocks of micro algal microbial assembly, which promotes the biomass harvesting efficiency. Combined consortium communities can perform functions that may be difficult or even impossible for traces or species of individuals. Lots of studies have been conducted for the performance of the mixed consortia for elimination of toxins from the wastewater. In case of bio-films, where the micro algae and bacteria activities are strong in the colony, no additional aeration is required as CO_2 and O_2 maintain a balance in this form.

There is a wonderful opportunity for microalgae-bacteria consortia to extract micro contaminants from waste water. Consortia-based treatment occurs in different configurations and sizes, along with algal ponds and photo bio reactors, as opposed to pure cultivation. However, most of the studies were performed in laboratory scale only; the pilot scale studies are absolutely meagre.

10.4 Combined systems

Technologies consisting of combination of the physical, biological and chemical approaches are in increased use for the removal of micro contaminants from waste water.

10.4.1 Physical process

It is possible to adsorb trace natural contaminants in water, by physical methods, like adsorption. A good variety of adsorbents are synthesized by researchers for adsorption of the emerging and other contaminants. Also, because of the molecular size of the contaminants, microfiltration and ultrafiltration are very effective for removing turbidity, although they are inadequate for removing micro pollutants. Some hybrid systems also consist of biochar-activated ultrafiltration, activated carbon-powdered ultrafiltration, and osmotic membrane bioreactor-microfiltration (Gao et al., 2015).

10.4.2 Advanced oxidation

The traditional physical, chemical and organic solutions are not enough to put off all the microtoxins, because of their continual and permanent structure. In such cases, ozonation and advanced oxidation may be helpful. In waste water treatment plants, ozone degrades pollutants by creating hydroxyl radicals, which can be durable and much less selective for growing compounds. However, oxidation do not provide an entire mineralization of certain compounds, so possibilities are there for formation of other by-products and metabolites (Dolu et al., 2017).

10.4.3 Combined technologies

A few combined methods for the elimination of pollutants in waste water are being suggested, recognizing the advantage of included techniques. For greater elimination of micro contaminants, microalgae-based technologies can be combined with advanced oxidation processes. The advanced oxidation process will convert resistant compounds into intermediates that are biodegradable. Then these by-products can be fully mineralized by the micro-algae procedures. The toxicity of incompletely oxidized materials can also be minimized by this combined system. The advanced oxidation method increases the efficiency. At some point in the design of the integrated structures, the configuration layout, pH, oxygen, and redox capacity are the essential aspects that need to be studied wisely (Wang et al., 2016).

10.5 Removal efficiency of various emerging contaminants by microalgae

Biodegradation by micro algae is a sustainable and eco-friendly technique for the elimination of emerging contaminants. However, the process is greatly dependent on the type, nature and the structure of the emerging contaminants, micro algae species and the environmental conditions.

10.5.1 Toxicity of wastewater

In particular, in large-scale real-life applications, waste water medium can be poisonous to a variety of microalgae species. This is the primary risk that the micro algae-based waste water system faces. The toxicity of waste water is based on the waste source and the type of waste water. These factors will target well-sized toxicity in urban waste water with high oxygen and ammonium concentration. When microalgae are exposed to excessive conditions, the development of poisonous degrading enzymes is stimulated (Abdel-Raouf et al., 2012; Bilal et al., 2018; Fomina and Gadd, 2014; Mahapatra et al., 2013).

10.5.2 Nutrient deficiency

Often the micro algae based bioremediation faces a shortage of essential vitamins in the waste water medium. The production of microalgae was limited by the availability of carbon, which attempts to eliminate the contaminants from the domestic waste water. In a specific case of the effluent medium of the palm oil mill, the biggest problem is that the nitrogen and phosphorus are consumed for the growth of microalgae, and ultimately bioremediation of the emerging contaminants are affected. It is very reasonable that micro-vitamin and essential mineral deficiencies will limit the growth in microalgae and worsen the bioremediation method in the same way (Smith et al., 1999).

10.5.3 Nutrient removal—nitrogen and phosphorus

Untreated or partially treated waste water has excessive vitamin ranges that produce eutrophication effects when dumped in water bodies. Eutrophication is an extremely critical danger to the long-term health of marine ecosystems and their equilibrium. In urban and agricultural waste water, the primary sources of anthropogenic nutrients that accumulate in these habitats are predominantly floor water and ecosystems. In wastewater treatment, the removal of nutrients to the applicable ranges for disposal or reuse is a major challenge, and some of the conventional solutions available, based on chemical and physical approaches, have high costs. Similarly, since they result in CO_2 emissions, do not eradicate N and P, and require the use of chemicals, these methods are environmentally unsustainable. Denitrification, in which nitrates are reduced into nitrogen fuel, is the most preferred method for nitrogen elimination. Phosphorus is periodically extracted with ferric chloride by chemical precipitation method. However, both phosphorus and nitrogen can be satisfactorily eliminated by the use of bacteria or algae, with the biomass subsequently destroyed. In autotrophic and heterotrophic environments, the assimilation of natural nitrogen by way of microalgae can occur. A few species of microalgae can assimilate a broad range of organic and inorganic nitrogen-containing compounds, specifically ammonium salts, nitrates and urea, for their bloom. Several parameters may have an effect on microalgae nutrient removal, including pH, temperature, nutrient concentrations, and lighting fixtures. Depending on waste water, these factors will vary and the removal ability will vary between algal species and water conditions (Chojnacka, 2010; Chong et al., 2000; Muñoz et al., 2006; Smith et al., 1999).

10.5.4 Xenobiotic compound removal

Industrial activities launch a variety of toxic chemicals into the biosphere. Bodily, chemical and biological remedies have been applied to eliminate dangerous compounds from business effluents. Due to their difficulty and cost, physical and chemical alternatives are deemed irrelevant. Then again, biological treatment marketers demonstrate capacity software and are economically attractive. Within the dispersion, chemical transformation and bioaccumulation of several toxic xenobiotic compounds, microalgae play a vital role. Micro-organisms like fungi, algae are able to remove natural contaminant by adsorption, absorption and biodegradation. The incomplete combustion of organic compounds forms polycyclic aromatic hydrocarbons (PAHs). Typically, they are chronic and highly poisonous, with carcinogenic and/or teratogenic properties, and are thus the subject of intensive control and surveillance. The metabolic profile of some microalgae species allows their survival in environments contaminated by PAHs and to even assimilate and degrade those contaminants beneath positive conditions. These organisms are able to bio-transform PAHs of low molecular weight (Bayramoglu et al., 2006; Peng et al., 2014; Torres et al., 2008).

10.5.5 Heavy metals removal

Contamination of water sources by heavy metals is a significant environmental situation, due to the fact that the pollutants are highly toxic and recalcitrant and show a bioaccumulation propensity. Almost, all of those contaminants originate from anthropogenic source, including the manufacture of batteries, paints, steel alloys; the production and use of fossil fuels; and mining etc. The presence of metallic ions, such as, cadmium, copper, lead, mercury, and chromium causes poisonous effects to plant life and fauna in aquatic medium and remains for a longer time. Those elements are incorporated into the food chain and collect along the distinct trophic levels (biomagnification) generating deleterious effects, mainly to purchaser organisms of the top of these chains. In urban waste water, a high concentration of heavy metals is located. The ability of treatment systems to do away with these toxic compounds is therefore extraordinarily crucial. Physio-chemical techniques together with precipitation and ion trade contain excessive funding and running expenses, hardly ever meet regulatory requirements, and produce problems of sludge disposal. Metal bioaccumulation with the aid of microalgae may be a possible approach to treat polluted wastewater. Maximum techniques for casting off metals from wastewater use algae, using dry biomass and useless forms, in which adsorption is the

Table 10.1 Performance of micro-algae towards heavy metal removal.

Name of the heavy metal	Micro algal species	Process conditions	Removal efficiency	Reference
Chromium	**Dried algae** Spirogyra condensate, Chlorella	Batch sorption equilibrium process	75–100 percent	(Onyancha et al., 2008; Han et al., 2008)
Chromium and Copper	Dried algaeminiata, Sargassum sp., Chlorococcum	Batch sorption process	>80 percent	(Jacinto et al., 2009)
Selenium	live algae; combined algal–anaerobic bacteria	High rate ponds	94–100 percent	(Gerhardt et al., 1991)
Nickel	Untreated and acid-treated dried algae Oedogonium *hatei*	Batch adsorption process	>75 percent	(Gupta et al., 2010)
Copper. Zinc, Cadmium, mercury	Live algae; Cladophorafracta	Batch process	97–99 percent	(Ji et al., 2012)

number one removal mechanism involved. More recently, methods have been studied to discover the ability to assimilate heavy metals into dwelling cells or into dried biomass. Microalgae are able to accumulate massive amounts of toxic heavy metallic ions from aqueous solutions; algae and some fungi produce polypeptides referred to as chelating dealers able to binding to heavy metals. In addition, these molecules, such as organo-metallic complexes, are divided inside vacuoles to allow the passage of the cytoplasmic concentration of heavy metal ions, thereby neutralizing the metal-associated toxicity. The subsequent step, known as chemisorption is slower and occurs intracellularly and is related to metabolic methods involving active binding groups. Evaluation of the cell distribution showed that huge quantities of metal ions bind to the cellular wall, whilst an insoluble fraction is amassed intracellular. The performance of micro algae towards heavy metals removal is presented in Table 10.1.

10.6 Biomass separation

The separation and healing of microalgae biomass from the growth medium is a crucial step within the production procedure. Microalgae cells range in length from 5 to 50 mm, form strong suspensions in culture media because of their bad surface charges, and secrete natural compounds that keep their balance within the dispersed state. The choice of separation technique relies on residences of the unique microalgae, along with density, length and value of the favoured products. Microalga cultures are commonly very dilute suspensions, with concentrations below 1 g L^{-1}. They can be perfect for mild penetration and can offer the essential biomass for manufacturing processes of by-products of interest to eliminate water and massive volumes of algal biomass, the ideal harvesting technique might also require one or more steps concerning chemical, bodily or biological procedures to acquire the preferred solid-liquid separation. The harvesting of algae may be divided into a two-step manner. First, big amounts of microalga biomass are separated from the culture volume. Subsequently, the wet biomass undergoes thickening, which requires a huge quantity of energy. Presently the most used techniques in microalga harvesting include centrifugation, flocculation, sedimentation, filtration and electrocoagulation (Ahmad et al., 2010; Ting et al., 2017; Wang et al., 2016).

10.6.1 Coagulation/flocculation

Flocculation is the coalescence of cells of suspended algae into large conglomerates which are weakly joined. To begin with, through the interaction between a coagulating agent and the surface charge of the cells, the suspended cells combine into larger debris. Ultimately, the aggregates combine into huge pellets, separating the from the suspension (flocculation). An effective technique for rapidly processing large quantities of biomass is found to be the flocculation technique. As flocculating agents, various chemical compounds have been studied, including a number of inorganic salts of polyvalent metals and organic polymers (polyelectrolytes). Micro-organisms are used to flocculate positive species of microalgae. Physical approach operations, consisting of sedimentation, flotation, centrifugation or filtration, usually involve an initial degree of flocculation as a means of cell grouping. The use of inorganic coagulants creates a large amount of sludge, causing the death of microalgae or the inhibition of growth. For some types of microalgae, aluminium is an efficient coagulant and it may be inappropriate to use metallic salts when the biomass may be applied in aquaculture, or as fodder or fertilizer. Flocculants should ideally be low-cost, non-toxic and solid at low concentrations.

10.6.2 Electrochemical methods

Physical/chemical processes that use electrical currents to remove sacrificial metal electrodes are electrochemical techniques to provide the necessary coagulation/flocculation ions. Electrocoagulation and electro flocculation are successfully used as alternative techniques for producing steel hydroxides for the separation of microalga biomass. Technology combining electrocoagulation and electro floatation have attracted significant interest for the separation of algal biomass.

10.6.3 Filtration

Filtration is the most aggressive technique of harvesting and it is viable to concentrate microalgae with efficiencies ranging from 10 to 20 percent (w), although the procedure requires relevant maintenance, along with the cleansing and substitution of filters. Different filtration mechanisms are micro, ultra, pressure; vacuum and tangential flow filtration. Microalgae lines with broad cellular dimensions can be concentrated using pressure or vacuum filters (Nurra et al., 2014).

10.6.4 Centrifugation

Through the use of centrifugal pressure, most of the microalgae can be retrieved from a dilute suspension. Centrifugation is a dewatering technique for concentrated biomass that has a higher life cycle strength intake (700 MJ/dry tone of algae), and equipment cost than other methods. But, it may still be considered a viable substitute, mainly within the production of microalgae to achieve high-value products or when utilized in an integrated manner collectively with different preliminary harvesting strategies. Energy-efficient methods need to be developed for the attention of biomass prior to centrifugation to permit the utility of this method in micro algae cultivation.

10.7 Harvesting of algal biomass

Microalgae are difficult to extract from a solution by nature. They are usually smaller than 30 lumen and have a density that is equivalent to water. The economic reaping of micro algae is very difficult, resulting in limited commercial use of algae. There are several strategies for microalgal harvesting and dewatering. A number of physio-chemical methods, such as sedimentation, flocculation, flotation, filtration, centrifugation etc. alone or in combination are used. A brief summary of the techniques of microalgal harvesting is given below (Lewis, 2013; Uduman et al., 2010; Van Den Hende et al., 2011; Van Wagenen et al., 2014; Vandamme et al., 2011).

10.7.1 Sedimentation

In flocculation process, algae cells present together in the floc spontaneously or due to the addition of a flocculant clump. The settling rate can be increased by increasing the size of the particles, which makes flocculation a viable pre-treatment step. Lime and other multivalent metal salts are widely used to remove suspended solids and are effective in extracting microalgae from waste water. Inorganic flocculants are not so effective and but the cationic polyelectrolytes are highly efficient in recovering microalgae.

10.7.2 Floatation

Floatation is a mechanism that moves microalgal particles to the surface using air bubbles. The system of bubble processing categorizes floatation processes. Dissolved air floatation (DAF) is a method that produces bubbles ranging from 10 to 100 lm, using pressurized water. This is the most common lagoon wastewater treatment system for microalgal biomass harvesting. In a reactor, continuous air is passed with a high-speed mechanical agitator, known as foam or froth flotation, is the phase through which algae float. Flotation is an efficient technique for extracting microalgae in conjunction with flocculation.

10.7.3 Filtration

Filtration is a highly effective algae harvesting technique. It is faster and a lower-cost option compared to centrifugation. However, due to factors like membrane replacement, fouling, and pumping, membrane filtration can be costly for large-scale processes. Microfiltration (0.1–10 lumens pore size) is usually most effective for algae harvesting, while macro-filtration can be used in the case of large cells. In the waste water industry, rotary vacuum filters and belt filters are commonly used in the harvesting of microalgae. However, the method may still be inefficient in terms of resources.

10.7.4 Centrifugation

In algae harvesting, centrifugation is rapid and effective and is considered to be one of the most practical methods of algal harvesting. On an industrial scale, several sizes and types of centrifuges are currently used. One of the most widely used is a disk stack centrifuge. Spiral plate centrifuges are considered as the best prototypes for algae harvesting.

10.7.5 Drying

Drying is required after harvesting. Due to the high enthalpy of water, drying can be an energy-cost method. Spray drying, drum drying, freeze drying, and sun drying are used commonly for drying micro algae. Solar drying is the most cost efficient, but needs wide areas of hydro-cyclone land decanter.

10.8 Necessity of pre and post treatment of micro algae

There are three types of pre-treatment techniques: physical (thermal and mechanical), chemical, and natural. Apparently, physical pre-treatments are the satisfactory approach for cellular disruption of microalgae. With the assistance of mechanical agitation (sonication and microwave), they work by breaking down the crystalline structures within the cellular wall. Thermal pre-treatment is more commonly studied among physical pre-treatments, but its efficacy is dependent on the species of microalgae. Pre-treatment has become critical as a physical separation and photolysis pre-treatment procedure to remove larger debris and reduce the organic load in case of bio-remediation. Similarly, post-treatment improves the effluent quality. Activated carbon adsorption, filtration, ozonation, and ultrasound treatments are some of the effective post treatment methods. For the wastewater treatment phase, each pre-treatment and post-treatment cycle may carry additional costs and energy consumption. Coupling the set-up of built wetland treatment with microalgae treatment or activated sludge approach. Such approaches can be cost-effective in contrast to other techniques (Sakarika et al., 2020; Serejo et al., 2020a; Serejo et al., 2020b).

10.9 Conclusion

In recent years, the micro algae based systems are widely studied for removal of the toxins from wastewater has it finds potential for the elimination of the emerging contaminants. Biosorption enables pollutants to be converted to less toxic products. However, degradation is dependent up on the algal species and the nature of the pollutant, so extensive research is needed for making a library of specific micro algae-contaminants grouping. Technologies using microalgae systems have become excellent alternatives to the physical, chemical and biological techniques used in conventional wastewater and gas effluent treatment. The method has some limitations due to secondary contamination by the algae. Microalgae have basic mobile structures and are capable of surviving in almost all environments. For their growth, they need optimum water, light and CO_2. A safe alternative for wastewater treatment has been found to be micro algal biosorption of heavy metals. A number of rising pollutants have been found to have the potential to flush out, extract or bio-transform microalgae. So, it can be concluded that the micro algae based systems are promising alternatives with some limitations for environmental protection.

References

Abdel-Raouf, N., Al-Homaidan, A.A., Ibraheem, I.B.M., 2012. Microalgae and wastewate treatment. Saudi J. Biol. Sci. 19, 257–275.

Agüera, A., Plaza-Bolaños, P., Fernandez, F.A., 2020. Removal of contaminants of emerging concern by microalgae-based wastewater treatments and related analytical techniques, in: Current Developments in Biotechnology and Bioengineering. Elsevier, pp. 503–5252.

Ahmad, A.L., Mat Yasin, N.H., Derek, C.J.C., Lim, J.K., 2010. Optimization of microalgae coagulation process using 27. Chojnacka, K. Biosorption and bioaccumulation – the prospects for practical applications. Environ. Int. 36, 299–307.

Ahmed, M.B., Zhou, J.L., Ngo, H.H., Guo, W., Thomaidis, N.S., Xu, J., 2017. Progress in the biological and chemical treatment technologies for emerging contaminant removal from wastewater: a critical review. J. Hazard. Mater. 323, 274–298. https://doi.org/10.1016/j.jhazmat.2016.04.045.

Aketo, T., Hoshikawa, Y., Nojima, D., Yabu, Y., Maeda, Y., Yoshino, T., et al., 2020. Selection and characterization of microalgae with potential for nutrient removal from municipal wastewater and simultaneous lipid production. J.Biosci. Bioeng. xxx.

Amenorfenyo, D.K., Huang, X., Zhang, Y., Zeng, Q., Zhang, N., Ren, J., et al., 2019. Microalgae brewery wastewater treatment: potentials, benefits and the challenges. Int. J. Environ. Res. Public Health 16 (11), 1910. https://doi.org/10.3390/ijerph16111910.

Bayramoglu, G., Tuzun, I., Celik, G., Yilmaz, M., Arica, M.Y., 2006. Biosorption of mercury(II), cadmium(II) and lead(II) ions from aqueous system by microalgae Chlamydomonasreinhardtii immobilized in alginate beads. Int. J. Miner. Process 81, 35–43.

Bilal, M., Rasheed, T., Sosa-Hernandez, J., Raza, A., Nabeel, F., Iqbal, H., 2018. Biosorption: an interplay between marine algae and potentially toxic elements—A Review. Mar Drugs 16 (2), 65. https://doi.org/10.3390/md16020065.

Chang, J.-S., Show, P.-L., Ling, T.-C., 2017. Photobioreactors. In Current Developments in Biotechnology and Bioengineering. https://doi.org/10.1016/B978-0-444-63663-8.00011-2.

Chojnacka, K., 2010. Biosorption and bioaccumulation – the prospects for practical applications. Environ. Int. 36, 299–307.

Chong, A.M.Y., Wong, Y.S., Tam, N.F.Y., 2000. Performance of different microalgal species in removing nickel and zinc from industrial wastewater. Chemosphere 41, 251–257.

Delrue, F., Alvarez-Dıaz, P., Fon-Sing, S., Fleury, G., Sassi, J.-F., 2016. The environmental biorefinery: using microalgae to remediate wastewater, a win-win paradigm. Energies 9 (3), 132. https://doi.org/10.3390/en9030132.

Dolu, T., Ates, H., Argun, M.E., Yel, E., Nas, B., 2017. Treatment alternatives for micropollutant removal in wastewater. Selcuk University Journal of Engineering, Science and Technology 5 (2), 133–143. https://doi.org/10.15317/Scitech.2017.77.

Escapa, C., Coimbra, R.N., Neuparth, T., Torres, T., Santos, M.M., Otero, M., 2019. Acetaminophe removal from water by microalgae and effluent toxicity assessment by the zebra fish embryo bioassay. Water (Basel) 11 (9), 1929.

Escapa, C., Coimbra, R.N., Paniagua, S., Garcıa, A.I., Otero, M., 2016. Comparative assessment of diclofenac removal from water by different microalgae strains. Algal Res. 18, 127–134. https://doi.org/10.1016/j.algal.2016.06.008.

Ferrando, L., Matamoros, V., 2020. Attenuation of nitrates, antibiotics and pesticides from groundwater using immobilised microalgae-based systems. Sci. Total Environ. 703, 134740. https://doi.org/10.1016/j.scitotenv.2019.134740.

Fomina, M., Gadd, G.M., 2014. Biosorption: current perspectives on concept, definition and application. Bioresour. Technol. 160, 3–14. https://doi.org/10.1016/j.biortech.2013.12.102.

Gao, F., Yang, Z.-H., Li, C., Zeng, G.-M., Ma, D.-H., Zhou, L., 2015. A novel algal biofilm membrane photobioreactor for attached microalgae growth and nutrients removal from secondary effluent. Bioresour. Technol. 179, 8–12. https://doi.org/10.1016/j.biortech2014.11.108.

Gerhardt, M.B., Green, F.B., Newman, R.D., Lundquist, T.J., Tresan, R.B., Oswald, W.J., 1991. Removal of selenium using a novel algal-bacterial process. Res. J. Water Pollut. Control Fed., 799–805.

Gruchlik, Y., Linge, K., Joll, C., 2018. Removal of organic micropollutants in waste stabilisation ponds: a review. J. Environ. Manage. 206, 202–214. https://doi.org/10.1016/j.jenvman.2017.10.020.

Gupta, V.K., Rastogi, A., Nayak, A., 2010. Biosorption of nickel onto treated alga (Oedogonium hatei): application of isotherm and kinetic models. J. Colloid Interface Sci. 342 (2), 533–539.

Han, X., Wong, Y.S., Wong, M.H., Tam, N.F.Y., 2008. Feasibility of using microalgal biomass cultured in domestic wastewater for the removal of chromium pollutants. Water Environ. Res. 80 (7), 647–653.

Jacinto, M.L.J., David, C.P.C., Perez, T.R., De Jesus, B.R., 2009. Comparative efficiency of algal biofilters in the removal of chromium and copper from wastewater. Ecol. Eng. 35 (5), 856–860.

Ji, L., Xie, S., Feng, J., Li, Y., Chen, L., 2012. Heavy metal uptake capacities by the common freshwater green alga Cladophorafracta. J. Appl. Phycol. 24 (4), 979–983.

Lewis, D.M., 2013. Harvesting, thickening and dewatering microalgae biomass. In: Borowitaka, M.A., Moheimani, N.R. (Eds.), Harvesting, thickening and dewatering microalgae biomass. Algae for Biofuels and Energy, 165.

Mahapatra, D.M., Chanakya, H.N., Ramachandra, T.V., 2013. Treatment efficacy of algae-based sewage treatment plants. Environ.Monit. Assess. 185, 7145–7164.

Matamoros, V., Gutierrez, R., Ferrer, I., Garcıa, J., Bayona, J.M., 2015. Capability of microalgae-based wastewater treatment systems to remove emerging organic contaminants: a pilot-scale study. J. Hazard. Mater. 288, 34–42. https://doi.org/10.1016/j.jhazmat.2015.02.002.

Matamoros, V., Uggetti, E., Garcıa, J., Bayona, J.M., 2016. Assessment of the mechanisms involved in the removal of emerging contaminants by microalgae from wastewater: a laboratory scale study. J. Hazard. Mater. 301, 197–205. https://doi.org/10.1016/j.jhazmat.2015.08.050.

Muñoz, R., Alvarez, M.T., Muñoz, A., Terrazas, E., Guieysse, B., Mattiasson, B., 2006. Sequential removal of heavy metals ions and organic pollutants using an algal-bacterial consortium. Chemosphere 63, 903–911.

Norvill, Z.N., Shilton, A., Guieysse, B., 2016. Emerging contaminant degradation and removal in algal wastewater treatment ponds: identifying the research gaps. J. Hazard. Mater. 313, 291–309. https://doi.org/10.1016/j.jhazmat.2016.03.085.

Norvill, Z.N., Toledo-Cervantes, A., Blanco, S., Shilton, A., Guieysse, B., Muñoz, R., 2017. Photodegradation and sorption govern tetracycline removal during wastewater 30 H. T. NGUYEN ET AL.treatment in algal ponds. Bioresour. Technol. 232, 35–43. https://doi.org/10.1016/j.biortech.2017.02.011.

Nurra, C., Clavero, E., Salvado, J., Torras, C., 2014. Vibrating membrane filtration as improved technology for microalgae dewatering. Bioresour. Technol. 157, 247–253.

Onyancha, D., Mavura, W., Ngila, J.C., Ongoma, P., Chacha, J., 2008. Studies of chromium removal from tannery wastewaters by algae biosorbents, Spirogyra condensata and Rhizocloniumhieroglyphicum. J. Hazard. Mater. 158 (2–3), 605–614.
Peng, F.-Q., Ying, G.-G., Yang, B., Liu, S., Lai, H.-J., Liu, Y.-S., et al., 2014. Biotransformation of progesterone and norgestrel by two freshwater microalgae (Scenedesmusobliquus and Chlorella pyrenoidosa): transformation kinetics and products identification. Chemosphere 95, 581–588.
Sakarika, M., Koutra, E., Tsafrakidou, P., Terpou, A., Kornaros, M., 2020. Microalgae-based remediation of wastewaters, in: Microalgae Cultivation for Biofuels Production. Academic Press, pp. 317–335.
Serejo, M.L., Farias, S.L., Ruas, G., Paulo, P.L., Boncz, M.A., 2020b. Surfactant removal and biomass production in a microalgal-bacterial process: effect of feeding regime. Water Sci. Technol. wst2020276. https://doi.org/10.2166/wst.2020.276.
Serejo, M.L., Morgado, M.F., García, D., Gonzalez-Sanchez, A., Mendez-Acosta, H.O., Toledo-Cervantes, A., 2020a. Environmental resilience by microalgae, in: MicroalgaeCultivation for Biofuels Production. Academic Press, pp. 293–315.
Smith, V.H., Tilman, G.D., Nekola, J.C., 1999. Eutrophication: impacts of excess nutrient inputs on freshwater, marine, and terrestrial ecosystems. Environ. Pollut. 100, 179–196.
Ting, H., Haifeng, L., Shanshan, M., 2017. Progress in microalgae cultivation photobioreactors and applications in wastewater treatment: a review. Int. J. Agric. Biol. Eng. 10, 1–29. https://doi.org/10.3965/j.ijabe.20171001.2705.
Torres, M.A., Barros, M.P., Campos, S.C.G., Pinto, E., Rajamani, S., Sayre, R.T., Colepicolo, P., 2008. Biochemical biomarkers in algae and marine pollution: A review. Ecotox. Environ. Safe. 71, 1–127.
Uduman, N., Qi, Y., Danquah, M.K., Forde, G.M., Hoadley, A., 2010. Dewatering of microalgal cultures: a major bottleneck to algae-based fuels. J. Renew. Sust. Energy 2, 012701.
Van Den Hende, S., Vervaeren, H., Desmet, S., Boon, N., 2011. Bioflocculation of microalgae and bacteria combinedwith flue gas to improve sewage treatment. N. Biotechnol. 29, 23.
Van Wagenen, J., Holdt, S.L., De Francisci, D., Valverde-Pérez, B., Plósz, B.G., Angelidaki, I., 2014. Microplate-based method for high-throughput screening of microalgae growth potential. Bioresour. Technol. 169, 566.
Vandamme, D., Pontes, S.C.V., Goiris, K., Foubert, I., Pinoy, L.J.J., Muylaert, K., 2011. Evaluation of electrocoagulation–flocculation for harvesting marine and freshwater microalgae. Biotechnol. Bioeng. 108, 2320.
Wang, Y., Roddick, F., Fan, L., 2016. Sunlight photodegradation of micropollutants in wastewater effluent. Ozwater '16 Water Liveable Communities Sustain Ind, 1–6.

CHAPTER 11

Bioaugmentation as a strategy for the removal of emerging pollutants from wastewater

Prathap Somu, Subhankar Paul
Structural Biology and Nanomedicine Laboratory, Department of Biotechnology and Medical Engineering, National Institute of Technology, Rourkela, Odisha, India

11.1 Introduction to bioremediation

Water is the prime requisite for life as the availability of clean and pure water for people, as well as industrialization, is vitally important for economic development depends. Although 71 percent of Earth's surface is covered with water, only 1 percent is suitable for human needs for both their personal and commercial application. Moreover, the release of untreated or sparsely treated wastewater into the water bodies lawfully or unlawfully further dwindles the water resources, thereby causing a water deficit. The world water council and several other organizations are working towards the documentation and amelioration of new plans to combat water scarcity. According to the United Nation's Water Development Report of 2020, there are about 2.2 billion people who do not have access to clean and safe drinking water (Hasan et al., 2019). The severe inadequacy of drinking water and the high level of contamination of water lead to multiple diseases like cholera, diarrhea, dysentery, hepatitis A, typhoid, and polio and thereby eventually affecting the healthcare system (Siddiqui et al., 2021). Therefore, it has become fundamental to purify and reuse the wastewater to combat problems related to water contamination and water scarcity in the water deficit areas.

However, the existing conventional methods of water purification faces a major challenge due to new emerging pollutants like micropollutants (Liu et al., 2020). The emerging pollutants may be either natural or synthetic compounds originating from Human activities such as agriculture, mining as well as industrial processes. The emerging pollutants may include pharmaceutical compounds, detergents, and surfactants such as triclosan, and alkyl phenols; fertilizers and pesticides; plasticizers such as phthalates; and fire retardants such as polybrominated diphenyl ether, etc., The other major pollutants include trace metals such as cadmium, iron, lead, and mercury, etc. (Gavrilescu et al., 2015). The major disadvantages of using current methods such as ozonization, chlorination, membrane filtration process for the removal of emerging pollutants is that they utilize a lot of energy, loss of water leading to secondary contamination, and high cost

Biodegradation and Detoxification of Micropollutants in Industrial Wastewater
DOI: https://doi.org/10.1016/B978-0-323-88507-2.00007-5

(Rathi et al., 2020). Hence, there is the requirement of much-needed up-gradation of existing technologies or replacement of new technologies, which is capable of encountering newly emerging contaminants.

Recently, the process of using living organisms to neutralize the harmful toxic substance known as Bioremediation is gaining importance (Dangi et al., 2019). Bioremediation involves the utilization of the beneficial microbes and plants where the contaminants are used as the food source to produce energy thereby assimilate these complex toxic contaminants to produce very simple harmless gases like methane and carbon dioxide (Azubuike et al., 2016). The reduction of the pollutants depends mainly on the nature of the pollutants which can be anything from heavy metals, sewage, hydrocarbons, plastics, agrochemical, and dyes. Apart from the nature of the pollutant, the site of application, the degree of pollution, location, cost, and environment are some of the criteria considered for the bioremediation processes. The most commonly used bioremediation techniques are Micro-remediation and Phytoremediation. Micro-remediation involves the application of microorganisms to transform hazardous pollutants into non-hazardous ones. For example, Extremophiles through their cellular metabolism detoxify the toxic compounds even in harsh conditions (Singh et al., 2008). Similarly, phytoremediation utilizes the uptake mechanism potential of plants, thereby reducing the mobility of pollutants in contaminating the soil and water. Macrophytic Plants are used for the removal of nitrogen and phosphorous from water (Yu et al., 2019).

Even though micro-remediation and phytoremediation are environmentally safe, but suffer greatly as the mechanisms involved is very unstable due to the absorbing capacity of the microbes and plants are quite limited and varies greatly based on the various physiological as well as environmental factors (Ojuederie and Babalola, 2017). The advancement in nanotechnology enables the application of these microorganisms and plants in bioremediation by providing better absorption and thus proving to be time and cost-efficient. For instance, nanomaterials such as Carbon Nanotubes, due to their great adsorption properties are used to adsorb various heavy metals such as cadmium, chromium, lead, zinc, and arsenic compounds. Nano Bioremediation in simpler terms can be described as an integration of Nanotechnology and Bioremediation to provide remediation that has double benefits than the individual processes.The process of inoculating the produced nanoparticle with the desired microorganism is proven to have a greater degradation capability (Galdames et al., 2017). Nano-stimulated phytoremediation and microbial remediation for the degradation of the resistant organic compounds into their simpler forms by the combined activities of plants, microorganisms, and nanoencapsulated enzymes. As each strategy has its advantages and drawbacks, the incorporation of remediation techniques could be considered as an answer to tackle remediation issues. Nanobioremediation is one such sort of strategy which got a great deal of consideration in recent years. Nano-bioremediation amalgamates the advantages of nanotechnology along with the favourable circumstances of bioremediation. This current section gives a brief record of nanotechnology and an assortment of nanostructured materials detailed for eliminating natural and inorganic impurities from ecological grids followed by the portrayal of nano-bioremediation procedure, its application cycles, and strategies.

11.2 Current bioremediation methods and their limitations

There are different types of bioremediation technologies that are used in the treatment of polluted areas such as phytoremediation, bioleaching, bioventing, composting, bioreactor, and biostimulation. There are two types of bioremediation treatment based on the removal of wastes namely, in situ and ex-situ bioremediation (Kamarudheen et al., 2020).

11.2.1 Microbial remediation

The limitations in physical and chemical remediation such as ineffectiveness, cost, time consumption, and various environmental factors led to the advent of Microbial remediation (Dangi et al., 2019). The remediation process involving microorganisms and their metabolic pathways either to inhibit the concentration of the pollutants or to convert the toxic compounds into an inert form. Microbial remediation is usually achieved either by introducing a large population of the bred microbes at the site of contamination or by fabricating favourable conditions like temperature, pH, food source, and oxygen in the polluted site for the ideal growth of the microbes. The microbial remediation processes generally work on the principles of oxidation–reduction reactions (Sharma, 2020). In these reactions either oxygen is commonly added as an electron acceptor to stimulate the degradation of reduced pollutants like hydrocarbons (Canak et al., 2019). or an organic substance is added as an electron donor to neutralize the oxidized pollutants such as nitrates, oxidized metals, perchlorates, chlorinated solvents, and propellants (Gu et al., 2018). Based on the type of microorganism used the microbial remediation is broadly classified into two namely the aerobic and the anaerobic microbial remediation. The aerobic microbial remediation is carried out by the microorganisms such as bacteria and fungi in the presence of oxygen thereby converting the highly toxic compounds into carbon dioxide, water, and some simple inorganic salts (Mikkonen et al., 2018). The anaerobic microbial remediation takes place in the absence of oxygen resulting in the end products such as methane, hydrogen gas, sulphides, and nitrogen gases (Logeshwaran et al., 2018). Based on the site of application of the degradants, there are two types of bioremediation, they are ex-situ and in-situ bioremediation.

11.2.1.1 Ex-situ bioremediation

The contaminants are removed from the contaminated site and are metabolized by the microorganisms is called Ex-situ bioremediation. There are two types of Ex-situ bioremediation namely, Slurry phase or bioreactors and Solid-phase ex-situ bioremediation which includes Land farming, Soil biopiles, and Composting.

11.2.1.2 In-situ microbial remediation

The microbial remediation are usually carried in-situ, a process where the microorganisms are added directly to the site of contamination especially in the cases of soil and water contamination. The in situ microbial remediation are known to be more advantageous than the ex-situ mode of remediation in terms of transport, efficiency, economy, and environment (Kuppusamy et al., 2016). The in situ bioremediation can be further classified as intrinsic in situ bioremediation and engineered bioremediation based on the microbial movement either towards or away from the chemicals in the pollutants, a condition known as chemotaxis. For instance, the halophilic bacterium *Halomonas anticariensis* FP35^{T} in salt concentrations degrades the aromatic compounds phenol and naphthalene as they are the chemoattractants of the bacterium (Tena-Garitaonaindia et al., 2019).

11.2.1.3 Intrinsic in-situ bioremediation

Intrinsic in-situ bioremediation is a process in which the pollutants in the contaminated sites are degraded by the addition of just the indigenous microbes without adding any stimulations. It can also be called Natural Attenuation or Passive Bioremediation. They depend only on the metabolic activities of the consortium of the native microbes present in the contaminated site which have degradation properties (Sharma, 2020). For any successful intrinsic in situ bioremediation, we must keep a check on the four basic parameters such as i) population of the microbes, ii) availability of nutrients for the microbes to grow, iii) optimal environmental conditions like the temperature, pH, and oxygen, and iv) sufficient time to be given for the natural process to take place (Sharma, 2019). The main advantage of the intrinsic mode of in–situ remediation is that it is cheaper than the other in-situ methods of bioremediation. As the intrinsic in-situ bioremediation involve no external stimulations, the process of bioremediation is very slow and time-consuming. Hence to enhance the effectiveness of the intrinsic in-situ bioremediation process we formulate the engineered in-situ bioremediation processes.

11.2.1.4 The engineered in-situ bioremediation process

Engineered in-situ bioremediation is achieved by the enhancement of the microbes introduced to the site of contamination and the physicochemical conditions of the environment. Generally, genetically modified microorganisms are used in these types of bioremediation. The engineered in-situ bioremediation process can be further classified into five types namely bioventing, biosparging, bioslurping, biostimulation, and Bioaugmentation.

11.2.1.4.1 Bioventing

Bioventing is a process in which either air or oxygen is delivered to the indigenous aerobic microorganisms to speed up the natural bioremediation of the volatile and semi-volatile organic compounds present in the vadose (unsaturated) region. It comprises a well and a blower for the even distribution of oxygen throughout the soil above the water level. Bioventing is well suited for drained, medium-coarse structured soil. Bioventing is one of the most common methods used in the biodegradation of hydrocarbons produced in the petroleum industries. It helps in the degradation of volatile organic compounds (Azubuike et al., 2016) and is also used in the treatment of non-aromatic hydrocarbons in petroleum polluted sites (Ossai et al., 2020) and crude oil (Speight, 2020).

11.2.1.4.2 Biosparging

Biosparging is a process in which air is induced in the water Table thereby causing the oxygen dissolved in the underground water to rise and enhance the biodegradation of the organic pollutants in the saturated zone (Zouboulis et al., 2019). This process prevents the accumulation of nitrates, formed as a result of the nitrification process and it also prevents the mobilization of the metals. It helps in treating fossil fuel contamination such as petroleum, diesel, and kerosene (Sharma, 2019).

11.2.1.4.3 Bioslurping

Bioslurping is the amalgamation of vacuum-enhanced pumping and bioventing to remediate the soil and groundwater concurrently. Bioventing ensures the aerobic degradation of the contaminants and the vacuum augmented recovery uses the partial vacuum formed due to the negative pressure to ensure the removal of the free products from the subsurface of the soil (da Silva et al., 2020). The main advantage of bioslurping is the conservation of groundwater, leading to a very cost-effective process and volatile organic compounds are also treated by this method (Azubuike et al., 2016).

11.2.1.4.4 Biostimulation

Biostimulation is described as a process to accelerate bioremediation by the addition of either nutrients or electron donors and acceptors to the site of contamination. They enhance the growth and metabolic activities of the indigenous microbes under a suitable physicochemical environment such as ideal temperature, pH, oxygen, and water conditions (Abatenh et al., 2017). The bioremediation of xenobiotic commonly uses biostimulation as their most feasible option (Landa-Acuña et al., 2020). Marine bacteria are also used in the remediation of certain pollutants such as *Pseudoalteromonas haloplanktis* (Dell'anno et al., 2020) and Deinococcus radiodurans (Manobala et al., 2020).

11.2.1.4.5 Bioaugmentation

The process in which additional microbes are added to the naturally occurring indigenous microbes to enhance the degradation process is known as bioaugmentation. The microbes are either genetically modified or added in a mixed consortium. Depending upon the site of contamination and the nature of the pollutants, the microbes or their metabolic pathways are engineered to take upon the additional stress they encounter at the contaminated sites. Usually, Biostimulation is incorporated along with bioaugmentation for the achievement of better results. Genetically engineered microorganisms such as *E. Coli, Vibrio harveyi*, Sulfate-reducing bacteria, Norcadia species are also used in the bioremediation of hydrocarbons and oil contamination (Singh et al., 2016). Schematic representation of biostimulation and bioaugmentation is shown in Fig. 11.1.

11.2.1.5 Limitations of microbial remediation

Though microbial remediation are environmentally safe, there are some limitations associated with them such as the contaminants are not fully transformed into their harmless forms, and they also may intermediates which are more toxic to the environment. Here the concentrations of the heavy metals and organic compounds also decline the activity of the microbes (Sharma, 2020).

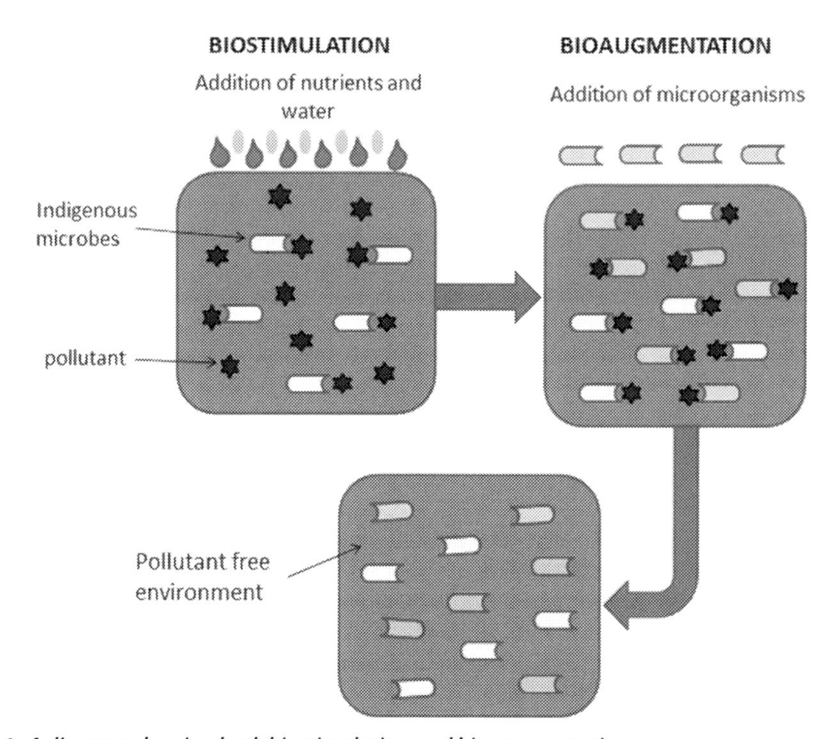

Fig. 11.1 ***A diagram showing both biostimulation and bioaugmentation.***

11.2.2 Phyto-remediation

The concept of phytoremediation was first observed by plants that showed resistance to heavy metals. Phytoremediation is the use of plants along with microorganisms to treat air, water, and soil contaminated areas. Plants absorb the metals with the help of roots and then transfer them to shoots. These plants are further harvested to produce bio-ore called Phytomining (Pirzadah et al., 2014). The plants were observed as metal excluders, metal indicators or bioindicators, and metal accumulators. Metal excluders prevent the entry of metals by changing the permeability of membrane Bioindicators or metal indicators helps in monitoring metals accumulated in estuaries (Phillips et al., 2015). Metal accumulators store metals in their aerial parts and hyperaccumulators have been used to store a large number of metals (Pirzadah et al., 2014). Hyperaccumulation also plays a major role in proving resistance to herbivores due to higher metal accumulation. Certain plants have been used to remove metallic components such as Zn, Co, Cu, Cd, Ni, Se, and other toxic metals (Pirzadah et al., 2014), organic and inorganic compounds. There are six types of phytoremediation which are Phytosequestration, Rhizodegradation, Phytohydraulics, Phytoextraction, Phytovolatilization, and Phytodegradation (Nwadinigwe and Ugwu, 2019; Bolan et al., 2011).

Phytosequestration or phytostabilization is the process of stabilization or immobilization of pollutants that are either absorbed by roots or adsorbed to the surface of roots or produce biochemicals that can sequester, precipitate or immobilize the pollutants. Both aquaticplants and terrestrial plants showed a high accumulation of metals such as Fe, Zn, Cu, Cr, Pb, Hg, As. Genetic engineering approaches have been employed to improve the phytosequestration (Yadav et al., 2018).

Rhizodegradation is the degradation of pollutants present in the soil with the help of plant roots by stimulating the microorganisms present in the soil by releasing organic compounds such as secretions, root exudates (Li et al., 2020). Polycyclic aromatic hydrocarbons (PAH) contaminated in the soil are efficiently removed by the combinations of biochar and rhizosphere environment. *Lolium multiflorum (Lam)* is commonly used for removing PAH due to its fibrous root system and many rhizosphere microorganisms are involved in the degradation of PAH such as Bacteroidetes. sp, Verrucomicrobia. sp, Actinobacteria. sp, Ascomycota. sp, etc. (Li et al., 2020; Kotoky and Pandey, 2020).

Phytohydraulics is the control of the movement of pollutants in the groundwater by the process of reverse osmosis or organic pumps by plants or trees to pull up the pollutants from the groundwater. Water-soluble organic and inorganic compounds were removed by Hybrid poplar, Willow, and cottonwood (Liu et al., 2017).

Phytoextraction is the process of uptake of pollutants from the wastewater or soil by the plant roots and accumulated in the aerial parts of the plant. The arbuscular mycorrhizal (AM) fungi help in the uptake of large amounts of pollutants by the plants. Genetic engineering is also manipulated to produced transgenic plants that show high resistance to heavy metals enhancing phytoextraction (Ali et al., 2017).

Phytovolatilization is the process of removing pollutants by absorption of organic pollutants and other metals by plants, modified into volatile form, and released into the atmosphere. It is used to treat both organic and inorganic compounds such as Se, As, Hg (Limmer et al., 2016). Arsenic compounds were effectively removed by Pteris vittata as vapors. The Se which is also found to be toxic is removed by this process by different plant species along with microorganisms (Chourasia et al., 2014).

Photodegradation is the breakdown of pollutants with the help of microorganisms that are present in the soil and water that are stimulated by the root zone. *Scirpus grossus* has been used to treat diesel contaminated water along with rhizobacteria (Nash et al., 2020).

11.2.2.1 Limitations of phytoremediation

The main disadvantage is that it takes more time and very slow process. It also has low efficiency and a limited number of targeted metals to be extracted. There is no control in the final destination of contaminants and not applicable for all components (Farraji et al., 2016).

11.2.3 Advantages and disadvantages of bioremediation

For bioremediation to be effective, the bioremediation strategies rely upon having the correct microorganisms in the ideal spot with the premium ecological factors for the degradation to happen. The privileged microorganisms are microscopic organisms or parasites, which have the physiological and metabolic capacities to eliminate the contaminations. Bioremediation offers a few points of interest over regular strategies, for example, land rolling or incineration. Bioremediation is possible on site, is regularly more affordable and site interruption is negligible, it dispenses long-term liability, disposes of long haul obligation, has more noteworthy public acknowledgment, with administrative support, and it tends to be combined with other physical or compound treatment techniques. Bioremediation also has its restrictions (Roychowdhury et al. 2019). Buildups from the treatment are normally not harmful substances like carbon dioxide, water, and cell biomass. The compounds which are considered toxic are converted to harmless products.

11.3 Nanoparticles for Bioaugmentation: Nanobioaugmentation

Nanotechnology application, in past decades, has taken up various sectors like electronics, medicine, sports, pharmaceutics, cosmetics, textiles, optics, and many. Even the area of environmental remediation is one among such sectors. All those ongoing researches and the number of published articles in this field of nanotechnology is evidence that it could efficiently take up remediation tasks and challenges efficiently (Singh et al., 2020). Sustainable remediation aims to reduce the concentrations of pollutants and increasing the additional environmental impacts.

Different types of Nanomaterials are used in bioaugmentation due to various reasons such as increased surface area per unit mass of a material is obtained at the nanoscale and hence, the contact is more between the material and surrounding material (Nanjwade et al., 2018). Another major reason is the lesser activation energy is required for feasible chemical reactions. The latter explains that NMs show quantum effect. The NPs also exhibits surface plasmon resonance using which toxic materials can be detected. Metallic and non-metallic NMs of various sizes and shapes can be used for environmental clean-up purposes (Rizwan et al., 2014). Since Nanoparticles can be easily entered into a contaminated zone or diffuse into the zones where microparticles are out of reach and the reactivity to redox-amenable contaminants is higher, different types of carbon base Nanomaterials, bimetallic Nanoparticles, single metal Nanoparticles are used. Pollutants like carbon tetrachloride (CT) can form weak complexes with oxide-coated Fe0. The reactivity is increased by oxide coating and hence, CT can be broken down into CH_4, CO or formate through electron transfer, whereas benzoquinone, by trichloroethene, and other chlorinated aliphatic hydrocarbons are broken down low toxic chemicals (Rizwan et al., 2014). Along with field applications, a photo-electrocatalytic reaction is followed by TiO_2 nanotubes to degrade pentachlorophenol (PCP) (Zhou et al., 2014).

For reductive dichlorination, single metal NPs can be used as biocatalyst. In a bio-reductive assay containing Pd (II), different substrates such as formate, acetate, and hydrogen are added to the *Shewanella oneidensis* as electron donors where the palladium, Pd (0) NPs are deposited inside the cytoplasm and on the cell wall of the microorganism so that charged with H* radicals. When a chlorinated compound is brought in contact with this charged Pd (0), which has been deposited on *S. oneidensis* cells, the chlorine molecule is removed from the chlorinated compound as a result of the H* radicals on the Pd (0) which can catalytically react with PCP. Nanoparticles can also be used in degrading specific chemicals by immobilizing microbial cells. Magnetic nanoparticles have been functionalized with ammonium oleate and coated on the surface of *Pseudomonas delafieldii,* which is much different from the conventional immobilization technique. These magnetic NP-coated cells are concentrated at a specific location on the reactor wall by applying an external magnetic field to these microbial cells which are separated from the bulk solution and recycled for the treatment of the same substrate (Sriplai and Pinitsoontorn, 2021). At high biomass concentration, these microbial cells are added into a bioreactor and were demonstrated to desulfurize organic sulphur from dibenzothiophene i.e., fossil fuels, as effectively as non-Nanoparticle coated cells (Rizwan et al., 2014). The specific NMs from here will be focusing on remediation of various types of waste which includes applications for groundwater and waste-water, solid waste, petroleum and petroleum products (hydrocarbon), soil, heavy metal pollution, and uranium remediation. The ability of NMs to reduce pollution is in progress and could potentially be the important revolutionary changes in the environmental field in the coming decades.

Nanoparticles with the size of 1 to 100 nm have specific properties such as high surface area, high adsorption, and high reactivity. Nanotechnology has been used in the treatment of wastewater, contaminated soil, and air. Nanomaterials used have different forms such as magnetic nanoparticles, carbon nanotubes, quantum dots, and metal nanoparticles (Singh et al., 2020). Nanoparticles can be synthesized as nanospheres, nanorods, nanocubes, nano-triangles, etc. Nanoparticles can be synthesized by either top-down method or bottom-up approach methods employing physical or chemical or biological methods (Singh et al., 2020). There are two groups of nanoparticles which include organic and inorganic nanoparticles (Nwadinigwe and Ugwu, 2019). Based on the site of synthesis, it is classified as natural NPs, incidental NPs due to human activities such as fossil fuel combustion and engineered NPs such as nanosilver, nanogold, etc. (Rai et al., 2018; Shukla and Iravani, 2018). The enormous amount of toxic substances such as metal contaminants like Ni, Fe, Pb, Cr, and industrial dye effluents, arsenite, tetracycline, sulphate, copper, phenol, atrazine, etc., can be removed by the nanoparticles. The different types of nanoparticles used are MnO_2, Pd, ZnO, TiO_2, CuO, CeO_2, nZVI, Al_2O_3, Fe/Ni bimetallic nanoparticles. Nanoparticles are used in bioremediation since it has high removal efficiency, highly economical and very less time (Singh et al., 2020). Fe NPs have been used to remove total petroleum hydrocarbons, heavy metals such as, Cr, Cu, Pb, Hg, trichloroethylene, brominated methanes, and many organic compounds, Zn NPs have been used to remove organic dyes, methylene blue, fuchsin, resorcinol, malachite green, etc. Cu NPs have been used to remove methylene blue, methyl orange and dichloromethane, Ag NPs have been used to remove organic dyes and textile effluents and Au NPs have been used to treat tertiary dye effluents (Chauhan et al., 2020).

Nano-bioaugmentation is an amalgamation of bioaugmentation and nanotechnology coming together to bring about remediation which is both time and cost-efficient, and environmentally safe rather than the individual process. A better degradation is achieved as the integrated approach overcomes the disadvantages of the separate technologies. For instance, the nZVI particles when coupled with microbial strains showed more efficiency in the remediation of the pollutants. In the case of recalcitrant compounds like Chlorinated Aliphatic Hydrocarbons (CAH), it was found that cannot be removed efficiently by the application of nZVI and organochlorine respiring bacteria (ORB) individually. The combination of nanoparticles along with the microorganism at an appropriate proportion showed effective removal of CAHs (Dong et al., 2019). The bacterial environment contains cysteine and vitamins for the regeneration of the spent nZVI. In another instance, the nZVI nanoparticle in combination with anaerobic bacteria such as Sphingomonas species in a reductive-oxidative strategy was found to be more effective for the remediation of Polybrominated Diphenyl Ethers (PBDEs). The aerobic bacteria effectively degraded the lower BDEs which are produced by the nZVI particles on reacting with PBDEs through reductive bromination (Zhao et al., 2018). The most important disadvantage concerning abiotic nitrate reduction namely the fluctuation of temperature remains unaffected by the biostimulation of nFEO nanoparticles with the nitrate-reducing microbial culture (Singh et al., 2020). Another study reported that the bimetallic nanoparticles like Pb/nFeo in integration with Sphingomonas sp. effectively degraded the carboxymethyl cellulose (CMC) pollutants.

Polychlorinated biphenyls (PCBs) treatment by approach was investigated by (Le et al., 2015) showed that PCBs can be effectively transformed to less toxic compounds by the sequential treatment of PCB with Pd/Fe nanoparticles followed by bioremediation with *B. xenovorans*. The toxic level of PCBs in E-coli DH5α before and after treatment were further analyzed using toxic equivalent values and reports shown that the cytotoxicity of residual PCBs toward E-coli was lower after treatment. (Němeček, Pokorný et al., 2016) observed that the Cr (VI) removal in an integrated system having nZVI and whey generated microbes up to 97–99 percent when groundwater contaminated with Cr (VI) were injected with both nZVI and whey. In addition to the removal of the contaminants, microbes were involved in regenerating the oxidized Fe0 nanoparticles which further reduced the dosage of nanoparticles by increasing the rate of remediation.

The application of the integrated nano-bioaugmentation process in the treatment system can be applied in two ways (Němeček et al., 2016). Initially, the contaminant is subjected to nanoparticles and then the further process is carried out by adding a bioagent. The above stated is one of the methods and the second one follows. It is a concurrent or combined method. In this method, simultaneously both bioagent and nanoparticle are added to the system. We have tabulated various reports were bioagents and nanoparticles were used in combination or concurrently for bioremediation of various pollutants in Table 11.1 as follows.

11.3.1 Nano-stimulated microbial bioaugmentation

Nano-stimulated microbial bioaugmentation is defined as the process of degradation of the pollutants by the nanoparticles synthesized with the help of microorganisms (Yadav et al., 2017). Nanotechnology-based remediation are more systematic because of their properties such as smaller size and large surface area to volume ratio (Baruah et al., 2018). The above-stated advantages of nanotechnology are integrated with the advantages of microbial remediation to ensure the most effective and environmentally safe methods of pollutant removal (Singh et al., 2020). The involvement of microorganisms like bacteria, fungi, and algae pave a safe pathway for the production of green nanoparticles (Bolade et al., 2020).

Microorganisms produce nanoparticles in two ways namely the extracellular and the intracellular mechanisms (Verma and Kuila, 2019). In the extracellular method, the metal ions are enzymatically reduced to produce nanoparticles that are widely dispersed and smaller in size in distribution. Here, the natural capping agents such as the proteins, peptides, and genes provide solidity and minimize the clumping of the nanoparticles (McClements, 2018). In the intracellular mode of action, the metal ions interact with

Table 11.1 Various nanoparticles used in combination with bioagents (Microbes, enzyme) for the nano-bioaugmentation process for detoxification of pollutant.

S.no	Bioagents	Nanoparticle	Contaminant	Mechanism	References
1.	Laccase derived from *Trametes versicolor*	Pd/nFe	Triclosan	Iron nanoparticles were used to remediate the triclosan along with the laccase produced from *Trametes Versicolor* which converts toxic byproducts into nontoxic compounds	(Bokare et al., 2012)
2.	Dehalococcoides spp.	nZVI	TCE	The metabolic activity of *Dehalococcoides spp.* is activated by nZVI to reduce the presence of chlorinating compounds. TCE was also removed by dechlorinating bacterial species by producing by-products.	(Xiu et al., 2010)
3.	Acinetobacter junii, Bacillus subtilis, E. coli,	nZVI-C-A beads	Cr(VI)	More than 90 percent of Cr(VI) was removed by nZVI-C-A beads which are covered by a thin biofilm. The combined effects of nanoparticle and biofilm provided enhanced removal of chromium	(Ravikumar et al., 2016)
4.	Shewanella oneidensis MR-1	Pd(0) nanoparticles	PCBs	Less harmful by-products are produced by the reduction of 90 percent of PCBs by the action Pd(0) nanoparticles formed from *Shewanella oneidensis* MR-1 species	(Windt et al., 2005)
5.	Pseudomonas delafieldii	Magnetic Fe_3O_4 nanoparticles	Dibenzothiophene	Dibenzothiophene was highly desulfurilated by Magnetic Fe_3O_4 nanoparticles coated *Pseudomonas delafieldii*than the free cells. It also has the advantage of reusability.	(Shan et al., 2005)
6.	Sphingomonas sp. strain XLDN2–5 cells	Fe_3O_4 nanoparticles/ gellan gum gel beads	Carbazole	Sphingomonas sp. strain XLDN2–5 cells coated with Fe_3O_4 nanoparticles/gellan gum gel beads showed the best reusability than increased degradation and nanoparticles can be easily separated from microbes using an external magnet	(Wang et al., 2007)
7.	Sphingomonaswittichii	Pd/nFe	2,3,7,8-tetrachlorodibenzop-dioxin (2,3,7,8-TeCDD	The successful degradation of highly toxic dioxin compound was done by the combination of Sphingomonaswittichii and Pd/ nFe nanoparticles	(Bokare et al., 2010)

the enzymes inside the cell by diffusion, to form nanoparticles (Ovais et al., 2018). A fungi *Aspergillus tubingensis* (STSP 25) fabricated with iron oxide nanoparticles was found to be more than 90 percent effective in the adsorption of heavy metals from waste-water (Mahanty et al., 2020) (Roychowdhury et al., 2019). In another study, the exopolysaccharides of the Chlorella vulgaris species were used in the co-precipitation of the iron oxide nanoparticles to degrade 91 percent of PO_4^{3-} and 85 percent of NH_4^+ present in the waste-water effectively (Govarthanan et al., 2020). Research states that the biogenic copper nanoparticles produced by a native copper-resistant bacterial strain Escherichia sp. SINT7 decolorizes azo dyes such as congo red, malachite green, direct blue-1, and reactive black-5 at 97.07 percent, 90.55 percent, 88.42 percent, and 83.61 percent respectively (Noman et al., 2020). A study further reports that a marine bacterium *Pseudoalteromonas* sp. CF10-13 degrades the metal complex dye – Naphthol Green B (NGB) into benzamide with the endogenous formation of black stable iron-sulfur nanoparticles which prevents the release of H_2S, exogenous addition of sulfur, and the build-up of metal sludge (Cheng et al., 2019). The production of biogenic nanoparticles either exogenously or endogenously for remediation purposes is a very superior technology. We have tabulated various nanoparticles used with microbes for remediation of pollutants is summarized as follows in Table. 11.2 as follows.

Table 11.2 Nanoparticles along with associated microorganism used in nano-bioaugmentation process for remediation of pollutants.

S.no	Nanoparticles	Microorganism associated	Advantages	The system where it was used	References
1.	Zirconia nanoparticles	*Pseudomonas aeruginosa*	Bioremediation of tetracycline	Waste-water	(Debnath et al., 2020)
2.	Silica nanoparticles	Actinomycetes	Decolorization of textile effluents	Textile effluents	(Mohanraj et al., 2020)
3.	Electrospun nanofibrous webs	*Pseudomonas aeruginosa*	Bioremediation of methylene blue dye	Water	(Sarioglu et al., 2017)
4.	Electrospun cyclodextrin fibers	*Lysinibacillus* sp. NOSK	Bioremediation of heavy metals and dyes	Waste-water	(San Keskin et al., 2018)
5.	Enzyme immobilized TiO_2 nanoparticles	*P. ostreatus*	Biodegradation of bisphenol-A and carbamazepine	Sewage	(Ji et al., 2017)
6.	nanoscale zero-valent iron (nZVI)	*Sphingomonas* sp. pH-07	Biodegradation of Polybrominated diphenyl ethers	Soil	(Kim et al., 2012)
7.	Fe_3O_4 nanoparticles	*Rhodococcus rhodochrous strain*	Biodegradation of chlorophenols	Aqueous phase	(Hou et al., 2016)

11.3.2 Nano-phytoremediation

Iron nanoparticles play a major role in the remediation of pollution by reduction. nZVI nanoparticles incorporated in soil have been used in the removal of 2,4,6-trinitrotoluene by *Panicum maximum* (Jiamjitrpanich et al., 2013) and it is also used in the removal of polychlorinated biphenyls (PCBs) present in the soil with e-waste contamination by combining with *Impatiens balsamina* (Gao and Zhou, 2013). The nanoparticles of *Noaea mucronate* and *Euphorbia macroclada* were used to remove heavy metals such as Pb, Cd, Ni, Zn, Cu by phytoremediation (Nwadinigwe and Ugwu, 2019). It also helps in the removal of heavy metal contamination in soil and organic components such as atrazine and molinate (Verma et al., 2021). Nano-encapsulated enzymes have been used to enhance phytoremediation which is used to remove large organic contaminants (Chauhan et al., 2020). We have summized the application of plant and nanoparticles in combination or concurrently for remediation of various pollutants in tabulated in Table 11.3.

11.4 Current technological barriers in using nanoparticles

The use of nanoparticles results in the accumulation in the environment which also affects humans and animals' health by incorporating them into water bodies. It also leads to the toxicity of the environment. The bioaccumulation of heavy metals by plants may result in affecting the food chain when other animals eat the plants (Cecchin et al., 2017). Only one percent of nanotechnological approaches have been commercialized (Dwevedi, 2018). The ability of plants to absorb pollutants depends on biochemical and

Table 11.3 Various plant and nanoparticles in combination or concurrently for remediation of various pollutants.

Nanoparticle	Plant	pollutant	References
nZVI	*Panicum maximum*	2,4,6-trinitrotoluene	(Jiamjitrpanich et al., 2013)
nZVI	*Impatiens balsamina*	e-wastes	(Gao and Zhou, 2013)
nZVI	*Alpinia calcarata*	endosulfan	(Pillai and Kottekottil, 2016)
Fullerenes	Populus deltoids	C_2HCl_3	(Ma and Wang, 2010)
Euphorbia macroclada Noaeamucronata	*Noaeamucronata*	Pb, Zn, Cu, Ni	(Mohsenzadeh and Rad, 2012; Nwadinigwe and Ugwu, 2019)
	Reseda lutea	Fe	(Mohsenzadeh and Rad, 2012)
	Marrobium Vulgare	Cd	(Mohsenzadeh and Rad, 2012)

physical characteristics (Kvesitadze et al., 2016). The nanoparticles have the persistent ability and tend to remain in the environment (Guedes et al., 2020).

11.5 Future prospective and conclusion

Nanobioremediation was considered an efficient technology for the treatment of heavy toxic metals that are harmful to humans and the environment (Kumar and Gopinath, 2016). Plant and microorganisms-assisted nanotechnology were found to be efficient and provided eco-friendly remediation of effluents. Enzyme encapsulated nanotechnology also will be stable and highly active to treat pollutants in a cost-effective manner (Wong et al., 2019). However, the risks possessed by the nanoparticles in the environment need to be accessed and new ideas should be raised to reduce the effects and enhance the nanotechnology in treatments of pollutants.

References

Abatenh, E., Gizaw, B., Tsegaye, Z., Wassie, M., 2017. The Role of Microorganisms in Bioremediation - A Review. Open J. Environ. Biol 2 (1), 038–046. https://doi.org/10.17352/ojeb.000007.

Ali, Z., Waheed, H., Gul, A., Afzal, F., Anwaar, K., Imran, S., 2017. Brassicaceae plants. In Oilseed Crops. John Wiley & Sons, Ltd, pp. 207–223. https://doi.org/10.1002/9781119048800.ch11.

Azubuike, C.C., Chikere, C.B., Okpokwasili, G.C., 2016. Bioremediation techniques–classification based on site of application: principles, advantages, limitations, and prospects. World. J. Microbiol. Biotechnol. https://doi.org/10.1007/s11274-016-2137-x.

Baruah, A., Chaudhary, V., Malik, R., Tomer, V.K., 2018. 17 -Nanotechnology Based Solutions for Wastewater Treatment. In Nanotechnology in Water and Wastewater Treatment: Theory and Applications. Elsevier, pp. 337–368. https://doi.org/10.1016/B978-0-12-813902-8.00017-4.

Bokare, V., Murugesan, K., Kim, J.H., Kim, E.J., Chang, Y.S., 2012. Integrated hybrid treatment for the remediation of 2, 3, 7, 8-tetrachlorodibenzo-p-dioxin. Sci. Total Environ. 435, 563–566.

Bokare, V., Murugesan, K., Kim, Y.M., Jeon, J.R., Kim, E.J., Chang, Y.S., 2010. Degradation of triclosan by an integrated nano-bio redox process. Bioresour. Technol. 101 (16), 6354–6360.

Bolade, O.P., Williams, A.B., Benson, N.U., 2020. Green synthesis of iron-based nanomaterials for environmental remediation: a review. In Environmental Nanotechnology, Monitoring and Management 13. Elsevier B.V, 100279. https://doi.org/10.1016/j.enmm.2019.100279.

Bolan, N.S., Park, J.H., Robinson, B., Naidu, R., Huh, K.Y., 2011. Phytostabilization. A green approach to contaminant containment. In Advances in Agronomy 112. Academic Press. https://doi.org/10.1016/B978-0-12-385538-1.00004-4.

Canak, S., Berezljev, L., Borojevic, K., Asotic, J., Ketin, S.S., 2019. (PDF) BIOREMEDIATION AND "GREEN CHEMISTRY." Fresenius Environ. Bull. 28. https://www.researchgate.net/publication/332318816_BIOREMEDIATION_AND_GREEN_CHEMISTRY.

Cecchin, I., Reddy, K.R., Thomé, A., Tessaro, E.F., Schnaid, F., 2017. Nanobioremediation: integration of nanoparticles and bioremediation for sustainable remediation of chlorinated organic contaminants in soils. Int. Biodeterior. Biodegradation 119, 419–428. https://doi.org/10.1016/j.ibiod.2016.09.027.

Chauhan, R., Yadav, H.O., Sehrawat, N., 2020. Nanobioremediation: a new and a versatile tool for sustainable environmental clean up-Overview.

Chauhan, R., Yadav, H.O.S., Sehrawat, N., 2020b. Nanobioremediation: a new and a versatile tool for sustainable environmental clean up-Overview. J. Mater. Environ. Sci. 2020 (4), 564–573. http://www.jmaterenvironsci.com.

Cheng, S., Li, N., Jiang, L., Li, Y., Xu, B., Zhou, W., 2019. Biodegradation of metal complex Naphthol Green B and formation of iron–sulfur nanoparticles by marine bacterium Pseudoalteromonas sp CF10-13. Bioresour. Technol. 273, 49–55. https://doi.org/10.1016/j.biortech.2018.10.082.

Chourasia, S., Khanna, I., Gera, N., Chinthala, S., 2014. (1) (PDF) Reduction of pollutants from RO reject using phytoremediation: proposed methodology. https://www.researchgate.net/publication/264547859_Reduction_of_pollutants_from_RO_reject_using_phytoremediation_proposed_methodology.

da Silva, I.G.S., de Almeida, F.C.G., da Rocha e Silva, N.M.P., Casazza, A.A., Convert, A., Sarubbo, L.A., 2020. Soil bioremediation: overview of technologies and trends. Energies 13 (18). MDPI AG. https://doi.org/10.3390/en13184664.

Dangi, A.K., Sharma, B., Hill, R.T., Shukla, P., 2019. Bioremediation through microbes: systems biology and metabolic engineering approach. Crit. Rev. Biotechnol. 39 (1), 79–98. https://doi.org/10.1080/07388551.2018.1500997.

Debnath, B., Majumdar, M., Bhowmik, M., Bhowmik, K.L., Debnath, A., Roy, D.N., 2020. The effective adsorption of tetracycline onto zirconia nanoparticles synthesized by novel microbial green technology. J. Environ. Manage. 261, 110235. https://doi.org/10.1016/j.jenvman.2020.110235.

Dell'anno, F., Brunet, C., van Zyl, L.J., Trindade, M., Golyshin, P.N., Dell'anno, A., et al., 2020. Degradation of hydrocarbons and heavy metal reduction by marine bacteria in highly contaminated sediments. Microorganisms 8 (9), 1–18. https://doi.org/10.3390/microorganisms8091402.

Dong, H., Li, L., Lu, Y., Cheng, Y., Wang, Y., Ning, Q., et al., 2019. Integration of nanoscale zero-valent iron and functional anaerobic bacteria for groundwater remediation: a review. Environ. Int. 124, 265–277.

Dwevedi, A., 2018. In Solutions to environmental problems involving nanotechnology and enzyme technology. Solutions to Environmental Problems Involving Nanotechnology and Enzyme Technology. Elsevier. https://doi.org/10.1016/C2016-0-04550-3.

Farraji, H., Zaman, N.Q., Tajjudin, R.M., Faraji, H., 2016. (PDF) Advantages and disadvantages of phytoremediation A concise review. https://www.researchgate.net/publication/306543535_Advantages_and_disadvantages_of_phytoremediation_A_concise_review.

Galdames, A., Mendoza, A., Orueta, M., de Soto García, I.S., Sánchez, M., Virto, I., et al., 2017. Development of new remediation technologies for contaminated soils based on the application of zero-valent iron nanoparticles and bioremediation with compost. Resource-Efficient Technologies 3 (2), 166–176. https://doi.org/10.1016/j.reffit.2017.03.008.

Gao, Y.Y., Zhou, Q.X., 2013. Application of nanoscale zero valent iron combined with impatiens balsamina to remediation of e-waste contaminated soils. Adv Mat Res 790, 73–76. https://doi.org/10.4028/www.scientific.net/AMR.790.73.

Gavrilescu, M., Demnerová, K., Aamand, J., Agathos, S., Fava, F., 2015. Emerging pollutants in the environment: present and future challenges in biomonitoring, ecological risks and bioremediation. New Biotechnol. 32 (1), 147–156. https://doi.org/10.1016/j.nbt.2014.01.001.

Govarthanan, M., Jeon, C.H., Jeon, Y.H., Kwon, J.H., Bae, H., Kim, W., 2020. Non-toxic nano approach for wastewater treatment using Chlorella vulgaris exopolysaccharides immobilized in iron-magnetic nanoparticles. Int. J. Biol. Macromol 162 (2020), 1241–1249. https://doi.org/10.1016/j.ijbiomac.2020.06.227.

Gu, T., Rastegar, S.O., Mousavi, S.M., Li, M., Zhou, M., 2018. Advances in bioleaching for recovery of metals and bioremediation of fuel ash and sewage sludge. Bioresour. Technol. 261. Elsevier Ltd, pp. 428–440. https://doi.org/10.1016/j.biortech.2018.04.033.

Guedes, M.I.F., Tramontina Florean, E.O.P., De Lima, F., Benjamin, S.R., 2020. Current trends in nanotechnology for bioremediation. Int. J. Environ. Pollut. 1 (1), 1. https://doi.org/10.1504/ijep.2020.10023170.

Hasan, M.K., Shahriar, A., Jim, K.U., 2019. Water pollution in Bangladesh and its impact on public health. Heliyon 5 (8), e02145. https://doi.org/10.1016/j.heliyon.2019.e02145.

Hou, J., Liu, F., Wu, N., Ju, J., Yu, B., 2016. Efficient biodegradation of chlorophenols in aqueous phase by magnetically immobilized aniline-degrading Rhodococcus rhodochrous strain. J. Nanobiotechnology 14 (1). https://doi.org/10.1186/s12951-016-0158-0.

Ji, C., Nguyen, L.N., Hou, J., Hai, F.I., Chen, V., 2017. Direct immobilization of laccase on titania nanoparticles from crude enzyme extracts of P. ostreatus culture for micro-pollutant degradation. Sep. Purif. Technol. 178, 215–223. https://doi.org/10.1016/j.seppur.2017.01.043.

Jia, S., Zhang, X., 2019. Biological HRPs in wastewater. In High-Risk Pollutants in Wastewater. Elsevier, pp. 41–78. https://doi.org/10.1016/B978-0-12-816448-8.00003-4.

Jiamjitrpanich, W., Parkpian, P., Polprasert, C., Kosanlavit, R., 2013. Trinitrotoluene and Its Metabolites in Shoots and Roots of Panicum maximum in Nano-Phytoremediation. Int. J. Environ. Sci. Dev., 7–10. https://doi.org/10.7763/ijesd.2013.v4.293.

Kamarudheen, N., Chacko, S.P., George, C.A., Chettiparambil Somachandran, R., Rao, K.V.B., 2020. An ex-situ and in vitro approach towards the bioremediation of carcinogenic hexavalent chromium. Prep. Biochem. Biotechnol. 50 (8), 842–848. https://doi.org/10.1080/10826068.2020.1755868.

Kim, Y.M., Murugesan, K., Chang, Y.Y., Kim, E.J., Chang, Y.S., 2012. Degradation of polybrominated diphenyl ethers by a sequential treatment with nanoscale zero valent iron and aerobic biodegradation. J. Chem. Technol. Biotechnol. 87 (2), 216–224. https://doi.org/10.1002/jctb.2699.

Kotoky, R., Pandey, P., 2020. Rhizosphere mediated biodegradation of benzo(A)pyrene by surfactin producing soil bacilli applied through Melia azadirachta rhizosphere. Int. J. Phytoremediation. https://doi.org/10.1080/15226514.2019.1663486.

Kumar, S.R., Gopinath, P., 2016. Chapter 2 Nano-Bioremediation Applications of Nanotechnology for Bioremediation, pp. 27–48. https://doi.org/10.1201/9781315374536-3.

Kuppusamy, S., Palanisami, T., Megharaj, M., Venkateswarlu, K., Naidu, R., 2016. In-situ remediation approaches for the management of contaminated sites: a comprehensive overview. In Reviews of Environmental Contamination and Toxicology 236. Springer, New York LLC, pp. 1–115. https://doi.org/10.1007/978-3-319-20013-2_1.

Kvesitadze, G., Khatisashvili, G., Sadunishvili, T., Kvesitadze, E., 2016. Plants for remediation: uptake, translocation and transformation of organic pollutants. In Plants, Pollutants and Remediation. Springer, Netherlands, pp. 241–308. https://doi.org/10.1007/978-94-017-7194-8_12.

Landa-Acuña, D., Acosta, R.A.S., Hualpa Cutipa, E., Vargas de la Cruz, C., Luis Alaya, B., 2020. Bioremediation: a low-cost and clean-green technology for environmental management. In Microbial Bioremediation & Biodegradation. Springer, Singapore, pp. 153–171. https://doi.org/10.1007/978-981-15-1812-6_7.

Le, T.T., Nguyen, K.H., Jeon, J.R., Francis, A.J., Chang, Y.S., 2015. Nano/bio treatment of polychlorinated biphenyls with evaluation of comparative toxicity. J. Hazard. Mater. 287, 335–341. https://doi.org/10.1016/j.jhazmat.2015.02.001.

Li, X., Song, Y., Bian, Y., Gu, C., Yang, X., Wang, F., et al., 2020. Insights into the mechanisms underlying efficient Rhizodegradation of PAHs in biochar-amended soil: from microbial communities to soil metabolomics. Env. Int. 144, 105995. https://doi.org/10.1016/j.envint.2020.105995.

Limmer, M., Burken, J.G., Limmer, M., Burken, J.G., 2016. Phytovolatilization of Organic Contaminants. Environ. Sci. Technol. 50 (13), 6632–6643. https://doi.org/10.1021/acs.est.5b04113.

Liu, Q., Zhou, Y., Lu, J., Zhou, Y., 2020. Novel cyclodextrin-based adsorbents for removing pollutants from wastewater: a critical review. Chemosphere 241, 125043. Elsevier Ltd. https://doi.org/10.1016/j.chemosphere.2019.125043.

Liu, X., Sun, J., Duan, S., Wang, Y., Hayat, T., Alsaedi, A., et al., 2017. A Valuable Biochar from Poplar Catkins with High Adsorption Capacity for Both Organic Pollutants and Inorganic Heavy Metal Ions. Sci. Rep. 7 (1). https://doi.org/10.1038/s41598-017-09446-0.

Logeshwaran, P., Megharaj, M., Chadalavada, S., Bowman, M., Naidu, R., 2018. Petroleum hydrocarbons (PH) in groundwater aquifers: an overview of environmental fate, toxicity, microbial degradation and risk-based remediation approaches. Environ. Technol. Innov. 10, 175–193. Elsevier B.V. https://doi.org/10.1016/j.eti.2018.02.001.

Ma, X., Wang, C., 2010. Fullerene nanoparticles affect the fate and uptake of trichloroethylene in phytoremediation systems. Environ. Eng. Sci. 27 (11), 989–992. https://doi.org/10.1089/ees.2010.0141.

Mahanty, S., Chatterjee, S., Ghosh, S., Tudu, P., Gaine, T., Bakshi, M., et al., 2020. Synergistic approach towards the sustainable management of heavy metals in wastewater using mycosynthesized iron oxide nanoparticles: biofabrication, adsorptive dynamics and chemometric modeling study. J. Water Process. Eng. 37 (2020), 101426. https://doi.org/10.1016/j.jwpe.2020.101426.

Manobala, T., Shukla, S.K., Rao, T.S., Kumar, M.D., 2020. Kinetic modelling of the uranium biosorption by Deinococcus radiodurans biofilm. Chemosphere, 128722. https://doi.org/10.1016/j.chemosphere.2020.128722.

McClements, D.J., 2018. Encapsulation, protection, and delivery of bioactive proteins and peptides using nanoparticle and microparticle systems: a review. Adv. Colloid Interface Sci. 253, 1–22. https://doi.org/10.1016/j.cis.2018.02.002.

Mikkonen, A., Yläranta, K., Tiirola, M., Dutra, L.A.L., Salmi, P., Romantschuk, M., et al., 2018. Successful aerobic bioremediation of groundwater contaminated with higher chlorinated phenols by indigenous degrader bacteria. Water Res. 138, 118–128. https://doi.org/10.1016/j.watres.2018.03.033.

Mohanraj, R., Gnanamangai, B.M., Poornima, S., Oviyaa, V., Ramesh, K., Vijayalakshmi, G., et al., 2020. Decolourisation efficiency of immobilized silica nanoparticles synthesized by actinomycetes, Materials Today: Proceedings. https://doi.org/10.1016/j.matpr.2020.04.139.

Mohsenzadeh, F., Rad, A.C., 2012. Bioremediation of Heavy Metal Pollution by Nano-Particles of Noaea Mucronata. IJBBB, 85–89. https://doi.org/10.7763/ijbbb.2012.v2.77.

Nanjwade, B.K., Sarkar, A.B., Srichana, T., 2018. Design and Characterization of Nanoparticulate Drug Delivery. In Characterization and Biology of Nanomaterials for Drug Delivery: Nanoscience and Nanotechnology in Drug Delivery. Elsevier, pp. 337–350. https://doi.org/10.1016/B978-0-12-814031-4.00012-X.

Nash, D.A.H., Abdullah, S.R.S., Hasan, H.A., Idris, M., Othman, A.R., Al-Baldawi, I.A., et al., 2020. Utilisation of an aquatic plant (Scirpus grossus) for phytoremediation of real sago mill effluent. Environ. Technol. Innov. 19. https://doi.org/10.1016/j.eti.2020.101033.

Němeček, J., Pokorný, P., Lhotský, O., Knytl, V., Najmanová, P., Steinová, J., et al., 2016. Combined nano-biotechnology for in-situ remediation of mixed contamination of groundwater by hexavalent chromium and chlorinated solvents. Sci. Total Environ. 563, 822–834.

Noman, M., Shahid, M., Ahmed, T., Niazi, M.B.K., Hussain, S., Song, F., et al., 2020. Use of biogenic copper nanoparticles synthesized from a native Escherichia sp. as photocatalysts for azo dye degradation and treatment of textile effluents. Environ. Pollut. 257, 113514. https://doi.org/10.1016/j.envpol.2019.113514.

Nwadinigwe, A.O., Ugwu, E.C., 2019. Overview of nano-phytoremediation applications. In Phytoremediation: Management of Environmental Contaminants 6. Springer International Publishing, pp. 377–382. https://doi.org/10.1007/978-3-319-99651-6_15.

Ojuederie, O.B., Babalola, O.O., 2017. Microbial and plant-assisted bioremediation of heavy metal polluted environments: a review. Int. J. Environ. Res. Public Health. https://doi.org/10.3390/ijerph14121504.

Ossai, I.C., Ahmed, A., Hassan, A., Hamid, F.S., 2020. Remediation of soil and water contaminated with petroleum hydrocarbon: a review. Environ. Technol. Innov. 17, 100526. Elsevier B.V. https://doi.org/10.1016/j.eti.2019.100526.

Ovais, M., Khalil, A.T., Ayaz, M., Ahmad, I., Nethi, S.K., Mukherjee, S., 2018. Biosynthesis of metal nanoparticles via microbial enzymes: a mechanistic approach. Int. J. Mol. Sci. 19 (12). MDPI AG. https://doi.org/10.3390/ijms19124100.

Phillips, D.P., Human, L.R.D., Adams, J.B., 2015. Wetland plants as indicators of heavy metal contamination. Mar. Pollut. Bull. 92 (1–2), 227–232. https://doi.org/10.1016/j.marpolbul.2014.12.038.

Pillai, H.P.S., Kottekottil, J., 2016. Nano-Phytotechnological Remediation of Endosulfan Using Zero Valent Iron Nanoparticles. J. Environ. Prot. (Irvine, Calif) 07 (05), 734–744. https://doi.org/10.4236/jep.2016.75066.

Pirzadah, T.B., Malik, B., Tahir, I., Kumar, M., Varma, A., Rehman, R.U., 2014. Phytoremediation: an Eco-Friendly Green Technology for Pollution Prevention, Control and Remediation. An Eco-Friendly Green Technology for Pollution Prevention, Control and RemediationSoil Remediation and Plants: Prospects and Challenges. Elsevier Inc, pp. 107–129. https://doi.org/10.1016/B978-0-12-799937-1.00005-X.

Rai, P.K., Kumar, V., Lee, S.S., Raza, N., Kim, K.H., Ok, Y.S., et al., 2018. Nanoparticle-plant interaction: implications in energy, environment, and agriculture. Environ. Int. 119, 1–19. Elsevier Ltd. https://doi.org/10.1016/j.envint.2018.06.012.

Rathi, B.S., Kumar, P.S., Show, P.L., 2020. A review on effective removal of emerging contaminants from aquatic systems: current trends and scope for further research. J. Hazard. Mater. 124413. https://doi.org/10.1016/j.jhazmat.2020.124413.

Ravikumar, K.V.G., Kumar, D., Kumar, G., Mrudula, P., Natarajan, C., Mukherjee, A., 2016. Enhanced Cr(VI) removal by nanozerovalent iron-immobilized alginate beads in the presence of a biofilm in a continuous-flow reactor. Ind. Eng. Chem. Res. 55 (20), 5973–5982.

Rizwan, M., Singh, M., Mitra, C.K., Morve, R.K., 2014. Ecofriendly Application of Nanomaterials: nanobioremediation. J. Nanopart. 2014, 1–7. https://doi.org/10.1155/2014/431787.

Roychowdhury, R., Roy, M., Zaman, S., Mitra, A., 2019. Bioremediation Potential of microbes towards heavy metal contamination. Int. J. Res. Anal. Rev. 6 (1), 1088–1094.

San Keskin, N.O., Celebioglu, A., Sarioglu, O.F., Uyar, T., Tekinay, T., 2018. Encapsulation of living bacteria in electrospun cyclodextrin ultrathin fibers for bioremediation of heavy metals and reactive dye from wastewater. Colloids Surf. B 161, 169–176. https://doi.org/10.1016/j.colsurfb.2017.10.047.

Sarioglu, O.F., Keskin, N.O.S., Celebioglu, A., Tekinay, T., Uyar, T., 2017. Bacteria encapsulated electrospun nanofibrous webs for remediation of methylene blue dye in water. Colloids Surf. B 152, 245–251. https://doi.org/10.1016/j.colsurfb.2017.01.034.

Shan, G., Xing, J., Zhang, H., Liu, H., 2005. Biodesulfurization of dibenzothiophene by microbial cells coated with magnetite nanoparticles. Appl. Environ. Microbiol. 71 (8), 4497–4502.

Sharma, B., Dangi, A.K., Shukla, P., 2018. Contemporary enzyme based technologies for bioremediation: a review. J. Environ. Manage. 210, 10–22. https://doi.org/10.1016/j.jenvman.2017.12.075.

Sharma, I., 2020. Bioremediation Techniques for Polluted Environment: concept, Advantages, Limitations, and Prospects. In Trace Metals in the Environment - New Approaches and Recent Advances. [Working Title]. IntechOpen. https://doi.org/10.5772/intechopen.90453.

Sharma, J., 2019. Advantages and Limitations of In Situ Methods of Bioremediation. Recent Advances in Biology and Medicine. https://doi.org/10.18639/rabm.2019.955923.

Shukla, A., Iravani, S., 2018. Green Synthesis, Characterization and Applications of Nanoparticles, 1st Edition. Elsevier. https://www.elsevier.com/books/green-synthesis-characterization-and-applications-of-nanoparticles/shukla/978-0-08-102579-6.

Siddiqui, F.J., Belayneh, G., Bhutta, Z.A., 2021. Nutritional and Diarrheal Disease and Enteric Pathogens. In Nutrition and Infectious Diseases. Springer International Publishing, pp. 219–241. https://doi.org/10.1007/978-3-030-56913-6_8.

Singh, M., Srivastava, P.K., Jaiswal, V., Kharwar, R.N., 2016a. Biotechnological Applications of Microbes for the Remediation of Environmental Pollution. In Book: Biotechnology: Trends and Applications. https://www.researchgate.net/publication/303330478_7_Biotechnological_Applications_of_Microbes_for_the_Remediation_of_Environmental_Pollution.

Singh, P., Kim, Y.J., Zhang, D., Yang, D.C., 2016b. Biological Synthesis of Nanoparticles from Plants and Microorganisms. Trends Biotechnol. 34 (7), 588–599. Elsevier Ltd. https://doi.org/10.1016/j.tibtech.2016.02.006.

Singh, R., Behera, M., Kumar, S., 2020b. Nano-bioremediation: an Innovative Remediation Technology for Treatment and Management of Contaminated Sites. In Bioremediation of Industrial Waste for Environmental Safety. https://doi.org/10.1007/978-981-13-3426-9_7.

Singh, R., Behera, M., Kumar, S., 2020a. Nano-bioremediation: an innovative remediation technology for treatment and management of contaminated sites. Bioremediation of Industrial Waste for Environmental Safety, Springer, pp. 165–182.

Singh, S., Kang, S.H., Mulchandani, A., Chen, W., 2008. Bioremediation: environmental clean-up through pathway engineering. Curr. Opin. Biotechnol. 19 (5), 437–444. Elsevier Current Trends. https://doi.org/10.1016/j.copbio.2008.07.012.

Speight, J.G., 2020. Remediation technologies. Natural Water Remediation. Elsevier, pp. 263–303. https://doi.org/10.1016/B978-0-12-803810-9.00008-5.

Sriplai, N., Pinitsoontorn, S., 2021. Bacterial cellulose-based magnetic nanocomposites: a review. In Carbohydrate Polymers 254. Elsevier Ltd, 117228. https://doi.org/10.1016/j.carbpol.2020.117228.

Tena-Garitaonaindia, M., Llamas, I., Toral, L., Sampedro, I., 2019. Chemotaxis of halophilic bacterium Halomonas anticariensis FP35 towards the environmental pollutants phenol and naphthalene. Sci. Total Environ. 669, 631–636. https://doi.org/10.1016/j.scitotenv.2019.02.444.

Verma, A., Roy, A., Bharadvaja, N., 2021. Remediation of heavy metals using nanophytoremediationAdvanced Oxidation Processes for Effluent Treatment Plants. Elsevier, pp. 273–296. https://doi.org/10.1016/b978-0-12-821011-6.00013-x.

Verma, S., Kuila, A., 2019. Bioremediation of heavy metals by microbial process. In Environmental Technology and Innovation. https://doi.org/10.1016/j.eti.2019.100369.

Wang, X., Gai, Z., Yu, B., Feng, J., Xu, C., Yuan, Y., et al., 2007. Degradation of carbazole by microbial cells immobilized in magnetic gellan gum gel beads. Appl. Environ. Microbiol. 73 (20), 6421–6428.

Windt, W.D., Aelterman, P., Verstraete, W., 2005. Bioreductive deposition of palladium (0) nanoparticles on Shewanella oneidensis with catalytic activity towards reductive dechlorination of polychlorinated biphenyls. Environ. Microbiol. 7 (3), 314–325.

Wong, J.K.H., Tan, H.K., Lau, S.Y., Yap, P.S., Danquah, M.K., 2019. Potential and challenges of enzyme incorporated nanotechnology in dye wastewater treatment: a review. J. Environ. Chem. Eng. 7 (4), 103261. https://doi.org/10.1016/j.jece.2019.103261.

Xiu, Z.M., Gregory, K.B., Lowry, G.V., Alvarez, P.J., 2010a. Effect of bare and coated nanoscale zerovalent iron on tceA and vcrA gene expression in Dehalococcoides spp. Environ. Sci. Technol. 44 (19), 7647–7651.

Yadav, K.K., K., G., N., K., A., R., L., M., Singh, N., et al., 2018. Mechanistic understanding and holistic approach of phytoremediation: a review on application and future prospects. Ecol. Eng. 120, 274–298. https://doi.org/10.1016/j.ecoleng.2018.05.039.

Yadav, K.K., Singh, J.K., Gupta, N., Kumar, V., 2017. (1) (PDF) A Review of Nanobioremediation Technologies for Environmental Cleanup: a Novel Biological Approach. J. Mater. Environ. Sci. 8 (2), 740–757. https://www.researchgate.net/publication/317970687_A_Review_of_Nanobioremediation_Technologies_for_Environmental_Cleanup_A_Novel_Biological_Approach.

Yu, S., Miao, C., Song, H., Huang, Y., Chen, W., He, X., 2019. Efficiency of nitrogen and phosphorus removal by six macrophytes from eutrophic water. Int. J. Phytoremediation 21 (7), 643–651. https://doi.org/10.1080/15226514.2018.1556582.

Zhao, C., Yan, M., Zhong, H., Liu, Z., Shi, L., Chen, M., et al., 2018. Biodegradation of polybrominated diphenyl ethers and strategies for acceleration: a review. Int. Biodeterior. Biodegrad. 129, 23–32.

Zhou, Z., Zhang, Y., Wang, H., Chen, T., Lu, W., 2014. The comparative photodegradation activities of pentachlorophenol (PCP) and polychlorinated biphenyls (PCBs) using UV alone and TiO2-derived photocatalysts in methanol soil washing solution. PLoS One 9 (9). https://doi.org/10.1371/journal.pone.0108765.

Zouboulis, A.I., Moussas, P.A., Psaltou, S.G., 2019. Groundwater and soil pollution: bioremediation. In Encyclopedia of Environmental Health. Elsevier, pp. 369–381. https://doi.org/10.1016/B978-0-12-409548-9.11246-1.

Index

Printed in the United States
by Baker & Taylor Publisher Services